# THE URBAN COMMUNITY

# THE URBAN COMMUNITY

## A World Perspective

NELS ANDERSON

Routledge
Taylor & Francis Group
LONDON AND NEW YORK

First published in 1959

This edition published in 2007
Routledge
2 Park Square, Milton Park, Abingdon, Oxon, OX14 4RN
Simultaneously published in the USA and Canada by Routledge
711 Third Avenue, New York, NY 10017
*Routledge is an imprint of Taylor & Francis Group, an informa business*

Transferred to Digital Printing 2007
First issued in paperback 2013



*British Library Cataloguing in Publication Data*
A CIP catalogue record for this book
is available from the British Library

ISBN13: 978-0-415-41840-9 (volume hbk)
ISBN13: 978-0-415-41931-4 (subset)
ISBN13: 978-0-415-41318-3 (set)
ISBN13: 978-0-415-86052-9 (volume pbk)

Routledge Library Editions: The City

Nels Anderson

Unesco Institute for Social Sciences
Cologne, Germany

# THE URBAN COMMUNITY:

## A WORLD PERSPECTIVE

ROUTLEDGE & KEGAN PAUL, LONDON

*First published in England in* 1960
*by Routledge & Kegan Paul Ltd.*
*Broadway House,* 68-74 *Carter Lane*
*London, E.C.*4



*Printed in Great Britain*
*by Lowe & Brydone (Printers) Ltd.*
*London, N.W.*10



**To Helen**

# Preface

BECAUSE OF CHANGES due to technological and other influences, new and wider approaches must be made in community research, particularly with respect to the urban community. We have entered an era of striding cosmopolitanism that is not confined to cities and that is both vigorous and challenging. Urbanism has always been vigorous and challenging but is now becoming more so, outstripping our efforts to understand it.

Precisely because the subject is so dynamic and complex, the study of the community becomes the more intriguing. It has become, in fact, much too complex for the old double approach of urban sociology on the one side and rural sociology on the other. For that reason I have not in this book tried to respect the areas claimed by the two sociologies.

Whatever may be claimed for ruralism, urbanism seems to be a border-crossing phenomenon. Every modern community is urban to a degree. One can study a city and limit the research to the municipal area, but urbanism itself is not so confined. One can study the rural community, using the city as a backdrop for rural-urban comparison, but a line cannot be drawn between the urban and the rural.

While recognizing the utility of both urban and rural sociology, it must be admitted that neither urbanism nor ruralism can be understood if the urban sociologist and the rural sociologist take back-to-back positions in the study of their respective subjects. Actually, we do not know the limits of ruralism on the urban side of the "line" or the limits of urbanism on the rural side. They are interpenetrating concepts, each being always a little of the other. Moreover, the interpenetration may assume a different character from time to time, although the urban-to-rural penetration tends ever to be the stronger.

This, of course, is not a new thought, although it might have sounded strange a decade or so ago. That the rural-urban dichotomy is one of unequal opposite factors is now being expressed by many writers, and here I must call attention to an article by Paul Meadows ("The City, Technology and History," *Social Forces,* Vol. 36, December 1957, pp. 141-147). In his view, urban sociology heretofore has been excessively "intra-urban," but it needs to be and tends to be more "inter-urban." He emphasizes the dominant position of urbanism and cites the city as the maker and shaper and the remaker of civilization. To him, urbanism and civilization are synonymous terms. To define our civilization is to define our urbanism, for our civilization is centered in the urban milieu of competitive creativeness—although it is not confined there.

In the pages that follow I have tried to distinguish between urban and rural but without venturing definitions. Nor have I tried to say, in terms of size, what is a village, a town, or a city. Rather, they must be visualized in respect to their way of life. I have been content with simple sociological distinctions: A village is a small place of primary group character, a city is a large place of secondary group life, and a town lies between the two. These places differ in the degree to which they are cosmopolitan in their way of life. (It is, of course, recognized that demographers, among others, need more exact definitions.) While large places tend generally to be more cosmopolitan and secondary than smaller places in their way of life, exceptions are very numerous. Towns and cities, especially, have an out-reaching and change-stimulating influence on their hinterlands. Places differ in the reach and the effectiveness of their influence: At the upper extreme, a city may be world-wide or global in its influence, and the number of places having a global reach in their influence tends to increase.

Although this global character of urbanism has been receiving increasing attention since World War II, a great deal more research is needed if we would understand communities as they are and as they are becoming. We see on every hand how industry becomes more global in reaching for raw materials and markets, how finance responds with great rapidity to far-flung economic changes, and how every type of secondary organization becomes increasingly world-wide.

Cities and towns must be dynamic and out-reaching if they would live, and whether they are regional or national or global in their reach depends on the volume and variety of goods and services they offer, as well as on the volume and variety of goods and services they need. It is in connection with these efforts that urbanism spreads. And now we find that governments are systematically giving aid in the extension of industrial urbanism to underdeveloped regions. The initiative in these efforts lies with the industrial urban countries of the West, particularly the United States.

Social scientists, and especially urban sociologists, can hardly escape the challenge of these global urban trends, whether their research interests are in New York City, Podunk, London, or Timbuctoo. Urban sociology must itself assume a more global perspective. That such wider interests are developing is seen in a great variety of scattered activities. I will mention only the series of world-region seminars being sponsored by UNESCO to study the problems of industrialization and urbanization. The first of these met in 1954 and was concerned with Africa south of the Sahara. The second seminar, in 1956, concerned industrialization and urbanization in the Far East. The third, to meet in 1960, will embrace the Near East and Mediterranean region, and the fourth seminar will meet in 1962 and will be in Latin America.

While being cognizant of a world perspective, one does well to orient his urban sociology primarily to his own culture and to his own home base. This does not negate its comparative value to urban sociology in other countries. This book is primarily preoccupied with urbanism as a way of life in the United States. Most of the sociological literature on urban and rural research now available is of American development. Community research has not only advanced further in the United States, but the special phases of such research have developed here more than in other countries. One can almost say that urban sociology is "at home" in the United States. A decade hence this may not be true.

It must be recognized, however, that urban sociology heretofore has been so very American in its content and focus that little attention has been paid to it by scholars in other lands. But it is also true that there are some notable exceptions to this observation, and the exceptions become more numerous. This trend has been gaining since World War II especially in areas of research that are very much identified with the phenomena of urbanism. I have in mind such specialties as industrial sociology, family sociology, urban demographic studies, and ecological research.

It will be noticed that I have not joined the chorus of complaints about the evils of cities—the dangers of anonymity, the vanishing neighborhood, and other effects of technological change. It seems that we can meet these problems much more realistically if we concern ourselves with forward-looking approaches rather than backward-turning solutions. The dynamics of urbanism as a way of life can hardly be other than socially disturbing. Too often the complainer wants the progress but he fails to see that the social disturbance may be a phase of this progress.

At this point a book that I have just reread comes to mind. The popular

American author of this little volume lives in a village and likes it, but his very entertaining and penetrating description of village life reveals his deep concern about its backwardness. He feels that something should be done to stop the growth of cities. If this is not possible, at any rate something must be done to prevent the city from destroying the village and little town.

While this popular writer sits in the village typing manuscripts that are sold in New York, he is, in fact, just as urban as if he lived in the city. He thinks that the natives should turn from their kerosene lamps, well water, outdoor toilets, and outmoded houses. He would have the village spruce up a little and be more alert. He recognizes that, while the villagers respect him, he has not been able to become one of them. What this writer would do is to take the village out of its traditionalism and complacency. He would urbanize it and that, of course, is exactly what the experts are trying to do when they are sent with their industrial urban "know how" to the underdeveloped countries.

If my view regarding the future of social change under urbanism seems to be optimistic, it is, at least, a considered optimism. It is based on the assumption that these social problems can be coped with more effectively if we understand better the processes and trends of urbanism. It seems to me, for example, that urbanism in the United States, as compared with urbanism in Old World countries, is better understood if we consider the role of the frontier in American life. This was not an influence that ended with the last of the homesteading; rather, it figured in the tone and tempo of town building and industrialism. American scholars, with a few notable exceptions, are still too close to that frontier to be impressed by it. We find almost no mention of the frontier, except historical references to the land frontier, in the works of American sociologists.

Another theme which appears frequently in this volume is the emphasis on work activity. With the emergence of industrial sociology, work is coming to be viewed as an essential sociological concept. I have attempted, on the one hand, to link it with leisure and, on the other hand, with the institutions of government. It seems important also to link work with the phenomena of welfare and security. Work, leisure, government, and security seem to be crucial elements in the presentation of urban sociology, although heretofore they have been somewhat neglected.

The urban sociologist must draw his grist from many mills. He must turn to experts on at least two different levels, beginning with the scholars: the historians, economists, political scientists, ecologists, psychologists, geographers, and demographers. At another level he must seek out the experts who are less preoccupied with the concepts and more intent on manipulation and action: the administrators, engineers, welfare workers, and city planners.

In preparing this book I am deeply indebted both to the wise men of the world of research and to the men of action. In every instance where thought has been borrowed consciously from any one of them, I have tried to give credit. In some cases I may have unconsciously used the thought of another as my own. Perhaps I read it in a book a long time ago, or in an article, or absorbed it in a conversation. If this has become part of my own thinking and I cannot name the sources, I am nevertheless grateful to them.

I have quoted a great many writers by name and for this I have asked permission, particularly for the longer quotations. The generosity of the editors and publishers has been gratifying. My thanks go to John W. Riley, Jr., for his critical reading of these chapters and for the frankness of his reactions. He gave the manuscript a severe and complete examination.

N. A.

Cologne, Germany
January, 1959

# Contents

# CHAPTER 1

# Urbanism as a Way of Life*

OUR TASK IS TO STUDY and understand the modern community and its way of life. In this opening chapter our attention is directed to a brief examination of modern community life. It is centered in the great cities and it is oriented to industry. It is often referred to as Western, and as a way of life is called urbanism. It is found in its most characteristic form in the United States, but tends to be present wherever a good share of a people live by industry and commerce and, concomitantly, at a tempo which is peculiar to urbanism.

Urbanism as a way of life is not confined to cities and towns, although it emerges from the great metropolitan centers. It is a way of behaving, and that means one can be very urban in his thinking and conduct although he may live in a village. On the other hand, a very nonurbanized person may live in a most urbanized section of a city.

## Characteristics of Urbanism

Urbanism as a way of life was once fairly restricted within the walls of the city and, within the city, it was limited to certain sections. The urbanized man remains oriented in the crowd. He is not disturbed by the coming and going of people, hence he is always making new acquaintances and forgetting the old ones; *transiency* is one of his characteristics. He cannot know all persons about him well and he may not wish to. Thus, again to use a term from

* The title of this chapter is borrowed from an early but still widely read article by Louis Wirth, "Urbanism as a Way of Life," *American Journal of Sociology*, Vol. 41, No. 1, July 1938, pp. 1-23. This author was a student with Wirth and is very much indebted to him and his pioneering in the urban sociology field.

Wirth, interpersonal relations are marked by *superficiality*. Since the urbanized man cannot know all people, and may not wish to, he acquires the ability to move in the crowd without caring who the people are about him, and he does not invite their approaches; *anonymity* is still a third characteristic.

Certain writers have questioned the transiency-superficiality-anonymity description of urbanism offered by Wirth. For example, Bascom studied the cities of the Yoruba in Africa. He found these cities were not well urbanized, and anonymity was less evident.[1] Sedky found that Alexandria, Egypt, also lacked these marks of urbanism. The people in Alexandria are still joined in extended family networks and they still live by the old tradition. She found little evidence of the emergence of individualism.[2] This writer does not know the cities of the Yoruba, but he does know Alexandria and other Middle East cities. Alexandria, much more isolated than Cairo, is much less urbanized. There is lacking in Alexandria the moving in and out of many strange people of varied races, religions, and nationalities. This may be said of all the cities of Iraq except Baghdad and it applies to all the cities in Iran except Teheran. But neither Baghdad nor Teheran is very industrial and neither could be described as being as urbanized as Cairo.

But urbanism is not merely a way of thinking and behaving. The urbanized man, wherever he may be, is ever adjusting to the new and changing. As he is congenial to initiative, he may also be intolerant of tradition if tradition stands in the way of getting things done. He is not only mobile himself, but he accepts the mobility of others. He may be loyal to his immediate family but he tends to lose contact with other relatives. As he tends to be more urbanized he is also more the individual than is possible in nonurban society.

A third characteristic of urbanism as a way of life concerns the standardizing influences which radiate from the cities. The farmer and the woodsman hear the same radio programs, view the same television programs, and visit the same movies as the most urbanized man. The American farmer, especially, uses machines made in the city and he has a city-made automobile. His children can receive the same education as do city children. He is connected with the world by his telephone and he is a newspaper reader. Not only does the newspaper come from the city, but his farm journals are edited and printed in cities. The farmer's wife buys the same packaged goods for her kitchen as the urban wife, and his children tell the same jokes and sing the

[1] William Bascom, "Urbanization among the Yoruba," *American Journal of Sociology*, Vol. 60, No. 5, March 1955, pp. 446-454.

[2] Mona Sedky, "Groups in Alexandria, Egypt," *Social Research*, Vol. 22, No. 4, Winter 1955, pp. 441-450.

same songs that happen to be in vogue among urban youth. A sort of network seems to exist by which all people tend to be mutually oriented.

This network is not merely one which extends urban influence outward. It has many other aspects. By it the thinking and creations of one country are transferred to other countries. Market news travels from one world city to another causing rapid responses in labor markets and industries.

Thus urbanism may be seen from a variety of viewpoints, three of which have been named above. It concerns the ability of people to behave in the urban setting, and it involves a sort of sophistication of the individual. It is also a kind of communication network by which people everywhere are knitted into a vast social system. Urbanism as a way of life is both complex and fluid, and tends to become more so.

## Degrees of Urbanism

Urbanism as a way of life in the United States tends to find its extreme form in cities like New York, Chicago, and Los Angeles. In Europe they would be London, Paris, Rome, and formerly Berlin and Vienna. But what of smaller cities and towns, hundreds of which are suburban to the most urban centers? Sometimes the outlying suburban village may be occupied almost entirely by people who work in the city. It may be a very urban population with respect to sophistication, but these people may foster a type of neighborhood community life which may resemble in many respects the neighborhood relations in a remote farm village.

To what extent is the nonurban place oriented in its daily life to the great urban centers, in its buying and selling, in its play and its work, in its family relations? To what extent do nonurban people join secondary organizations which have branches in different cities, towns, and villages? If such tests were applied to American towns and villages it would be difficult to find any place that is not urbanized to some degree. The rural person, often depicted in the mass media as a comic stereotype, hardly exists today, but he did exist in many parts of the United States a few decades ago, although never as a comic character.

The same observation holds for some European countries. The United Kingdom is quite urban, the influence extending from the great cities to all of its shores. Efforts are being made to revive and restore some of its rural way of life. In Holland the peasant with his wooden shoes and wide pantaloons has disappeared, except in areas most visited by tourists and here the old garb is used as a trade uniform. At Arnhem, Holland, is a well-known outdoor museum in which the various farm houses and farm villages have been recreated, and even the modern Dutch farmer visits this place to see what farm life was like a generation or more ago. While the peasant is still found in France, he hardly exists any more in Germany or in tightly inhabited Belgium.

This does not mean that the farmer as a worker in agriculture is disappearing, although his numbers tend relatively to diminish in the industrial countries. It means only that ruralism as a way of life tends to disappear, and is replaced by urbanism as a way of life. As will be seen in other chapters, certain characteristics of that older way of life tend to linger and may be found even in urban communities.

## The Process of Urbanization

More and more people in most countries are becoming city or town dwellers. This flow of population from land-bound occupations to other types of work is something new in human history. Such an urban trend could not have happened prior to the industrial revolution, and since the industrial revolution the trend has gone forward even though there have been efforts here and there to curb or guide it.

In a document on the causes and implications of urbanization prepared by the Economic Commission for Asia and the Far East (ECAFE), the term was defined in these words: "In its most simple and demographic sense, urbanization can be defined as the process whereby population tends to agglomerate in clusters of more than a designated size."[3] This definition reflects the definition of the demographer Warren S. Thompson writing in the *Encyclopaedia of the Social Sciences* who called urbanization the "movement of people from communities concerned chiefly or solely with agriculture to other communities, generally larger, whose activities are primarily centered in government, trade, manufacture or allied interests."[4]

This definition cannot be accepted without reservations. Assuming for the moment that urbanization is simply a one-way process, it need only be added that the rate of the process varies from time to time as well as from

[3] United Nations Economic Commission for Asia and the Far East, Document E/CW.11/URB.2, 28 June 1956, p. 4. Seminar on urbanization at Bangkok, 8-18 August 1956. The definition was adapted from article by Kingsley Davis and Hilda Hertz, "Urbanization and the Development of Pre-industrial Areas," *Economic Development and Cultural Change*, Vol. III, No. 1, October 1954, p. 8.

[4] "Urbanization," *Encyclopaedia of the Social Sciences*, Vol. XV, p. 189.

place to place. During the pioneering settlement period in the United States, for example, it was necessary for governments to encourage the location and building of towns. Before a half-century had passed, town-building was moving at a rapid rate and by 1900 many Americans feared their country was becoming too urban too fast. Yet with each decade since 1900 the United States has become increasingly urban. This trend has been frequently described in terms of the relative percentages of rural and urban population. Moreover, the rate of urbanization, as a one-way process, has been more rapid in all countries since the industrial revolution. It would, in fact, be difficult for some countries to survive were most of the people not living in urban agglomerations.

Such a definition of urbanization, however, tends to exclude too much. It is more than a shifting of people from country to city and from land-bound work to urban types of work. Merely moving a man to the city does not necessarily urbanize him (although it helps), while another rural man may be very much urbanized and never leave his rural work or habitat. Urbanization involves basic changes in the thinking and behavior of people and changes in their social values. It is not merely a matter of an individual or group changing from one kind of work to another, but involves changes in attitudes toward work, and it means entering a new and ever-changing division of labor. Thus Karl Mannheim used the term, "urbanization and its ramifications."[5]

Urbanization may assume the form of "idea migration" from the most urban to the less urban places. In the case of the American dairy industry we find an example of a rural occupation which has become quite urbanized. Most of the dairy farms are located in the most urban parts of the United States. The fact that this type of rural work is often called an industry is seen in the timing of the dairyman's work with the tempo of the city. Raper notes that as cities have grown, farm people have been "brought into closer and closer contact with urban centers." He then names some characteristic influences of urbanism on dairy farming and mentions as a second form of urbanization, the "continued growth of part-time farming and rural residences within commuting distance of urban or other non-farm employment."[6]

Thus it would be a mistake in our study of modern community life to

[5] Karl Mannheim, *Freedom, Power and Democratic Planning,* New York, Oxford University Press, 1950, p. 62.

[6] Arthur F. Raper, "The Dairy Areas." See Carl C. Taylor and Associates, *Rural Life in the United States,* New York, Alfred A. Knopf, 1950, p. 431. On p. 491 Raper observes that the degree of urbanization varies from one rural region to another, but as urbanization increases "living and thinking of rural and urban populations become more and more alike."

think of urbanization as merely a one-way process; the urban way of life also moves into the nonurban and less urban areas. Urbanization also means a changeover from one way of life to another. It corresponds in the variety of its application to "detribalization," a term used by students of social change in Africa. Some Africans move from the tribal villages to industrial places and there enter the struggle of becoming adjusted to the urban way of life. But the ways of doing and thinking and the things characteristic of the urban way of life also invade the tribal villages, and there the struggle for adjustment may take another form. However the term is used, the end product of urbanization is urbanism as a way of life.

## Ruralism

In most countries urban areas of habitation are distinguished from rural areas of habitation in terms of the number of people living there. In the United States if the number is under 2,500 the aggregate is designated as rural. Larger places are divided into classes: 2,500 to 5,000 population, 5,000 to 10,000 population, and so on to cities of 1,000,000 people or more. These are arbitrary distinctions which must be accepted in order to have comparable population statistics for comparison through time, and each country establishes its own arbitrary classifications.

The ambiguities inherent in this system of distinguishing rural from urban are too well known for discussion here. Nor do we need here to go into the various efforts of population statisticians to meet the problem. This much must be said: It is generally recognized that rural population and rural occupations are found in the United States in agglomerations far in excess of 2,500 population, while urban population and urban occupations are found in places of less than 2,500 inhabitants. Thus the size-of-population yardstick, however useful for many purposes, is not very helpful in measuring the presence or absence of a rural or an urban way of life.

Allan Nevins, in a preface to a book on great cities, having in mind the American situation, noted that "quite different faculties of mind and character, broadly speaking, appertain to the city and to the country. . . . The rural outlook is the more serene and conservative; the urban outlook the more volatile, alert and radical." In his discussion he mentioned differences in wit and humor, and the fact that in some important respects rural and urban characteristics tend to be mutually opposite. He implied, however, that the urban are dominant: "But we must make up our minds now to be a predominantly urban civilization, and try to nurture rural virtues in a citified environment."[7]

Without discounting in the least the statistics on population and occupations, we must also recognize that before referring to a person or an aggregate of people as urban or not urban we must take account of their thinking and behavior, which may be very urban, very rural, or some urban-rural mixture. The tendency in most countries is for rural places to assume a more urban way of life. This process of urbanization is easily to be seen, as Nelson has observed it in certain isolated Mormon villages:

> Urbanization, or as some sociologists call it, secularization of life, is proceeding at a rapid pace. Communication and transportation devices which characterize contemporary life place the remotest corners in instantaneous contact with the world. The diffusion of urban traits to the countryside is everywhere apparent. Farmers are declining in numbers and farms increasing in size. Life becomes more impersonal, mutual aid declines, and contractual forms of association increase. Formal organizations multiply as new interests arise—economic, social, recreational, educational. New occupations come into being as specialization and division of labor grow more elaborate. Homogeneity of the population gives way to increasing heterogeneity. Attitudes change. The sense of community suffers as cleavages develop around special interests. These developments are clearly evident in the Mormon village today, as they are in the communities of the United States elsewhere.[8]

Nelson's closing observation applies to rural villages in most countries, but in particular in the Western countries. The global trend toward the urban way of life touches every community.

The trend of rural change is away from what we call ruralism. The term is not one of depreciation. Communities dominated by ruralism—and they are not numerous—are likely to be described as backward. Raper, quoted earlier, noted that the degree of urbanization (our term urbanism) differs from one region of the United States to another, and this also means that the degree of ruralism differs. The same could be said of regions in other countries, although country-to-country comparisons are difficult to make for we must

[7] Allan Nevins, Preface. See Ernest Barker and 12 Associates, *Golden Ages of the Great Cities*, London, Thames and Hudson, 1952, p. xiv.

[8] Lowry Nelson, *The Mormon Village*, Salt Lake City, University of Utah Press, 1952, pp. 276-277. In a more recent book he makes the observation that it "seems inescapable that rural and urban segments of American society are drawing closer together, becoming more alike. This is quite likely due more to the 'urbanization' of the countryside than to the 'ruralization' of the city, although the influence of one upon the other is reciprocal." *American Farm Life*, Cambridge, Harvard University Press, 1954, Preface. For a report which describes rural change in Holland, see E. W. Hoffstee, *Rural Farm Life and Rural Welfare in The Netherlands*, The Hague, Ministry of Agriculture, Government Printing Office, 1957.

take into account each country's historical development and traditions. On this basis the United States, for instance, is a comparatively new country and lacks the firmly rooted traditions and long history of Europe.

While emphasizing the global spread of urbanism, it must not be forgotten that more of the world's population lives in a rural than an urban way of life. Many more people live in villages than in cities, but the rate of urbanization is increasing in momentum. In some parts of the world so many people are crowding into the cities that the labor markets are glutted. Industry cannot be developed fast enough to meet the demands for employment and so living standards remain low. On the other hand, urban influence is moving to the rural areas at an equally rapid rate. This is another phase of urbanization—the city goes to the country.

## The Rural-oriented and the Urban-oriented

While it is true that more of the world's population is in the rural-oriented category than in the urban-oriented, the relative proportions are changing rapidly, and have been, particularly since the beginning of this century. The rate of changeover has been especially fast since World War II, due in large measure to programs of aid to underdeveloped countries instituted and continued by the Western countries, in many cases, because of the growing interest of Communist countries in these same underdeveloped areas. These efforts are various, sometimes short sighted and often ambiguous, but the end results are that more and more of the world's population, on the initiative of others, is becoming culturally uprooted by being lured or forced out of the old tribal ways of life. The urban-oriented segment of the population is increasing rapidly.

This changing over from old forms of life and work to new ones, ruthless as far as traditional values are concerned, gives rise to many complex and baffling social problems. Our interest here, however, is not in the problems as such—they will be discussed in other chapters—but in the fact that urbanism is spreading. As a way of life, urbanism is becoming global. This growth has always been painful and problem creating, and it continues to be so, although the problems may not be the same. It is, nevertheless, a trend that must be accepted. By many, it is called progress. In Communist terminology this urbanization process is called liberation, and the Communist countries have their active programs for systematically pushing and guiding the process.

At this juncture we must point out that urbanism in its "pure" form would be as difficult to find in most of the Western countries as the "pure" form of ruralism. Any country has, of course, population agglomerations that are the most rural and those that are the most urban, but between these extremes are found various mixtures of urbanism and ruralism; the one diminishing in the urbanward direction and the other diminishing in the ruralward direction. There are no sharp dividing lines.

While the most urban area or the most rural area may be recognizable within the United States, we are likely to find that the characteristics of "most rural" and "most urban" are not the same from country to country. These differences notwithstanding, the relationships between the rural and the urban in all of these countries, particularly the industrial countries of the West, are in a continuous process of change. Change in every country is in the direction of the less urbanized places becoming more urbanized.

The change tends to be a steadier process in the more industrialized and more urbanized countries such as the United States, the United Kingdom, Belgium, Germany, and others than it is in the least industrialized countries. India is an outstanding example of a country where the strains of urbanization seem to be ever at the breaking point. The urban labor market is continuously glutted with hordes of untrained rural job seekers. To quote Deshmukh, "A vast majority of the rural exodus is forced to keep on floating from town to town and from town to village in search of employment, and to earn a precarious living during this migration process."[9]

## Urbanism and Industrialization

The urban segment of any population lives indirectly from the land. It must provide goods and services which can be marketed in rural areas in return for the products of the land. Towns and cities have grown, some more than others, as urbanites have found more ways of producing more goods and services. Many cities, besides making things for sale, rendered services by bringing goods, such as raw materials and food, from far places in ships and then redistributing them to other places by other means of transport. However, before the emergence of power industry there tended to be growth limits which were attained by certain urban places and not exceeded for centuries. As we shall see later, growth-limiting factors are ever present, but they are subject to change.

For generations it seems to have been the accepted order in cities that they produced goods and services under conditions of continuous depression.

[9] M. B. Deshmukh, "A Study of Floating Population, Delhi," *The Social Implications of Industrialization and Urbanization*, Calcutta Unesco Research Center, 1956, p. 150.

Hand work was the rule, and is so much the rule still in some towns of the Far East that modern textile mills have difficulty in getting started because of resistance from the hand-loom tradition. In these towns the situation is further complicated by the fact that there is not sufficient other industry to relieve labor market pressure. Most European industrial cities have also lived through this struggle between hand work and machine work. Fortunately, the United States did not have to face such a problem because the westward-moving frontier drained off the labor surplus while industry was getting started.

Industry in the Western countries has not relieved cities of growth problems or of occasional deep unemployment, but the population no longer lives in a condition of continuous depression and drudgery. Industry enables more people to find work than was possible before, and while work has become increasingly productive, it has also been lightened and working hours shortened. It was rare that a pre-industrial city in the West exceeded half a million inhabitants living by hand production and commerce, for that system of production and exchange had its limits. Industry has enabled many of the same cities to approach or exceed a million inhabitants, and the people in the new industrial cities enjoy living standards which could not be imagined previously.

An industrialized type of urbanism is much more than a condition involving the industrial cities, for the industrial city has been forced to develop a new kind of relationship with the outside. The old city was a seat of wealth, power, and influence to which people came to sell and to buy. The new city does not wait; it goes out to sell and to buy. The new city had to take the initiative in making roads where they hardly existed before, for the outside had little interest in roads, and often resisted their construction. This is one of the basic differences between ruralism and urbanism: Ruralism conserves its isolation, while urbanism, especially industrial urbanism, encroaches upon isolation.

In order to grow, the industrial city had to extend its networks to improve transportation and communication and to extend its contacts over wider areas. Thus industrial urbanism attains markets and access, but in this process it also becomes more easily accessible to more people, and so circulation and the movement of people parallels the movement of goods.

Industrialization has not only introduced new work ways in the urban centers, new and ever-changing divisions of labor and new work standards, but these innovations tend to be exported into the outside areas, often with disturbing results. Here, again, urbanism and ruralism differ fundamentally. This is another outward-extending urban influence that gathers momentum.

It is true that the factory first made its home in towns and cities, and is still most at home there. However, for reasons of competition and economic advantages, industries may be established in areas from whence the raw materials come—factories near the beet fields, canneries in regions of special farming, sawmills in the forests, and metal-reduction plants near the mines. To a large extent the current programs of aid to underdeveloped countries are concerned with establishing industries in these remote places. Several Western countries, for example, are now engaged in various efforts to industrialize Africa, thereby bringing urbanism to the Dark Continent.

To introduce urbanism to a new region through the development of industry, however, calls for much more than the building of factories. Say Thompson and Woodruff:

> The economic development of an area means the use of more capital, or more labour, or more natural resources, either more of all or more of any of them, to raise the real incomes per head of the inhabitants. In the underdeveloped countries of Africa the emphasis is heavily on the use of more capital; whether social or economic, development involves the direction of resources in the present to uses which will only yield benefits in the future. But in backward countries the people have a low standard of living; the level of national income per head provides little scope for large savings. Left to themselves, therefore, the progress of backward countries would be slow and the attainment of a rapid rate of development is dependent on an inflow of funds.[10]

Bringing industry into a nonurban area in the United States may create some problems, but even the most rural American communities have had some experience with new developments. They have their civil organizations for coping with new and unexpected difficulties; ways are found within the local aggregates for achieving adjustment. But the African natives know nothing of this type of collective living; nor have they been trained in the skills of industry, for the facilities for such training are absent. Community facilities are also lacking, and, moreover, the people know little about the public services a community normally should have, such services as water supply to schools. Civil organization in many areas is still in the hands of outsiders (Europeans) and years may pass before these people acquire the civic consciousness needed for urban living.

Nevertheless, through industry urbanism is being taken to nonurban places, even to peoples emerging from tribalism. The processes of adaption will be long and painful. But it has been long and painful in the developed countries.

[10] C. H. Thompson and H. W. Woodruff, *Economic Development in Rhodesia and Nyasaland,* London, Dobson, 1953, p. 23.

## Urbanization and Responsibility

President Truman in his inaugural address of January 20, 1949, proposed the so-called Point IV doctrine of extending aid to underdeveloped countries. This was in fact an extension of the earlier Marshall Plan for the postwar recovery of European countries. The adoption of this doctrine was the basis of a national policy of aid and guidance. It was an expression of interest in the peoples of the underdeveloped countries, not merely in the exploiting of the natural resources of these lands. This idea of national responsibility came to be adopted later by the Organization for European Economic Cooperation (OEEC). A study of the situation south of the Sahara was made and a report appeared in 1951. The following quotation reflects the viewpoint of this interest:

> The countries responsible for the territories have already made a great contribution toward helping the inhabitants in their efforts to achieve better living conditions and enabling the territories to occupy a significant place in the world economy. These joint efforts must be continued and widened in scope, for it is a matter of general concern that the most should be made of the world's resources. It is clear that any expansion of the economy of the area and any rise in the standard of living of its inhabitants will be to the mutual benefit of both the peoples of the African continent and of other continents as well, and that any growth in the trade of the territories is bound to lead to a further expansion of world trade.[11]

The principal OEEC countries with African territories are Belgium, France, and the United Kingdom. It should be mentioned that each of these countries has now established its social science institutes in Africa. The support of these efforts seems to indicate an attitude of public responsibility for the underdeveloped areas. Whatever else may be said of these efforts, this is pertinent to the present chapter: Industrial urbanism is being systematically introduced into Africa, and Western governments are accepting the responsibility for coping with the social problems involved. The expansion of urbanism to global dimensions is being supported by national policies.

## Specialization of Work and Place

The city has always been the haven of specialists, but never to such a high degree as today. This springs in part from the complex division of labor in the urban center and partly from the complexity of its social life. Although

[11] *Investments in Overseas Territories, In Africa South of the Sahara,* Paris, OEEC, 1951, p. 8.

there is much in the urban situation to stimulate specialization, it tends also to level people in their occupations. Yet this same "level-down" worker is able to range forth and back between a variety of related jobs. Individual specialization, however, is less pertinent to the theme of this chapter than is the specialization of aggregates of population. Cities, towns, and smaller places, because they are so interrelated in their market relations, tend to influence each other and to specialize.

For reasons of cost and market advantages, cities tend to specialize in products for which the raw materials are most conveniently available. Thus it was natural at one time that Minneapolis, located in the wheat belt, should become a flour-milling city. It was just as natural that Milwaukee and St. Louis, near the same wheat belt, should become leading brewery cities. However, it was also natural for other cities in the Middlewest to embrace both industries, as they have done. Gary, Indiana, has a natural location for the kind of city it is, a steel center. Other factors may also be influential in determining the work of a place—the type of labor available, the distance to markets, costs of transportation, and so on.

Specialization in some cities may be due to a chain of coincidences or to some type of human initiative. Thus, Los Angeles is the film capital of the world; Detroit is the automobile capital; New York is coming to be, if it is not already, the publishing capital of the world; and Paris holds the title of the world's fashion capital.

Older cities also specialized, but for very limited markets. The modern city cannot operate this way, and precisely because the markets are not limited, the principal work of an urban center may suddenly change. Here we have still another difference between modern urbanism and ruralism: The economic base of a rural aggregate is usually fairly constant. For example, Iowa and corn are about as permanently associated as are Kansas and wheat, or as California and citrus fruits. Thus the rural way of life tends to be more secure with reference to work than is possible with the urban way of life; urbanism is much more beset with hazards.[12]

[12] Of the Yoruba cities in Africa, Bascom reports that many inhabitants are farmers. Others perform other work, "weaving, dyeing, ironworking, brass-casting, wood-carving, ivory-carving, calabash-carving, beadmaking, and leather-working, as well as drumming, divining, the compounding of charms and medicines, and certain other activities or crafts whose techniques are known only to a small group of specialists. These specialists, who are organized into guilds, supply all other members of the community with their particular goods or services. Formerly these occupations tended to be more hereditary within the clan or lineage, but the apprenticeship system provided a method by which individuals from outside the kinship unit could be taught a craft." This is the ancient type of hand work specialism, quite remote from that of industrial urbanism. *Ibid.*, p. 449.

Unfortunately, the industrial-urban way of doing things has operated in many ways to introduce this kind of insecurity into large sections of agriculture. Farmers may be induced to neglect general farming in favor of single crops—sugar beets, for instance, in one area to keep the sugar factory going, and other crops in other sections. Compared with other parts of the world, however, industrialized farming in the United States is hardly a source of rural insecurity. The OEEC report, already quoted, makes this observation about Africa:

> The vulnerability of the territories is all the greater for the fact that the economy of most of them is based on a limited number of commodities. For example, groundnuts represent 99% by value of the exports of Gambia; metals and ores 95% of those of Northern Rhodesia; tobacco and tea 78% of those of Nyasaland; cocoa 69% of those of the Gold Coast and 45% of those of the French Cameroons; and sisal 55% of those of Tanganyika.[13]

## The Money Economy

It would be difficult to find areas in the United States today that are not operating wholly on the money economy, although this was not true a century ago when large sections of the frontier population were getting on quite well with various exchange economies. The changeover into the money economy, however, involved no adjustment problems. This is not the case when industrial urbanism enters certain underdeveloped countries where the introduction of money into the native groups may be the source of breakdown in the social order. The native wife, for example, who is able to sell produce for cash is likely to be less obedient to her husband. But usually it is the young male who is soonest influenced by having money in his pockets. He is encouraged to be mobile to a degree never possible under the subsistence economies. Although he may have much bitter experience learning to use money, at least in the learning he has the adventure of feeling emancipated. He may return home later, if only to visit.

Textor made a study of samlor drivers in Bangkok, mostly young men from the remote villages of Thailand. They leave home to earn money and they enjoy a kind of freedom while doing so, but their primary interest is to return home one day with their pockets full. For this they work long hours, make sacrifices, and save, and become somewhat urbanized in the process. Finally the young man is ready for the home journey:

[13] *Ibid.*, p. 24.

Very commonly the returner will devote much of his savings to building a new house or house addition. Besides treating friends and relatives to refreshments, he is likely to devote some of his money to merit-making at the temple. A particularly devout returner might spend many hundreds of *baht* to purchase a Buddha or some other improvements for the local temple. Another common type of expenditure, difficult to classify as exclusively a consumer or capital item, is the bride price. Local maidens and their parents are far from completely unimpressed by the new opulence of the young bachelor returner, and the youths in turn regard the stint at samlor-pedaling in Bangkok as a promising route to finding a bride possessed of beauty, wealth or both.[14]

Such young returners bring back to the village some flavor of urbanism and hence, indirectly, the isolation-invading medium of money facilitates the accumulation process of urbanization.

## Urbanism and Individualism

We are told in many ways that the individual enjoys greater freedom under the urban way of life than is possible under ruralism. There are more choices to make under urbanism, and they are choices which the individual himself must make. In addition, one is less obligated to the demands of a wide kinship group.

Americans are regarded by other peoples as being individualists, and they often describe themselves as such; in fact, individualism is sometimes regarded as a type of American cult. It was perhaps necessary during the frontier period which is only now coming to a close and during which rule by tradition was not very useful for getting things done. People had to learn to stand alone if necessary and to make their own decisions, but they took the consequences of their decisions. If the individual's adopted course led to failure, according to the rules of the game, he would try again, or try something else. Each of the new countries where European civilization was planted developed some brand of individualism, but it took a different course in the United States because the opportunities here were more ample, and perhaps because there was a greater mixture of peoples among the pioneers. Because

[14] Robert B. Textor, "The Northeastern Samlor Driver in Bangkok," *Social Implications of Industrialization and Urbanization*, Calcutta, Unesco Research Center, 1956, p. 33. The samlor is a three-wheeled vehicle pedaled as a bicycle, with a seat for two passengers behind the driver. It is an invention which began in Bangkok, but is spreading. Sometimes the samlor is equipped with a motor. It is more suitable than taxis in cities having great street congestion.

the frontier situation did not encourage much carry-over of their Old World traits and traditions, the pioneers were forced to operate as individuals.[15]

This peculiar background may explain the extreme individualism of the American brand of industrial urbanism, but another cause is found in the fact that urbanism generally is conducive to encouraging in the person the development of unique qualities and capacities, usually designated as "individuation." Urban individualism, to quote Hayek, "does not deny the necessity of coercive power but wishes to limit it—to limit it to those fields where it is indispensable to prevent coercion by others and in order to reduce the total of coercion to a minimum."[16] It favors rational and immediately realistic control rather than merely traditional control, and these rules must be subject to change.

The individual under urbanism participates in a changing manner to the life about him, but he is an individual and so regarded in the labor market. This does not mean that he operates without social obligation to family and friends; rather, it means that he is not represented by his family. He votes as an individual[17] and, after reaching adulthood, he is responsible as an individual to the law. He usually enjoys the right to choose his own occupation and his place of work, although for some social groups family control may be very strong still even in great cities, especially for families of wealth. With the same reservations the urban individual is free to choose his marriage partner, and they are free to live separate from their extended families. The fact that these characteristics of urban individualism are today not peculiarly urban in the United States tends to indicate how much the values of urbanism have extended outward to the hinterland.

By the rural as well as the urban American it is generally accepted that the individual who works is entitled to keep and to expend his earnings. This was not always the case even among Americans two generations ago, and is certainly not true today in the less urbanized countries. Hoselitz speaks of the industrial worker in underdeveloped countries:

> Similarly, any compensation he may get for his effort is regarded by him not merely as a reward to be employed for his individual purposes, but to be

[15] For a presentation of the American idea of individualism, see Abbott P. Herman, *An Approach to Social Problems*, Boston, Ginn, 1949, Chapt. VIII, "Our Values of Individualism."

[16] F. A. Hayek, *Individualism and Economic Order*, London, Routledge and Kegan Paul, 1949, p. 27.

[17] Despite the clearly important influences on how people vote. See, for example, Bernard R. Berelson, Paul F. Lazarsfeld, and William H. McPhee, *Voting*, Chicago, University of Chicago Press, 1954.

> shared with other members of the collectivity of which he forms a part. In many societies there are still recognized traditional arrangements for the control of money income which secure to the older members of the community the major handling of what money income is secured. Laborers must share their income with chief, parents, fathers-in-law or other relatives and persons of respect.[18]

When such a native goes to a town for work he finds himself in an agglomeration of people in the process of being urbanized. He settles with a group from his own village and area, usually with members of his own family group present. He retains his old loyalties, although they may weaken a little. His children born in the town may become less attached to the village. He and they gradually become "detribalized." They find about them increasing numbers of former villagers who have broken completely with the village. They begin to assume the roles of urbanized individuals which the situation demands, for even these places become increasingly urban in thought and conduct.

## Individualism and Uniformity

The idea of individualism as a characteristic of urbanism can be overemphasized. It is generally true both that the urbanite has more opportunities for choice and for expression than has the person under ruralism, and that it is therefore necessary for him to make individual and often unusual choices. While he is expected to behave as an individual, however, more so than would be expected under ruralism or tribalism, his freedom to behave as one is not entirely complete. He still must conduct himself according to rules, and many of the rules are secondary and impersonal.

The urban man must learn to live with groups of people and to associate with others on levels of intimacy ranging from friendship to anonymity. He must be able, as Wilson observed, to "convert people into things," but there is method and control even in this.

> Clearly we cannot treat the thousands of people whom we meet daily with the intimacy and emotional tone characteristic of *human* relationships. People come to have no particular significance in and of themselves, in their own rights. They become pre-eminently means to our ends (and, of course, vice versa). We relate ourselves to others insofar as they are of service to us, and shrink from contact beyond a kind of contractual relationship. . . . Thus we are likely to deal with people in terms of those symbols which define the utility of the relation-

[18] Bert F. Hoselitz, "The City, the Factory and Economic Growth," *American Economic Review*, Vol. 45, No. 2, May 1955, p. 182.

B

> ship. Uniforms, badges, headgear, labels—these all tell us that we are dealing with the laundryman, the milkman, or mechanic. We deal with patients, clients, customers. Beyond whatever these labels imply we need not know—nor do we want to know. The relationship is purely instrumental.[19]

The same urban situation that stimulates individuation also tends to create uniformities and conformities. The individual must see large groups in stereotype terms; he finds stereotyped patterns of behaving, which he accepts and helps to maintain. By accepting these uniformities and conforming to them, he becomes at once a person under mass control and a factor for maintaining that control. Without such adjustments, varied and global, the individual would be in a state of confusion. These ways of adapting are productive of order, and although it is a changing order, the individual is able to maintain equilibrium within it.

The uniformities peculiar to urbanism as a way of life in thinking, behaving, and conducting social relationships tend to acquire mass dimensions. Thus people by the millions eat certain foods, wear certain types of clothes, live in houses of a strikingly uniform kind furnished in much the same way (although varying with social class), and enjoy their leisure in fairly similar activities. Although these uniformities may vary from country to country and everywhere they may be in process of change, urbanites acquire a generally uniform capability for adjusting to change. The urbanite, then, lives within a framework of uniformities but is at the same time afforded a multitude of choices in which he may act as an individual.

## The Problems of Social Change

These mass uniformities are ever-changing and, for competitive commercial reasons, they must change. Detroit changes car models each year and Paris brings out new fashions. Interior decorators, competitive and creative, are constantly working on new designs for house furnishings. These continuous change-working efforts regarding what consumers will buy reflect the spirit of urbanism. All the media of mass communication are used to spread ideas and to induce the desired conduct. These manifestations of the urban way of life are wide-reaching. All people are reached but different groups of people respond differently.

Changes in the things with which man works or uses in his free time

[19] Everett K. Wilson, "Some Notes on the Pains and Prospects of American Cities," *Confluence,* Vol. VII, No. 1, Spring 1958, p. 9.

lead to changes in his thinking and behavior. Changes in man's means of transport which facilitate his movement, or changes in his means of communication, affect his relationships with others, for such changes permit a man to move about more freely and communicate with greater facility. Changes in the house or in house furnishings as well as changes in the individual's place of residence have their effect on his social values, and his conduct may be in some way modified. This is the essence of social change.

All people are affected by these changes in the material environment, whether they live in the densely urbanized eastern states or in the more open regions of the West or Middlewest, but their reaction patterns may be very different. Some people are not able to maintain social and psychological balance in making their adjustments. At the other extreme are those who never seem to lose balance, however varied the adaptations required. Many who seemed unable to adjust during their youth may be wholly adjusted, urbanized, in their later years. Still others who were very much oriented to urbanism during their early years may become very much out of balance in their later years. Years may pass before some persons of rural origin are able to feel adjusted to the urban situation, much as some urbanized people are unable to feel at home if it is necessary to change from a more urban to a less urban place of residence.

While adaptation is largely individual, it is not entirely so. Reactions vary among different groups or categories of people. Country or place of origin, age, education, economic class, and religion are factors which influence the group-adaptation process.

Because change in the urban community is continuous, the necessity for adjustment is continuous. The rate of social change may vary with times and conditions, and tends to move more rapidly under certain circumstances than under others. And just as there are different rates of change, so there are different rates at which people and groups adapt to change. The community of more rapid change is likely to be more the scene of social conflict than a community of less rapid change. Such an area is also likely to be one with some degree of social discord and various acute social problems.

The urban milieu is of its nature an ever-changing center of work and living, and technological creativity. Social change in the less urban places cannot help but be stimulated by the dynamic influences which radiate from such centers. Here more than elsewhere people work and create and compete sharply, and while at it seek new ways of working and competing to greater advantage. Here the people of many origins meet and mix and, whether willingly or not, change old ways of living for new ones.

## Urbanism and Mobility

We have noted that industrial urbanism abhors isolation. Its nature is to expand. In many respects it is a market-finding process, which means that it is naturally occupied with finding or making transportation routes, and with gaining access to those markets. Urbanism's effort to expand stimulates people to move from one place to another, generally from smaller to larger places. This stimulation to move is of a kind that does not permit rest, and the individual, once having moved, may be stimulated to move again. That is one type of mobility which must be associated with urbanism. There are others.

Work must necessarily also change in response to the creation of new tools and machines and the making of new products. Hoselitz called attention to work changes that follow the migration of natives from the tribal villages to industry where they must use tools and perform tasks previously strange to them. Each of these migrants had his fixed place in the village, which was usually determined by some combination of family status and seniority. These criteria for status have no place in the factory where villagers meet under disturbing conditions.

> Older men or chiefs or other persons with high traditional status may be subordinate in the factory to younger men, and this subordination may express itself not merely in a lower wage but in actual positions of inferiority in the factory hierarchy of a work crew. The disruptive effects not only on traditional structures but also on the psychological security of the persons involved, are obvious if a man with high traditional status must obey orders of someone who in the traditional ranking is far below him simply because the new distribution and valuation of skills makes the new relation mandatory.[20]

When people move physically from one situation to another and change from one work to another, professional or occupational mobility may result, or one may remain in the same locality and still move from one occupational or professional level to another. Or, this professional mobility may take place between generations—the father may be a laborer, while his son may become a skilled worker and his daughter a private secretary. Not only is the urban way of life conducive to physical and professional mobility, it is also tolerant of mobility from one social class to another.

[20] Hoselitz, *op. cit.*, p. 182.

Such changes often take place under urbanism, and the urban man does not regard them with wonder. These types of mobility may be found in American rural society, but to a lesser degree. They are even less evident in Old World rural society, which is more tradition bound than rural society in the United States and less under the influence of urbanism. Such changes, however, would hardly be found in the tribal society, even within walking distance of a factory.

## The Two Sociologies

There are two community sociologies, rural and urban, and each is prominently recognized by that discipline in the United States. The field of rural sociology is rural society and rural living while that of urban sociology concerns society and living in towns and cities. The pages of this opening chapter could be interpreted as a challenge to the existence of two separate sociologies, or as the basis for a single community sociology which is at once both rural and urban.

Much that has been called rural sociology in the past was either a plea for the old basic values of rural life or was on guard against the "evils" of urbanism. As Caplow writes, "Much of what passes for sociological literature consists of sermons of protest against the uprooting of old beliefs and the emancipation of the secularized urban dweller from the tight control of the primordial village community in which most of the men who ever lived have passed their lives."[21] Observations of this type do not apply to either the literature of rural or urban sociology today, but these sociologies have other limitations.

One of these limitations is the fact that many rural sociologists stand with their backs to the city. The reverse is also true—many urban sociologists feel somewhat less self-conscious if their work keeps them out of the hinterland. Each avoids the possible embarrassment of being caught in the territory of the other. The fact is that neither knows exactly where his territory ends and that of the other begins. The time has come, it seems, when we must recognize that it really makes no great difference where the line between rural and urban is drawn. In fact, it might be just as well if we stopped thinking about the dividing line. Rural and urban tend to take on more meaning if

[21] Theodore Caplow, *The Sociology of Work*, Minneapolis, University of Minnesota Press, 1954, p. 284.

seen in terms of social organization and interpenetrating social processes, even in relation to a division of labor pattern.

The approach in this book is essentially that of urban sociology, but it is urban sociology that draws heavily on rural as well as urban data. It must be recognized that the urban is not only dominant but is becoming increasingly so as this way of life becomes global in its reach.

Such trends do not render obsolete the need for rural sociology. On the contrary, rural work and rural living are vital to society. The people who do this work in the United States are no less rural even though they become more urban in their way of life, nor are they less a special segment of society even though they come increasingly to be regarded as a specialized occupational group. While this segment of the American population is percentagewise small, it has its social needs and its social problems just as the nonrural people have theirs. The difference is in the spheres, one of which is primarily the field of rural sociology and the other field of urban sociology. Each penetrates the area of the other but does not obviate the work of the other.

## Summary

Urbanism as a way of life is primarily, although not entirely, associated with living in towns and cities. It differs from ruralism mainly in ways of work, habits of thought, and with respect to traditional controls. The tempo of life and work is faster under urbanism. Urbanism is the more dynamic of the two, while ruralism is the more self-isolating. Urbanism is more given to creativeness and to extending its influence outward. Ruralism is the more passive and negative, although the use of such terms do not involve negative implications.

We have distinguished between those characteristics of urbanism which concern living in large cities and those which relate to urbanism as an outward-extending, change-stimulating influence emanating from urban centers. In the city, urbanism has been described as superficial, anonymous, and transient. But this way of life as it extends outward may not be anonymous, although it tends to be contractual and secondary. The superficiality, so much a part of urbanism in the city and outside, is not always superficial, for urbanism is also profoundly creative. Transiency is a mark of urbanism wherever its influence extends, and transiency tends to increase with nearness to urban centers.

We noted that different elements of urbanism may be present in different degree, depending on the relative mixture of ruralism and urbanism in particular areas. We took notice of the term "urbanization," commonly used to describe the moving of people from less urban to more urban places and from nonurban to urban types of work. We recognized that urbanization must also mean the changeover from rural to urban ways of living, and that this may take place without migration to the city. The influences that affect people and promote urbanization may be exported by the city. Wherever people live in the hinterland, they may be described as more "rural-oriented" or more "urban-oriented"—that is, urbanized to a lesser or greater degree.

We have noted that, within limits, cities can grow and be important without being industrial. Yet, urbanism as we know it has been made possible by the expansion of industry and commerce, the second being largely an auxiliary activity of the first. Just as industry must be urban-centered, so it must extend outward for markets and raw materials. It disturbs old work ways, old consuming habits, and established social relationships. It requires new and ever-changing divisions of labor. It encourages specialization, but it also reduces mass work to common levels. It raises levels of living.

Urbanism tends to "individuate" people, liberating them, putting them in a position favorable to making choices and being creative according to their ability. The individual and not the family is the unit in the labor market and before the law. But urbanism also imposes uniformities on great masses of people over wider areas. Uniformity applies to machines, tools, clothing, houses, means of transport and communication, commercialized leisure activities, and so on. These mass uniformities are continually changing. As they change they effect changes in ways of thinking, behavior, and relationships, giving rise to problems of social change.

Urbanism as a way of life encourages mobility, and could not be possible without increasing mobility. This is not limited to the moving of people from place to place; it also means mobility from one kind of work to another and the movement of people from one social class to another. Mobility-connected social changes may also give rise to social problems, yet mobility is essential to urbanism.

The approach to the understanding of the urbanized community which this book undertakes may be described as urban sociology. It is not an encroachment on rural sociology. Even though the rural way of life changes to become more urban-oriented, *rural* as a way of work and life remains and that means a continuing need for rural sociology.

CHAPTER 2

# The Nature of the Community

WITHOUT DEFINING THE TERM, we considered the community in the previous chapter. As we shall see presently, a localized society, whether urban- or rural-oriented, is a community. While our interest is primarily in the cosmopolitan community, we need a term that will include the most urban as well as the least urban segments of society, whether found in a more advanced or a less advanced country. Since "community" serves so many uses for many different people, we must try to make clear what it means in this book.

## Uses of the Term

Sometimes the social scientist avoids a term if it is very much used in common conversation. It tends to be "loaded" with different meanings, and that is true of "community," as of other useful words such as "group," "society," "country," "culture," and many more. "Community" is a term which, if not used, is difficult to set aside; a substitute is hard to find.

The term "community" has been appearing in recent years with a variety of applications. For example, the North Atlantic Community is the group of countries joined together in the North Atlantic Treaty Organization (NATO). Similarly, Belgium, France, Germany, Holland, Italy, and Luxembourg together comprise the European Coal and Steel Community.

The term is also used often to indicate "community of interest," which relates to some matter of special concern in terms of which groups or formal organizations may find a basis for cooperation. Thus, several states have a community of interest with respect to the Tennessee Valley Authority. A group of villages may find a community of interest when a common problem arises that they all must recognize and act upon. A community of interest may exist without any formal understanding, as when families having cottages along a stream refrain from littering the stream banks with rubbish.

Americans use the term in a variety of special ways. A "community center" is a place, often in a residential area, where groups of people have their meetings or social gatherings, or where children come to play and where young people may hold their dances. It may be privately owned and supported by groups in the vicinity, or it may be a public property. "Community organization" is a term used by recreational leaders, welfare workers, and others to describe local programs for organizing and leading community life. The "community chest" is a relatively recent creation, where social agencies join together in a single fund-raising organization rather than conducting individual drives to secure funds. It is not found in other countries.

We need only to add that the term has still other recognized uses. The natural scientists, for example, speak of a "plant community," meaning a common habitat occupied by a variety of plants. In a similar sense an area occupied by animal species of different kinds, down to insects, may also be called a community.

## Multiplicity of Definitions

Hillery undertook a collection of definitions of community and was able to bring together 94 of them. Some had special application to urban and others to rural communities. As might be expected he found that those descriptive of rural communities were most in agreement.[1] Aside from other values of this endeavor, the isolation of so many definitions indicates what an important place the concept has in American sociology.

René Koenig undertook also to find the common elements in a variety of definitions of community. His comparison included writers in France, Germany, Italy, Switzerland, the United Kingdom, and the United States. He proposed this composite definition: "A community is above all a global society of a kind that has a local unity, with an indefinite number of institutions, social groups and other phenomena within it, and besides a great variety of association forms which operate within the mentioned groupings, and also there are the obviously essential outside organizational contacts (social, economic, legal,

[1] George A. Hillery, "Definitions of Community, Areas of Agreement," *Rural Sociology*, Vol. 20, No. 2, June 1955, pp. 111-123.

administrative, etc.)."[2] This defines community in terms of complex relationships within a variety of social organizations, but this definition might also be applied to a society, or to any self-sufficient aggregate of people.

Koenig would accept the specification that community must be identified with place, a location in which the individual maintains certain fairly habitual relationships. To Parsons, it is the base of operations for a group. In his words, "A community is that collectivity the members of which share a common territorial area as their base of operations for daily activities."[3] Another observer might argue, however, that it is important to think of community as a place where a collectivity shares common experience, where the interests of people are localized.

For example, Suranyi-Unger thinks of the community as an economic creation. Before there can be a community there must be some economic reason for people being there, or wanting to be there. People must have common wants which are satisfied by residence in that particular place, and "all communal ties and activities are only justified if and so far as they arise from corresponding common wants. The communities grow with the growth of the underlying common wants, and they decay with the decadence of these wants, which are virtually the motive force of the communal development."[4]

The community, in short, may be thought of as a global social unity in which exist various types of social organization; it is also a location, and it is also a place where people find the means to live. It is a place not only of economic activity and of human association, but it is also a place where memories are centered, both individual and "folk" memories. Moreover, the community has the quality of duration, representing an accumulation of group experiences which comes out of the past and extends through time, even though the individuals making up the community are forever coming and going.

## Ambiguities of the Term

One very often quoted definition of community is that offered by MacIver:

> Any circle of people who live together, who belong together, so that they share, not this or that particular interest, but a whole set of interests wide enough

[2] René Koenig, "Die Gemeinde in Blickfeld der Sociologie," *Handbuch der kommunalen Wissenschaften und Praxis,* Berlin, Springer, 1956, p. 23.

[3] Talcott Parsons, *The Social System,* Glencoe (Illinois), The Free Press, 1951, p. 91.

[4] Theo Suranyi-Unger, "Individual and Collective Wants," *Journal of Political Economy,* Vol. 56, February 1948, p. 18.

> and complete enough to include their lives, is a community. Thus, we may designate as a community a tribe, a village, a pioneer settlement, a city or a nation. The mark of a community is that one's life *may* be lived wholly within it.[5]

This definition recognizes variety in types of communities, ranging from a primitive face-to-face village to a cosmopolitan city. Community is not static, but is subject to change, like the social organizations within it. In addition to being dynamic and changing, and in addition to having different aspects, as the term is applied to different sorts of human aggregates, community may have different meanings for different people. In other words, the nature and extent of one's community is largely a matter of individual definition.

The youth who goes from the village to the city and who later returns to the village not only moves from one type of community to another, but, on returning home, finds that the village becomes a different community to him than it was before his migration. His associations after returning to the village may be wider and his work relations may be different because his horizons have been widened. Members of his family may not have changed and they may not be in sympathy with the changes in him.

How individual the idea of community may be can be illustrated by assuming the case of a big-city banker. He might be a Chicagoan, for example. His home is in a North Shore suburb, but he keeps an apartment in the city. In winter he usually goes to a cottage in Florida. In summer he spends a few weeks at a lodge in the Canadian woods. But the suburb is his home and he calls that his community. His wife is prominent in Chicago club life and is a social leader in the North Shore community. His daughter is a debutante with a social life of her own. His son is a sportsman who moves in another circle. It may be said that these four members of the same family live in four different worlds, but each tends to center in the family home in the North Shore suburb. Each of these four persons lives in an individual type of community which differs from the others as much as all of them differ from the community of the family cook, the chauffeur, the gardener, or the secretary in the banker's office. Each belongs to the greater Chicago community, but each belongs within a special frame of reference.

An American who meets another American in Paris may say with some pride (especially if the other American comes from a small inland town), "I am a New Yorker." But the same American when in New York may

[5] Robert M. MacIver, *Society, Its Structure and Changes*, New York, Richard R. Smith, 1932, pp. 9-10.

explain that he does not live in the city although he has his work there. He lives in Suburb X on Long Island, just outside the city limits. He may say, "I wouldn't live in this town if you gave it to me." Actually, as a suburbanite he is an anti-New Yorker. But Suburb X is a city of more than 20,000 and, much to his dislike, is growing. However, he still takes pride in saying, "I live at the East End," and he complains about how "the riffraff is crowding in at the West End." This man's definition of his community changes as he uses the term in one frame of reference or another, and who can say that these different uses are not proper ones, even though they seem to contradict one another?

The farmer, the rancher, or the fisherman, on the other hand, may not be confronted with such complexities with respect to identifying or defining his community. The geographical borders are more clearly known and so is the approximate center, and for this reason we may expect to find that definitions of rural communities are in much more agreement.

## Validity of the Term

We would encounter similar difficulties with respect to the word "society," which also has different applications depending on the frame of reference, but we do not put the word aside. A community is a unit of society in place. In some of its uses, community refers to a primary face-to-face collectivity, and may be called a neighborhood, a term to be considered later. Or, the community may be a secondary collectivity made up of a mixture of people, a mixture of work activities, and a mixture of formal and informal groups, with a good share of its life anonymous.

Community in the secondary sense may have a formal or civic identity which is quite separate from the many types of groups in terms of which its life is organized. In other words, we may think of Denver or Des Moines as aggregates of people living in habitual relationships, or we may visualize either place in terms of layout and structures. But each of these communities is also a *corporate body,* a responsible civic organization which operates quite impersonally in the name of all the people. Thus, community in one use may have reference to an informal organization—a community of scattered farmers, for example; or it may have reference to a formal organization.

## The Roots of the Community

Before leaving for the moment the matter of definition, a word must be said about the antiquity of the community. This is perhaps not so important in the United States where so many communities, even large ones, have very little history behind them, compared with villages, towns, and cities in the Old World. The modern communities known to our generation are the heirs of many centuries of agglomerate living, and it is hard to think of any other way for humans to live. Keith makes this observation:

> There is nothing new in postulating that early mankind was divided into an exceeding large number of small communities; what is new is that this mosaic of humanity endured throughout the entire period of man's major evolution and provided the most favorable circumstances for bringing about rapid changes in brain and body.[6]

It is suggested that because man lived in communities he was better able to bring about social organization and to evolve culture. This resulted in various types of interdependence and division of labor which were impelled by survival needs. Because it was needed for such living together, language came into use. Cole sees the evolution from one stage of community living to another as being, in large part, a result of various basic inventions. The use of animals and metals led to use of the plow, man made use of the winds to sail his boats, and, further:

> Crafts and trades brought people together in large numbers. Media of exchange were developed to do away with the necessity of barter. Many craftsmen did not provide their own food, but became dependent on the rural hinterland for food. Artisans of special skills had to be trained to build the towns, and special occupational groups had to be trained to operate and maintain them.[7]

Cole calls attention to technological and other changes which accompanied the Copper Age and the Bronze Age, and which attended the introduction of iron. Inventions certainly have had their influence on the evolution of community life, but community life, in turn, stimulated inventions. In this process community life was evolving and people were learning how to live more effectively in larger agglomerates. Thus community living was in a stage-to-stage development over many centuries and has moved from relatively slow to relatively rapid rates of change. Inventiveness has increased at a corresponding rate, but the acceleration of change has without doubt stimulated that inventiveness. The centers of such stimulation were the towns and later the cities.

[6] Arthur Keith, *A Theory of Human Evolution,* New York, Philosophical Library, 1949, p. 169. William G. Sumner observed that the beginnings of community life are "lost in mystery" and it is only by analysis and inference that we can learn about origins, *Folkways,* New York, Ginn, 1940, p. 310.

[7] William E. Cole, *Urban Society,* Boston, Houghton Mifflin, 1958, pp. 14-15. Cole cites in this connection Gordan V. Childe, *Man Makes Himself,* New York, New American Library of World Literature, 1951.

Much of this community evolution took place before such new countries as the United States were known; but in the evolution of community life, particularly during the past century, the United States has participated fully. Actually, much that is called progress in the structure, the work, and the life of urban communities is of relatively recent creation, but even the most modern of urban places shares in a long history of community evolution. There is, therefore, a certain kinship among all urban communities.

## Community and Neighborhood

"Neighborhood" is often used synonymously with "community," and not improperly when the term is descriptive of small compact places. Wurzbacher and Pflaum made a study of a German *Gemeinde,* a political subdivision about the size of an American county. In it there were 25 villages and hamlets, most of which they called neighborhoods, and the villages included two or more neighborhoods. They found that during three generations the character of these neighborhoods had changed greatly, although in size they remained about the same. Formerly they were more "closed"—that is, they were seldom visited by outsiders and the inhabitants did not go often to other places. On the other hand, each house was more "open" than today, in the sense of neighbors visiting back and forth. Today these places are much more in contact with other localities—more strangers come and go, there are branches of outside organizations, and there is contact with the outside by papers, radio, movies, bus lines, etc. The neighborhoods are now said to be "open," while the homes are more "closed" than formerly. The neighborhoods still exist but the neighborly lack of privacy has been replaced by more discrete relationships and restraint. These places have come into the influence area of industrial urban life.[8]

West, in describing *Plainville, U.S.A.,* noted that people in that locality talked much of their neighbors and distinguished between "real" neighbors and nearby families. Thus "my neighborhood" conveyed a selective meaning; one chooses to designate who his neighbors are, and he may change his mind about some of them and later change his mind again. West reported that the party telephone line which enables scattered families to be very much in contact, often with several housewives talking at the same time, may create a type of neighborhood. West concluded that "the important function of neigh-

[8] Gerhard Wurzbacher and Renate Pflaum, *Das Dorf im Spannungsfeld industrieller Entwicklung,* Stuttgart, F. Enke, 1954, Chapt. 5.

borhoods is to provide neighbors." As with the German villages described by Wurzbacher and Pflaum, West found that *Plainville* included a number of neighborhoods which differed from one another in important respects.[9]

West doubtless would not argue that a number of families living in close proximity is any less a neighborhood because all the families are not on speaking terms (perhaps good evidence that the area is a neighborhood). The test of neighborhood is that the neighbors know one another, or that they be mutually aware of each other's presence and behave accordingly. As an area of mutual acquaintance, the neighborhood is small. If the aggregate breaks into two or more neighborhoods, then it would seem that the name community must be given to the larger group. But an isolated hamlet of a few families may be at the same time a neighborhood and a community.

The neighborhood has always been a primary social control area which, as Dawson and Gettys say, sets the "standards of behavior which are expected," and exerts pressures "upon those who deviate from these norms."[10] In connection with the control function, the neighborhood may be very rural at one extreme or very urban at the other. Most people in human history have lived in such neighborhoods, and do so now. But in the urban community certain areas that might be called neighborhoods do not have the primary group character, or it is found there only to a limited degree. Such areas, however, may still be called neighborhoods. Let us, therefore, consider the two types: neighborhoods of primary participation and neighborhoods of secondary participation.

## *Neighborhoods of Primary Participation*

Natives from the villages who come to new African towns to work in the industries tend to settle in prepared housing areas, called compounds. These people have no other conception of community life than that of the primary contact neighborhood of the village, and they behave accordingly when they enter the town:

> The fact that the compounds are wide open to neighbors (even when fenced off), that the dwellings are made up of several houses between which people can come and go, and that owing to the influence of the climate and of tradition, people live much out of doors, imparts a rural air to the streets and street life. People see what their neighbors are doing, they talk to each other from compound to compound, they are of necessity in unbroken contact.

[9] James West, *Plainville, U.S.A.*, New York, Columbia University Press, 1945, p. 71.
[10] Carl A. Dawson and Warner E. Gettys, *An Introduction to Sociology*, New York, Ronald Press, 1948, p. 17.

> From this it follows that every newcomer is spotted, observed, catalogued, located with reference to his tribe, the village he came from, the relatives or neo-brothers who are putting him up, his trade, etc., by the neighbors who converse with each other in their compounds, in the street, at the bicycle repair shop, at the tobacconist's or fishmonger's, in the bar, etc. According to the number of ties discovered between him and this or that group, the new arrival will be more or less quietly assimilated; at least he will be involved in neighborhood relationships.[11]

The people described here by Clément are in process of being urbanized. They have already moved from nonurban work and they live part of their time in relation to an incipient type of urbanism. In their habitations, however, they have recreated the village type of primary group life, although some of them later escape to live in other types of residential areas where neighborhood relations are less conspicuously present. Immigrants in American cities have established such concentrated culture areas and try to prolong the relationships known to them in their Old World villages.

However, one must belong to that kind of primary group neighborhood, and one must be accepted. A case to the contrary is that of a writer who moved from New York City to a hill village in western Massachusetts because he liked "the direct simplicity" of village life. He had his books and his work much as in the city. He spent hours over his typewriter. He read books and wrote manuscripts which he sent by post to the city. When not doing this he chopped wood or dug in the garden, or he engaged neighbors in conversation. Nevertheless, he could not quite become one of them. If they came to his house they were clearly as self-conscious as when he went into their houses. He was highly respected and the village was proud to have him there, but he was not an accepted neighbor in the same sense that the villagers were neighbors to one another.

## *Neighborhoods of Secondary Participation*

The more urbanized a community, especially a community that is growing, the more mobile its population tends to become. Although the neighborhoods of such a community may seem to be reasonably stable, some people are moving in and others are moving out. Some are moving up and others down in the social scale. Neighborhood acceptance in such a situation, as Park

[11] Pierre Clément, "Social Patterns of Urban Life," in *Social Implications of Industrialization and Urbanization in Africa South of the Sahara,* Paris, UNESCO, 1956, p. 375.

described it, becomes selective and relationships become more flexible.[12] Some people in any urban neighborhood may *suffer* isolation and are not accepted as neighbors.[13] Others *seek* isolation regardless of the neighborhood in which they live; thus even in a quiet suburb there are people who have no desire for neighboring and like to be left to themselves. Others have a passion to be good neighbors, and to those who like isolation such people can be a nuisance.

There seems to be a tendency in urbanized communities for people to think of themselves more as *belonging* to neighborhoods and less as *being* neighbors. Thus one may feel very satisfied with his neighborhood and yet know (or want to know) few people there. This is one of the manifestations of urban secondary life and anonymity. It is hardly possible for a person in the rural-oriented neighborhood to live there in the anonymous way he can live in an urban neighborhood.

An urbanite may think of his neighborhood as being intersected by X Street and Y Avenue where he has lived in the same apartment building for several years. He has a "nodding acquaintance" with a number of other tenants in the same building, but he has no close friends there. He knows by first name only the elevator operator and the girl at the desk. He knows who the janitor is but has rarely spoken to him. He knows (first names only or last names only) the clerks in the grocery store, the meat shop, the bakery, and the flower shop, and the tailor who also does cleaning, and the newspaper dealer at the corner. For the most part, these are the people he talks to in what he calls his neighborhood. Most of his habits of going and coming are timed and habituated to this one locality. He has many friends in the city but they are widely scattered. They know his address and his telephone number. Most of them have been in his apartment, some of them often.

This is an extreme example, but it is not unique, of the anonymous urban neighborhood where the measure of being a good neighbor is that one "minds one's own business." This would be the mark of an unfriendly or antisocial neighbor in the traditional neighborhood in a rural community. But our urbanite may be a very social person. He has more good friends than he has time to see during the year, but he sees them under circumstances that are

[12] Robert E. Park and others, *The City*, Chicago, University of Chicago Press, 1925, pp. 6-7.

[13] During the Great Depression in the United States many argued that the problem of relief was a neighborhood problem. Bakke made a study of unemployment in a Connecticut city and found little evidence of neighborhoods accepting responsibility for jobless neighbors. Rather, the jobless worker tended to lose status and to avoid the neighbors. They would "hold more to themselves in unemployment than otherwise." E. Wight Bakke, *Citizens without Work*, New Haven, Yale University Press, 1940, p. 7.

related to a variety of special interests. Some he meets in a downtown restaurant or at a bar in his neighborhood.[14] There is no lack of social activity for either him or his wife; it is, however, selective social activity.

## Networks of Acquaintance and Friendship

At the conclusion of a study of neighborhoods in an urban community, Riemer expressed the view that there is a need for a new conception of neighborhood. Associating neighborhood with residential proximity does not fit the geographical realities of urban friendship and acquaintance: "Our thinking about social relationships in the city has been dominated by the spatial dimension." We have seen how town planners and community organizers have tried in vain to impose the old traditional concept of neighborhood on urban people. Thus, says Riemer, "A small town culture trait has been superimposed upon the urban environment." This works appropriately in the small town and village, but it does not work in the city: "In the city man has gained the freedom of making social contacts with little regard to geographical distance."[15]

Proximity of residence may have some part in the formation of urban acquaintance among children or among parents who have small children. Urbanites living in the same apartment block may strike up an acquaintance merely by walking their dogs in the same park. But proximity of residence may not lead to a great degree of acquaintance and friendship for urban adults. They are more likely to come together in connection with their work or leisure activities. Or, friendships may hold over from school days, or result from chance contacts, such as a meeting on a train.

London families, according to Bott, have acquaintance and friendship networks. She made a long-term study of twenty "ordinary" families ranging from working class to upper middle class. Each of these families had its scattering of acquaintances, friends, and relatives, who constituted contact points with the anonymous millions in the city. These she described as net-

[14] Much has been written about the "pub" in the English community, a place where people meet to drink their beer. The usual "pub" has a clientele of the same people meeting about the same time almost every day. One habitue recalls a "pub" in the 1890's. "Cronies as we were none of us knew anything of one another's lives. We knew one another only by nick-names and where each went after leaving for the night was known only to the individual. We had no curiosity in the matter." Quoted by Roger Fulford, "Jubilee London," article in a symposium by Ernest Barker and others, *Golden Ages of the Great Cities,* London, Thames and Hudson, 1952, p. 287.

[15] Svend Riemer, "Villagers in Metropolis," *British Journal of Sociology,* Vol. II, 1951, p. 40.

works. When a couple gets married, it means the joining of their networks and, while the married pair may continue to have a joint network, each partner will have a network somewhat different from the other. Some persons in the individual's network may be socially closer than others. Some networks include many points, while others may include only a few. Some persons in an individual's network may be there because of work interests, which may also become social interests. If the urban couple invites a number of persons for an evening party they select from their respective networks the persons who will be congenial to one another. For another evening the selection might include an entirely different list of persons.

Bott found that a worker family living many years in a working-class area had a somewhat local network, and the network was also "close-knit." Persons known to the family were also known to one another. The network is neighborhood-like. Lines radiating from the family to its contacts (individuals or families) may be indicated by points *A*, *B*, *C*, *D*, and so on, but these points (*A-B*, *A-C*, *B-D*, *A-C-D*, and so on) may also be connected. The network of another family in another urban area may be "loose-knit" with little network-connectedness between the persons known to the family. The more close-knit the network of the family or individual (the more contact between *A*, *B*, etc.), the more it becomes a primary group social control relationship. The individual or family with a loose-knit network may be directly influenced by the persons in the network but, since the persons in this network seldom meet, they do not constitute a concerted influence on the individual or family.[16]

The two types of networks may be illustrated thus:

1. Loose-knit network

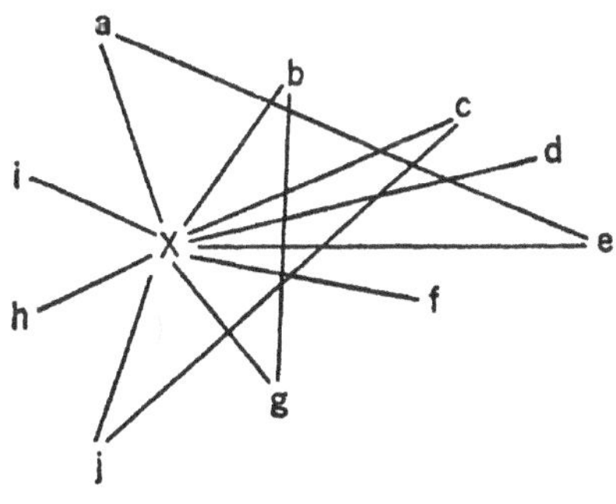


2. Close-knit network

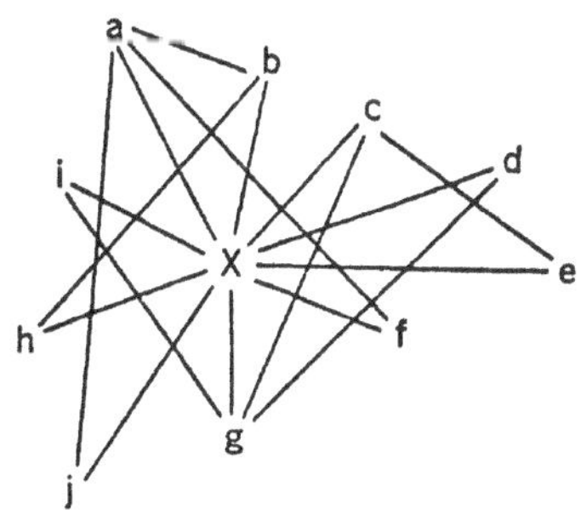


[16] Elizabeth Bott, *Family and Social Network*, London, Tavistock Publications, 1957, primarily Chapter III. In her report Dr. Bott does not attempt such a general application of the network idea as is used here. She applied the idea only to the twenty families.

*X* is the individual or family. The other letters stand for other families or individuals of contact. The number of such contacts may be many or few. The control implications of the close-knit network are clear. It should be clear, too, that as the individual is the center of his own network, he is also a point of contact in the networks of others, perhaps many others with different degrees of social distance. It is easily recognized that networks exist in the most intimate neighborhoods as well as in the most anonymous ones. The social relationships portrayed by the two network diagrams are not new, but the diagram enables us to see "loose-knit" or "close-knit" networks in comparable terms. The network is an abstraction that can be used with or without the spatial dimension so necessary to the concept of neighborhood.

## Urbanism and Community Integration

Our primary task is to understand the nature of the community in its most complex form, not in the little places where small aggregates of people live their lives in full view of one another, but in large aggregates where great numbers are visible. The setting in this modern community has been aptly described by Hauser:

> One of the most striking aspects of the modern era has been increasing urbanization, that is, the increasing proportion of the world's population residing in cities. This phenomenon is the physical manifestation of the urban way of life which is so characteristic of today, and of the many new types of problems incident to what has become mass society. Moreover, the urban mode of life extends far beyond the boundaries of cities and even beyond those of metropolitan areas.[17]

Man is learning to live in mass society, and this way of life, regardless of the numbers in the urban agglomerations, produces very different patterns of order. People tend to become integrated into webs of relationships through which they are able to function as individuals and as groups with a minimum of friction. Whatever the elements are that enable great agglomerations of people, most of them strangers to one another, to find satisfaction in urban living, these are the elements that render the urban community a reality.

Let us look at an urban agglomerate in which certain of these integrative elements seem to be lacking—the city of Jinja, a place of less than 10,000 in Uganda, Africa. Here, as reported by Rhona Sofer, the population is 45

[17] Philip M. Hauser, "World Urbanism," Editorial, *American Journal of Sociology*, Vol. 6, No. 5, March 1955, p. 427.

percent Asian, 53 percent African, and 2 percent European. Of the Europeans, most of them British, a few have brought their families. Almost all are in Jinja on working contracts, and when these are completed they will leave. Almost none have any interest in Jinja and almost none would consider making a home there. Yet these Europeans hold most of the power, governing Jinja according to European conceptions. They possess most of the technological skill for managing Jinja's industry. The Asians, mostly Indians, either work for the Europeans or are engaged in business. They are socially excluded by the Europeans and feel themselves above the Africans. As for the Africans, they are divided into tribal groups, and many of these have only a transient interest in Jinja, their loyalties being to the villages from whence they came. There are types of social groups and associations, but few of these are oriented to life and problems in Jinja. Some of these groups are internally integrated. Sofer concluded:

> Jinja is not a European town, nor an Asian town, nor an African town. It is an urban area whose characteristics are largely a function of an inter-racial structure. The parts played by the different racial categories in producing the present situation are important for the immigrants mainly in so far as they facilitate or impede their participation and their formation into coherent groups in the existing and evolving social structure.[18]

Without doubt Jinja is in process of becoming an urban community. What seems to be needed for theoretical integration, at least, is more people with permanent social and economic interests, more shared experience on the part of the inhabitants, more rivalry and competition for advantages that can be enjoyed in Jinja, and a wider sharing of the responsibility of citizenship; in short, more participation by more people in the common life. It does not matter that the people are of different races and social classes if the individual, whoever he may be, is able to make a place for himself there and to have a part in the common life. The community takes form as the people there "take root." This idea was expressed by Rosencrance, who was not thinking of a new city in Africa, but industrial communities in the United States:

> Therefore, it is proposed to examine the effect of our industrial civilization: 1) on the individual's sense of community in various kinds and sizes of community structural organizations and 2) in specific interest groups found therein. A sense of community involves not only an individual's feeling of identification, but a feeling of being respected by others, a willingness to adjust personal pur-

[18] From summary of a study in Uganda by Rhona Sofer, "Social Survey of Jinja," in *Social Implications of Industrialization and Urbanization in Africa South of the Sahara*, Paris, UNESCO, 1956, p. 616.

poses to desirable group purposes, sharing and decision-making, participation in group endeavors, and a commitment to work for common ends.[19]

The term "consciousness of community," used elsewhere by Rosencrance, is one that will be returned to in later chapters. In its most elementary application it concerns the feeling the individual has of belonging to a place (even though temporarily) as he occupies himself there with making his living and gaining some degree of status. Whether he works for his own interests alone or also tries to serve others, even though he is no more than a unit in the mass, if he is integrated in the work and life about him he makes his contribution to that kind of community life we call urbanism.

## The Community in Practical Terms

In sociopsychological terms, a community exists in the minds of people living in an identifiable place. If all the people were to move away, however, the community would disappear, except as the record of their way of life might be later envisioned from the artifacts and structures left behind. In our time it is quite unlikely that such a total evacuation could take place, but to imagine such an eventuality serves to remind us that the modern community, especially the town or city, exists in terms other than the sociopsychological. It is also an impersonal civic organization and a network of physical structures. These we need to consider briefly.

To the political scientist the civic community is an area of public administration in which a large variety of public services are performed. If it is a large city it may be divided into wards or districts, each being a type of administrative unit. Whether or not the subadministration districts exist, the community is a political subdivision which can be seen on a map. Moreover, the entire community—its geographical form, its streets, and various locations—also exist on maps with all the limits and boundaries specifically indicated. Within this area civil government is empowered to exercise certain types of authority: the health service, the regulation of traffic, fire protection, public education, water supply, sewage, etc. In such terms the community as a civil corporation is very real. As such it has its records and its history, its folkways and its laws.

But the community is also a visible reality of buildings, streets, parks,

[19] Francis C. Rosencrance, contribution to discussion on nature of the community in symposium edited by Eugene Staley, *Creating an Industrial Civilization*, New York, Harper & Brothers, 1952, p. 284.

and other properties. Some of these may be private properties: houses, stores, office buildings, factories, and institutions. Others are the public properties: the streets, bridges, and tunnels; the network of drains and sewers; the water-supply system; the many kinds of public buildings, the monuments, the parks, sport places, and so on.

These visible properties of the community, all of which tell something about it, are interrelated into a single structural pattern. They are the operating facilities, the "plant" of the community. Many of them, like the great office building or the modern apartment block, are complex mechanisms. There tends to be, in fact, a technical or mechanical relationship in the operation of most of these structural properties. For the urban man, part of being oriented to his community involves adjustment to the many mechanisms in its structural equipment.

Any effort to define and understand the modern community must, in short, take account of the visible equipment, what one sees and orients himself to as he goes about the place. We may say, however, that while the "real" community exists in the minds and memories of people, the basis for the community in the mind is the community "on the ground." The latter community is an arrangement of places and facilities whose locations and utilities become incorporated in the habits of community living of the people, as the parks become common facilities and the monuments become objects of common pride. These are the elements of the community as a physical phenomenon in relation to which the human aggregate integrates itself into a type of functioning unity.

## Complexity of Communities

It is often said, and with good reason, that the way of life of the very urban community is more complex than the way of life in a rural place. But such a generalization is not without hazard. Students of preliterate peoples have called our attention to the remarkable complexity of simple societies. Benedict, for example, reported the rules of living in the matriarchal Dobu community where the outer lines for each habitation area are strictly drawn and the rules for cross-border relations piously respected. The mother-dominated family groups are protected against one another and linked with one another by codes which are precise in their application. How complicated a very simple activity may be is illustrated in the growing of yams. From planting to eating the process is involved in a network of incantation and folklore. Seed yams descending through the husband's family must never be mixed with those descending through the wife's family. "Yams are conceived of as persons and are believed to wander nightly from garden to garden. The vines remain behind but the tubers are gone. Toward the middle of the morning they normally return."[20] Thus yams must not be dug too early in the day, and they resent being harvested too soon. All aspects of Dobu life are surrounded by traditional taboos or equally strict requirements which must be correctly learned and precisely obeyed.

Loomis and Beegle made an extensive comparative study of various types of rural community life, designated by them as rural social systems. In the study they compare the different American rural social systems with certain Old World systems. While the subject matter is not presented in terms of relative social complexity, it is clear that the rural community, more than the urban community can be, is oriented to the farm family, and the farm family is a unit for work and living. "The fact that the life and occupation of all members of a family are inextricably intermingled so that all members consider themselves a part of a going concern affects the thinking and attitudes of rural people as much as or more than any other feature."[21] As they describe each rural social system, it becomes evident that each in its way is complex and each sets its requirements for learning to do the right things in the right way. They also recognize that American rural families tend to come increasingly under urban influence.

Ways of work and living in the more rural communities have a season-to-season continuity, much as the tribal ways of work and living have a generation-to-generation continuity. One learns what he needs to know and it is not necessary to unlearn much of it or to learn new ways and to acquire new attitudes. The American rural community, however, tends to lose much of that former isolation as it is encroached upon by urban ways of work and living.

To say that the rural or the primitive community's ways of work and life are less complex than the urban ways calls for additional explanation, for each type is complex in its own way. The complexities in the nonurban com-

[20] Ruth Benedict, *Patterns of Culture,* Boston, Houghton Mifflin, 1934, p. 146.

[21] Charles P. Loomis and J. Allan Beegle, *Rural Social Systems,* Englewood Cliffs, N. J., Prentice-Hall, 1950, p. 32. This book introduces the *Gemeinschaft* and *Gesellschaft* concepts of Ferdinand Toennies to American rural sociology. *Gemeinschaft* is presented as the familistic rural type of community life, while *Gesellschaft* is seen as characteristic of the contractual type of life of the urban community. Each represents a particular kind of complexity of social organization and social relationships.

munities are more traditional and rendered sacred in many respects by folk beliefs. The urban complexity is less traditional and more oriented to rational thinking, to mechanisms, and to the acceptance of technological change. It takes no longer for a person to learn the complexities of the urban community than those of the rural or the tribal community. The difference is that in the urban community the learning process never ends. Change is ever present.

## Functions of the Community

To Bardet, the community comes into existence through the "fusion des consciences et des activités" and it finds its unity through a sort of internal consensus. It has a material base which is somewhat independent of the volition of the inhabitants. This base, like the social agglomerate, is rooted in history and geography. Above all, the community is an assemblage of organizations which have their purposes and relationships and which perform diverse functions.[22] This thought about the functioning of organizations and groups in the community is pertinent to any effort to understand the meaning of community life.

Whether the community is small and rural or large and urban, its existence is explained only if we understand the functions essential to community living which are performed there. It does not matter whether these functions are performed by public bodies or agencies, by private groups or organizations, or by some combination of the two. They may be performed differently in different communities. Although these essential functions are numerous, their meaning can be indicated by the few following examples.

### *The Function of Governing*

Keeping order is perhaps one of the oldest functions of community life. It may be performed in the primitive community by the council of old men and supported by community opinion. In the more modern community having some type of civil government, order is the final responsibility of the police and the courts. But these alone could not safeguard life, protect property, and maintain order without the support of the various private groups and nongovernmental organizations. These private groups, sometimes called secondary organizations, are supported by their respective memberships.

The function of government has been evolving not only in structure but

[22] Gaston Bardet, *Mission de l'Urbanisme,* Paris, Les Editions Ouvrières Economie et Humanisme, 1949, pp. 149-151.

in its "why and wherefore" aspects. It has become urbanized just as have those groups which help manage community life. On this theme Hauser wrote:

> There is no doubt that the complex of technological, economic and social changes which constitute "urbanism" is the major factor in the rapidity with which governmental functions have proliferated, often despite the express intent of administrations. The urban way of life, the increasing interdependence of the elements of the social order and the increasing inability of traditional and inherited social institutions to cope with the new problems of urban living have led inexorably to the multiplication of government functions, powers and personnel; and the process is still under way.[23]

In this process of being urbanized, government is but one phase of an organizational network which may be complex or not, depending on the community. It includes organizations of different kinds with different interests, some competitive, but, in the wholeness of the community, all recognize their subordination to this central civic authority which is based on law.

### *The Function of Supply*

The community is also an aggregate of consumers. The people must eat. They must be supplied with water. They must be safeguarded against health hazards. Some of these functions are performed by private organizations, but the civil authority has the final responsibility for keeping the supply lines open. The community must be accessible to its hinterland and it must have continuous access to its hinterland of supply.

### *The Rearing of Children*

In the child, the past and the future of the community meet. The child must be kept in health and educated in the culture of the community. In all societies, simple and complex, this is primarily the responsibility of the family, but in the modern urban community the task is too complex for the family alone. The community must share the responsibility for safeguarding and educating the children, or at least for setting the minimum standards for their rearing and education. Increasingly in the urban community civil authority shares the work of rearing the young and also aids and guides parents in this important responsibility.

### *Individual Rights and Responsibilities*

Whatever the developmental level of the community, from the most rural

[23] Philip M. Hauser, "On the Impact of Urbanism on Social Organization, Human Nature and Political Order," *Confluence*, Vol. VII, No. 1, Spring 1958, p. 66.

to the most urban, it is assumed that the individual has certain rights, as he is also assumed to have certain recognized responsibilities. The ultimate authority, whether it is a familial hierarchy or a form of civil government, is looked to for safeguarding the individual's rights, and it, in turn, looks to the individual to assume his proper responsibilities. This may be regarded as part of the community function of keeping order, but the implications of protecting rights and keeping the individual aware of his responsibilities are somewhat broader.

The rights and privileges of the individual concern the owning and using of property and thus create the need for continuing public records. The individual has certain rights to buy and sell in the market and the organized community is the watchman of the market. The integrity of weights and measures must be safeguarded; perhaps prices need to be regulated and the quality of the merchandise inspected. When such functions, as well as others equally essential are performed, they evidence the reality of the community as an integrated organization.

## Communities and Outside Relationships

As we noted in the previous chapter, communities do not live in isolation. For the modern community, isolation means extinction. Once it was possible for communities with subsistence economies to survive indefinitely, but that is impossible today under the impact of global urbanism. It is, in fact, the existence of other communities that gives each a sense of self-identity. The internal consensus, the self-awareness that the population acquires, becomes the more readily articulate in relation to the proximity of other populations. These relationships with the outside not only contribute to the formation of a "consciousness of community," but they contribute in many ways to making a community the sort of place it is.

The work a community does depends in large part on its market relations with other places, as well as on the kind, the amount, and the source of raw materials. The fact that the community is a small aggregate or a large one may be in part due to the initiative of the inhabitants living there, but its size is likely to be more the result of the number and kind of contacts with the outside. Thus Washington becomes one kind of a community and Pittsburgh another; Liverpool becomes a community of a very different type from London. Each is a different kind of a workplace, engaged in different tasks and performing in a somewhat special role, while each may differ from the others, it may be like them in many respects.

Not only is a community in large part the creation of its outside contacts, but its character must change as the nature of these contacts change. Mention of community interdependence was made by Mitchell in his study of certain areas in England. He concluded that a community tends to be rejected if it does not maintain contacts, whether they be competitive or cooperative. The community must change as other communities change, and if it is unable to change it will be left behind.

> In such a community, to reestablish satisfactory relationships, change must take place, and this change takes on the aspect of threat; this gives rise to reaction. If the community possesses elasticity, then the reaction will be to facilitate the change, if not, then the change will be resisted. Where resistance takes place it will be for two reasons. First, because it involves a change to some new and unknown form, or one imperfectly known. Secondly, because the new form is known to be inferior to the one it seeks to replace. These reasons may be conscious among some members of the community, but for the most part they tend to be unconscious.[24]

Whatever stand the community may take in response to pressures for change, if it fails to adapt it may very well find itself isolated to some degree. Thus its status as defined by other communities may decline, as its status may rise if it effectively takes advantage of contacts with the outside. In other words, the self-identity a community acquires is only in part due to internal consensus; it is also defined in part by outside relationships and outside recognition.

## Summary

The term "community" is much used by Western peoples and, like other useful words, it tends to have a variety of applications. Partly because of this wide and diverse use, social scientists have had difficulty in finding commonly accepted definitions. No attempt is made in this book to evaluate or reconcile these many definitions. However, in our effort to arrive at a suitable definition it was necessary to take account of the diverse and ambiguous character of the elements that must be included in *any* definition. In one situation community has a different meaning than in another situation, and there are, obviously, many situations.

[24] G. Duncan Mitchell, "Social Disintegration in a Rural Community," *Human Relations,* Vol. III, No. 3, 1950, p. 303.

The neighborhood, as a closely related concept, is understood to be a small area occupied by a limited number of people living in close proximity and in frequent contact, a primary face-to-face group. A community, even a small village, may include a number of neighborhoods. The boundaries may overlap and some families may belong to adjoining neighborhoods. The concept of neighborhood is more applicable to rural than to urban communities. Although urbanites may live in the same building, they may prefer anonymity. An urban neighborhood, geographically, may be occupied by people who have no wish for "neighboring."

It is being increasingly recognized that urbanites live their social lives less in terms of residential proximity and more in terms of acquaintance and friendship networks. Each individual has his own network which may be in part the network of his family. The basis for the network is some common interest and one may have different networks according to different interests. If the friends and acquaintances within the circle are known to one another, the network is "close-knit" and the individual is much more under its influence than another individual whose network is "loose-knit," whose friends and acquaintances are in less frequent contact with one another.

A community takes form and character as the people within it acquire common experiences and memories. Feelings of "consciousness of community" may be the result of rivalries and competition or of cooperation. People become identified with the whole as they develop reasons for being there and wanting to remain, as they "take root." Integration takes place as individuals and groups, consciously or not, become identified with the processes of community life.

A community is also a place "on the ground"; it is a geographical fact. The sociopsychological community in people's minds has for its base the array of physical structures, streets, monuments, and so on which can be seen and experienced.

It is generally recognized that urban communities differ from rural communities in complexity. Each has its own complexity, however; the difference is in kind. The urban child learns the urban way of life as quickly as the rural child learns the rural way of life and work. But the rural way of life tends to repeat itself season by season, except as it must change in response to urban influences. The urban community is one in which the way of life continually changes. As it is less traditional, it is also more rational and technical. The complexities continue to change and hence learning must continue.

Whether urban, rural, or primitive, the organized community serves the people within it by performing certain basic functions. It maintains law and order. It exercises its authority in emergencies in order that the people may eat and drink. It rears and educates the young and it safeguards the individual in various recognized rights. The performance of such essential functions constitutes the ultimate test of the community.

While the community acquires self-identity in terms of internal consensus, this self-identity is also the result of outside relationships. Community work, character, and size are in large measure determined by contacts with other places. In the industrial urban society a community may decline or even vanish if it is isolated. Because its growth and strength depend largely on its importance to other places, a community must change as changes occur in surrounding areas.

CHAPTER 3

# Communities and Their Natural History

THE PURPOSE OF THIS CHAPTER is to consider the evolution of communities. Man has been a long time learning to build villages, then towns, and now great cities. He lived in houses for many generations before he learned that fire can be used safely to heat room space, that chimneys can be used to take the smoke away, that glass can be used to let in the light. Learning to live collectively has been a slow process, particularly in ancient times, but it goes forward much faster today. It goes faster in urban than in rural places, but whatever the type of community, innovations are received with caution. This caution is understandable, whether the innovations relate to institutions or tools, group organization or beliefs, food or clothing or houses. Community evolution involves many kinds of change, learning in different areas. So much of this evolution is outside the embrace of history that the term "natural history" seems more applicable.[1]

## The Idea of Progress

We are reminded by Ayres that the term "progress" is generally in bad academic odor because of its association with transcendental notions about preconceived goals or future states of perfection. Yet the term undoubtedly has a valid place in our vocabulary. It can be used, for example, in relation to tool making and in tool combinations, in work assembly arrangements, or in other connections involving movement from one level of development to

[1] David Bidney (*Theoretical Anthropology*, New York, Columbia University Press, 1955, p. 283) would not accept the idea of describing community evolution as natural history. It is "cultural history" and concerns the cultural evolution of mankind.

another. However, in Western thinking, according to Ayres, we still "retain the conviction that if we are to think of social progress at all, we must do so in terms of transcendental 'ends.' "[2]

The idea of progress is new and it is Western. It had no place in primitive cultures where the present in every detail was inherited from the past. If change did take place, it was incidental, and its purpose was not to reach envisioned goals. In comparing underdeveloped with advanced peoples we speak of progress in terms of leaving old ways behind and taking up new ones, acquiring new skills, making and using new tools or materials, or making new contacts. It is a way of looking at the way of life which has been evolving since the beginning of the industrial revolution. Men have been motivated to aspire to achievement of one sort or other; in fact, it is now quite the expected thing. Ayres says of the advance:

> What the evidence shows is that humbug, cruelty and squalor have been decreasing for the population as a whole throughout modern times, as they have been decreasing throughout the history of the race. No one seriously advocates turning back the clock to the day when Plato dispensed sweet wisdom to a few disciples while all the rest of the world lived in fear of evil spirits, or to the day when theology was most angelic and the clergy lived in open concubinage, lords enjoyed first-night rights with every bride, and no man was safe from violent molestation, or from small pox, typhus and starvation.[3]

Man has managed to get from there to here, from voodooism to atomic science, from a circle of grass huts to the steel-ribbed skyscraper or the suburban community, from tools of bone or stone to precision instruments, from grubbing with sticks to modern agriculture, from isolated community life to global urbanism. This story, if it could be told, is natural history.

Each special element of the material phase of modern civilization has been developing over centuries. The facilities of transportation, for instance, have developed very fast in the past century, but they were emerging slowly long before. Toynbee, considering the evolution of ships, calls attention to the long learning process in which ship construction and seamanship evolved together,[4] but this development was stimulated by other developments, and it stimulated still others. A chain relationship existed between many evolving elements, as it still does.

[2] C. E. Ayres, *The Theory of Economic Progress,* Chapel Hill, University of North Carolina Press, 1944, p. 122.

[3] *Ibid.,* p. 244.

[4] Arnold Toynbee, *A Study of History,* London, Oxford University Press, 1955 printing, Vol. IX, pp. 366-372.

A look at the "processes of becoming" as they relate to communities is too much of a task for this chapter, but we can give at least passing notice to some of the more pertinent aspects of community evolution.

It has been recognized, for example, that this movement from one way of life and work, or one level of civilization, to another may be both wasteful and painful. People may look forward with uncertainty, if not fear, as they also look wistfully to the past, seeing what is most pleasing (the fine fronts of some monumental buildings but not the narrow filthy streets). Mumford wrote in 1944 of unmitigated confusion and disintegration. Although ". . . behind all these phenomena of physical destruction we can detect an earlier and perhaps more fundamental series . . . ," in between he saw only breakdown in stability and a "loss in form and purpose in many of the arts."[5]

Sorokin wrote in 1941, reiterating what he had said in 1929, "We are living and acting in one of the epoch-making turning points in history" which he said would be marked by revolution, bloodshed, anarchy, "social, moral, economic, political and intellectual chaos." While it might be said of any decade in modern times that it is "one of the epoch-making turning points in history," Sorokin argued that civilization has moved through different stages, but now it is being purged and may move into some sort of "super-system" which he designates as "ideational-motivated."[6]

These viewpoints are expressive of an overlook at our changing civilization. But our look must be more concerned with the particulars, with step by step developments in the growth of communities from earliest villages to modern cities. And in examining this growth we must recognize that it was indeed both wasteful and painful. Man's fumbling efforts before he learned to live in large aggregates and then learned how to keep himself clean by the use of sewers and systems of piped water, his learning how to erect tall buildings that would not fall down or burn down, how to carry on long-distance trade, how to secure and preserve quantities of food—these are typical of the aspects involved in the more detailed approach of the natural history of the community.

## Cumulative Community Experience

We are reminded by Mumford in his *Technics and Civilization* that man was mechanism conscious long before he made the machine and the "will-to-order" has been present in communities for centuries, as in the organization

[5] Lewis Mumford, *The Condition of Man*, New York, Harcourt, Brace, 1944, p. 14.
[6] Pitirim A. Sorokin, *The Crisis of Our Age*, New York, E. P. Dutton, 1941, p. 12.

C

of military campaigns. Most of the initial elements of the industrial revolution —clocks, printing presses, water wheels, and so on—were already in long use, and there had also been a long "preliminary period of ideological and social preparation."[7] The idea that culture is an ever-cumulative process is not new, but it needs frequent repeating.

What Mumford wrote about the machine applies with equal force to the community, of which the machine and the factory are only phases. There are other phases and each is cumulative from the earliest villages to the metropolis, since all of the essential elements of work and life in the metropolis were present in the early village. Goodfellow was interested in the Bantu village, a group of what he called "homesteads in close neighborhood," although he recognized that many people might not be impressed:

> They are by no means mere technical structures; they are erected to meet the needs of the people, their spiritual as well as their material needs. An understanding of why the huts are placed as they are gives at once a deep insight into the people's culture, and when we learn not only of the form of the huts but of their purposes, we probe into social life and begin to appreciate the economic aspect of the life.[8]

As Goodfellow recognized, these Bantu villages were built somewhat in circles, and there were reasons for that, as there were equally practical reasons for the size and shape of the houses. Here a group of families, ecologically oriented to their work and physical surroundings, had learned to live in a self-contained manner. Their arts, their foods, their beliefs, their family organization, their divisions of labor, were not only interrelated as of the moment but drew support and confirmation out of the past, and were expected to continue. All that can be said of the Bantu village for being a socially and economically self-contained unit would also apply to the more advanced but isolated Iranian village described in Chapter 6. The Bantu village has no protective walls, as many Iranian villages still have, but this does not mean that the villagers have always lived at peace with their neighbors. It is in such original villages that much of man's experience in community living and intercommunity relations has been gained.

In different parts of the world the same basic learning process has been going on in the primary communities. It has involved learning to live to greater advantage with the local environment and its resources, finding

[7] Lewis Mumford, *Technics and Civilization*, New York, Harcourt, Brace, 1934, p. 4.
[8] D. M. Goodfellow, *Principles of Economic Sociology*, London, Routledge and Sons, 1939, p. 148.

the techniques for doing work, and developing various aspects of the social and cultural facets of community life. The other phase of learning has concerned relationships with other communities, first the near ones and then the far ones. It was in relation to these outer contacts that some communities evolved faster than others and some achieved domination over others, through trade or by force, or both. In these places of dominance, where people were doing new kinds of work and encountering new needs, the processes of change moved faster. They grew in size and diversity of activity, and they became holders of power. In the most elemental sense, the dominant places became urban. According to Turner:

> The specialization of social classes within the urban culture group occurred in a relatively clear order. Economic, social, political, and intellectual aspects of urban cultural development were in the main correlative; each changed as the others changed and each contributed to the development of the others. Interaction between the urban cultural group and out-groups followed a sequence more or less parallel to the order of specialization of urban social classes within the culture group. Lastly, the degree of integration of urban cultures varied with the interaction set up by these external and internal movements; in other words, at certain points in the development of urban cultures the patterns were loosely formed, while at others they were far more rigid.[9]

These places that acquired some degree of ascendancy over others, where a type of dominance prevailed, either "loosely formed" or "more rigid," were centers of innovation and change. It is to be assumed that these first cities were in spots of greater advantage, economically and geographically. They became central points in networks. A favorably located village in a small valley might have a dominant relationship with a group of nearby villages in the same valley, a relationship which may have been held firm at times by force or maintained loosely by mutual consent. The dominant community, rather than the others, would have relations with another dominant community in another part of the valley. Understandably, there would be competition for supremacy between dominant communities, with wars, the emergence of kings, the extension of commerce, and so on over centuries.

Elemental and crude as the first cities must have been, they marked the entry of man into a rudimentary form of urbanism which grew out of a transition through many centuries. During this period people had to learn many new kinds of work. Perhaps at first the life of people on the land was not greatly affected, but there emerged in the towns certain groups which

[9] Ralph Turner, *The Great Cultural Traditions*, New York, McGraw-Hill, 1941, Vol. II, p. 1291.

were wholly or in part detached from the land—workers in leather, metals, and wood, makers of cloth, practitioners of magic and, in time, various types of intellectuals. These groups became fixtures of the first cities, and have been there ever since. Among them must certainly be included the builders whose record of learning can be traced in the structures they erected. Although we cannot trace back to the very earliest cities, we can go back to some of those that stood five to seven thousand years ago. From then to now the story is one of continuous evolution.

## The Ancient Cities

Wallbank and Taylor define civilization as "a type of advanced human life based usually on city living and an involved pattern of activities made up of such forces as writing, law, government, commerce, and refined concepts of religion."[10] Civilization, according to this concept, could not have developed except in cities, which afford stimulating human relationships. The city is a way of life that began in the valleys of the Nile about 5,000 B. C. and later continued in the valleys of the Tigris and Euphrates. For our purposes, cities like Heliopolis, Memphis, and Babylon, or the later city of Ur, are of interest mainly in terms of their contribution to the initial evolution of urbanism—and they did develop some of the essentials.

Various estimates about the populations of certain ancient cities seem a bit fantastic, but it is true that large numbers of people did live in these cities. Thousands of people living in agglomerations detached from the land was something new in human experience, and they consequently had to learn how to live in this new relationship. Rules and regulations had to be found for governing people en masse. A method of communication was needed, and hence the alphabet and writing evolved. Time had acquired a new meaning, and the calendar was invented. It was necessary to measure the time of the day, something never needed by man before, and so the sundial was developed. The needs of the market gave rise to money, and since numbers were needed to count money and to convey ideas quantitatively, it indirectly led to the development of mathematics.

With possible exceptions, the ancient city was primarily a center of authority sustained by the military, and military considerations usually determined the location of the city. Trade was usually an incidental consideration,

[10] T. W. Wallbank and A. M. Taylor, *Civilization, Past and Present*, Chicago, Scott, Foresman, 1950, Vol. I, p. 11.

but military and trade interests were not necessarily at variance, since both had to take into account food supply, fuel, water, and convenience of access to other places by land or water. There were no roads but there were routes over which armies or caravans might travel. If supplies were not carried by boats, they were carried on the backs of camels, donkeys, or humans. This meant that the area of sustenance was limited, and this in turn limited the size of the city. There was little movement of goods from the city to the hinterland. What the historians describe as trade was largely in luxury goods, light in weight but high in value. It is hard to think that such luxury trade figured greatly in the economy of the city.

However, these early cities brought into existence the idea of a public authority and a public economy which much later was to be crystallized by Rome, and earlier to some extent by Greece, into a system of public administration under law. Although this public authority functioned unevenly in the more ancient cities, it was used to get things done, to build the walls of the city and its palaces, to organize the military, to organize systematically the work of people so that the army might be supplied and armed, and to administer by one means or other the collection of taxes and tribute.[11] This, then, is another type of collective experience which could be gained only in cities and, however fumbling it was, it was a cumulative experience.

These early cities had to develop new kinds of leadership. The situation required people with different skills and functions—judges, artisans, tax collectors and money changers. Here, too, the poets, musicians, philosophers, and magicians found refuge in a protective relation to the king and ruling families. Such groups, functioning in stimulating roles, are still at home in the city, but with each passing century their numbers have increased and new kinds of idea groups have been added. Theirs is a type of cumulative experience that lies close to the heart of urbanism.

## Old Cities on the Mediterranean

The cultural influence of the cities of the Middle East roved westward rather than eastward. Contacts with the Far East were few and difficult to make, as can be gathered from the legends recorded by Richard Francis Burton in "A Thousand Nights and a Night," particularly the fantastic unabridged narrations of Sinbad, the sailor. In China were some very old

[11] Arthur Keith, *A New Theory of Human Evolution*, New York, Philosophical Library, 1949, p. 293.

cities and some of these, as Gutkind reports, were built by plan.[12] There were very old cities in India, but these were in some cases established by the mandate of some king.[13] Apparently the evolution of Far Eastern cities paralleled in many respects that of Middle Eastern cities, although in ancient times neither had much influence on the other because of the wide natural barriers between.

The Mediterranean cities were the primary inheritors of the cumulated experience of the old cities on the Nile, the Tigris, and the Euphrates. These newer cities on the Mediterranean had been stimulated by open-sea long-distance commerce. They could exchange foods, animal products, woods, metals, and other products with far places. They learned boat building and seamanship. Banking grew along with commerce. Craftsmen were finding new markets for their wares. In no phase of the new urban life was greater advance made than in building, for the builders had access to a wider variety of materials. Most of these activities were very old, but on the Mediterranean they evolved and expanded. The Mediterranean situation offered the stimulation of variety, a new invitation to venturing and discovery. It was an invitation which beckoned not only from over the inland sea, but from mysterious inland regions.

As there had to be new ways of work, so were there improvements in the arts of war, and these activities were carried on with vigor. Organizing armies, waging military compaigns, and bringing other places under domination had now become a major type of public enterprise. This naturally opened up new markets and stimulated commerce, but, even more important for the evolution of urbanism, these early city states on the Mediterranean faced the need of developing effective methods for organizing and managing large groups of people, whether an army on the march or the masses in the city. Law and administrative regulation, only feebly present in the very old cities, were now becoming firmly established. While such developments, stimulated either by military or commercial purposes, were leading toward goals not then visualized, they have all the characteristics of progress.

Few rulers of the earlier ancient cities concerned themselves with the problems of access to their subject territories. Because they tended to be larger,

[12] E. A. Gutkind, *Revolution and Environment*, London, Kegan Paul, Trench, Trubner, 1946.

[13] For some observations on the old cities of India, see Robert L. Crane, "Urbanism in India," *American Journal of Sociology*, Vol. 60, No. 5, March 1955, p. 470. Also see article in the same issue of the *Journal* by Morton S. Ginsberg, "The Great City in Southeast Asia," p. 459.

however, this access to supply was more important for the later cities. It was left to Rome to pioneer in road building—public works of a new kind—the easier to extend its authority to subordinate regions and to bring supplies to the capital. These roads opened the way for cultural influences to be diffused.

The Mediterranean cities advanced far over their predecessors in structural ventures, and the remains of many of these achievements in grand buildings and monuments can still be seen. Rome made prodigious efforts to provide herself with potable drinking water and to keep the city clean by means of sewers, although Carcopino states that in spite of the greatness of these construction projects, the benefits of flowing water and sewers were available only to ground-floor occupants. Upper-floor occupants had to use public latrines and buy their water from street vendors as had been done through the ages. And Carcopino also reminds us that Romans did not have chimneys in their houses, which meant that space heating was almost absent.[14]

The fact that cities had already had about five thousand years of history before Rome was building sewers and bringing in a flow of fresh water should serve to remind us that the common people—at least 70 to 80 percent of the urban population—lived in poverty and squalor, a condition which hardly improved through thirty or more centuries. It is hard to see that their levels of living were any higher, if as high, as those of rural people, except for a period when Rome had a state security system for giving food to the unemployed. The advances in art, architecture, literature, philosophy, mathematics, finance, public administration, and other branches concerned only an inner circle of the population in the ancient cities. There is little reason to believe that the mass of people knew about such developments, or cared or were affected by them except indirectly in ways they could not understand.

## Urban Evolution in the West

While the emerging cities in Western Europe inherited much from the cumulating stream of urban experience that flowed from the cities of the Middle East and subsequently on to Rome, these European cities also had a cultural background of their own. From the folkways and folk thinking of their own ancestors, they had accumulated experience in community living with their own traditions and social values. They were, moreover, vigorous people and, for one reason or other, they were more susceptible to indi-

[14] Jerome Carcopino, "Rome under the Antonines." See Ernest Barker and others, *Golden Ages of the Great Cities,* London, Thames and Hudson, 1952, pp. 35-36.

vidual venturing than were the masses in the older cities. In a limited way they had developed types of intercommunity trade, but they had not become city builders. The idea of urbanism came from the outside, brought in through the military might of the Romans who occupied large sections of Western Europe for about four centuries.

Through these four centuries of Roman tutelage, the West began inheriting from Rome not only culture traits that originated in Rome, but traits that Rome had inherited from the East. But Rome had more to offer than the "barbarians" were ready to receive, or even able to comprehend. With the defeat of Rome, the famous network of roads fell into disuse. With a few exceptions, Roman structures disappeared, but various nonmilitary Roman influences, notably that of the church, remained.

A further influence was that of the traders, with centuries of experience of their own. Although their bases had long been the Mediterranean cities and their trade was carried on mainly over sea routes, they had been able during the Roman occupation to venture inland to such river-route places as Cologne, Frankfurt, Paris, and Vienna. Ships from the Mediterranean cities found their way to such places as Amsterdam and London.

Although all of these stimulating influences were at work, urban life in Europe fell into several centuries of decline which were called the Dark Ages. Wallbank and Taylor describe them as follows:

> While the new and the old were being blended, there was a decline in civilization, and trade and commerce suffered a setback. For nearly five hundred years after the great invasions of the fifth century, European civilization was unsettled. It retrogressed rather than advanced. But the period of the so-called Dark Ages was one of preparation, in which a new civilization, even more fruitful than the old, compounded of both Germanic and Roman elements was being evolved.[15]

Following these centuries of reorientation, as well described by Pirenne, urban life in Europe began to take form.[16] Cities grew on the more solid basis of industry and commerce, not on the basis of military might and tribute as in former times. But again, since law and order prevailed in the cities and not in the intervening areas, and because defense was needed against authorities centered in other cities, these new cities, like the ancient ones, were enclosed in walls.

[15] Wallbank and Taylor, *op. cit.*, I, p. 238.

[16] Henri Pirenne, *Mediaeval Cities, Their Origin and Revival of Trade*, Princeton, Princeton University Press, 1925.

It developed with the passing of time that industrial and commercial cities could scarcely be contained in walls. Thus we are able to see in various European cities that the first ring of walls later had to be replaced by a wider ring of walls, and perhaps later by a still wider ring. With the crowding in of people, as the cities afforded more work, it was necessary to build houses higher. To save space, streets were kept narrow. Workers coming into the city had to leave their families in outside villages and so the percentage of single persons of both sexes found within the walls increased. As population pressure increased, necessitating the building of wider rings of walls, there was a higher percentage of families. When removed, the old walls usually made space for wide streets, the "ring streets" found today in cities like Cologne and Vienna. With new methods of warfare the walls proved useless and were rapidly removed.

When European cities began removing their walls, mostly during the first quarter of the nineteenth century, it was symbolic of the new phase into which the urban community had already entered. It needed the combination of gunpowder and cannon balls to shock people into recognizing that the land occupied by walls could be better utilized for streets—streets that led to roads and roads that led to markets. It has since become increasingly clear not only that cities cannot be contained within walls, but that they cannot be contained by the political boundaries with which they are generally surrounded.

## The Emerging of Modern Urbanism

Any effort to identify the characteristics of the modern city, and thus to place the finger on the approximate time of its emergence, must take into account at least two and perhaps three aspects of its evolution, and even these are difficult to pinpoint. One of these is the evolution of public services: sewers and piped water, street lighting, police and fire protection, compulsory education, public welfare, health and sanitation and professional civil service. We find that each of these services has a long history. Each is a special kind of cumulative community experience, although each in the modern community is assumed to meet certain minimum standards that did not exist one or two centuries ago.

A second group of characteristics concerns the work of the community and the level to which the work ways and organization of the people have attained. We might say, for example, that the modern city began with factories and power-driven machines, but we must remember that machines driven by wind or water power have been in use three or more centuries. Or the point at which the modern city began might be identified with certain developments in commerce that enabled special craftsmen to speed the production of their wares for distant markets.

Or the city may be considered as having become modern in terms of the widening participation of the people in the political life, or in terms of the presence or absence of secondary organizations such as political parties, trade unions, social clubs, professional associations, and the like.

Actually, whether one test or several tests are used, it would be difficult to fix a point in time when the modern urban community came on the scene. It differs with countries as well as with cities. The same applies if we try to fix the approximate dates for the beginning of the industrial revolution; while it began in England after 1750, it did not begin in Germany until after 1850, and those dates also concern the modern city.

Mumford, interested in the social implications of the machine, found it necessary to divide its evolution into three periods. The first period he named *eotechnic,* a very long period dating from about 1750 back to about the tenth century. This period was essentially Western European, a period of occasional important discovery or invention. He names, among many other inventions, movable type and printing, paper, the magnetic compass, spectacles, cannon, the water-power mill, and the spinning wheel—all of which had a preparatory relation to the second period, the *paleotechnic.* This second period, extending from about 1750 to about 1832, was closely identified with the emergence of modern cities. Workshops of craftsmen were becoming factories. Machines were being invented and production employment reorganized. Cities were attracting crowds of new labor. Work was being degraded and so were workers. Mumford calls this period a "rubbish-strewn avenue between the *eotechnic* and the *neotechnic* economies." The neotechnic period, about a century under way, is still with us.[17]

Both the eotechnic and the paleotechnic periods were a preparation for the neotechnic period, but the preparation was mainly Western and the direction of the march was toward industrial urbanism. In terms of this development, the modern city is essentially Western, although other cities elsewhere were also moving toward their own type of modernism. Bendix observed that one of the preconditions for industrial urbanism in the West was a "relative

[17] Lewis Mumford, *Technics and Civilization,* New York, Harcourt, Brace, 1934, pp. 211-214.

independence of economic activities from government regulations." Continuing, he noted:

> This independence was enhanced when commercial as well as legal transactions became increasingly secure against arbitrary interference by the king or by members of the landed aristocracy. Such increasing security involved a century-long process, in which the earlier identification of the national government with the interests of the royal household and the identification of judicial and administrative functions with the prerogatives of the aristocracy were very gradually abandoned. This process of disentanglement began in the early Middle Ages with the separation of one after another government office from the corresponding office in the royal household. And it eventuated in the establishment of a professional civil service and a professional judiciary during the eighteenth and nineteenth centuries. . . . The end result of this process was the development of the "state as a service-rendering organization for the protection of rights and the enforcement of duties."[18]

Western European cities, many of them much older than Western European states, and most of them enjoying a high degree of independence, apparently contributed much to this condition of independence-for-enterprise described by Bendix. When such enterprise was beginning to expand in a world-wide way, when industries were getting settled and their productivity was beginning to increase, when society in the cities was being oriented in a dynamic way to expanding industry and commerce—that is the point at which we may say that the modern city was beginning to take form. In general, that point would be somewhere between 1750 and 1800. It differs among European countries, and in the United States the emergence was not under way until 1830.

## New Phases of the City

Primarily, the modern city is oriented to trade and industry, and in this orientation it has a dynamic character of its own. In spite of the fact that it "belongs" legally to the state, it acquires a self-identity which, although it cannot be described as antistate, is certainly not confined to the state. Its financial function may operate in any region where markets can be developed or in any region from whence it may draw supplies. Not only does it develop its own way of life but, often deliberately by the media of communication

[18] Reinhard Bendix, *Work and Authority in Industry*, New York, John Wiley & Sons, 1956, p. 118. The inner quote used by Bendix is from Ernest Barker, *The Development of Public Services in Western Europe*, London, Oxford University Press, 1944, p. 6.

centered in the city, it spreads its way of thought and living to nonurban places, both within the state and beyond. It takes the initiative in opening or enlarging transportation routes and lines of communication. It is dynamically occupied with growing and with finding the means for sustaining itself. Surviving means being ever-modern and ever-resourceful.

It may be said that the modern city puts itself in a missionary role with respect to contact with its hinterland, but it does so for reasons that are more practical than moral. As Hoselitz observed, it takes to the hinterland its "style of life, in outlook and in behavioral norms. The culture of the agricultural sectors of an overwhelmingly industrialized country finally becomes assimilated to that of the city; the rural countryside becomes 'citified.' "[19]

The modern city does not wait on the slow processes of social change, but tries to stimulate backward people out of their tradition-bound isolation. Carey's report of his efforts as an official of the British government to speed the evolution of the Africans offers an example of this process:

> But the chief hated the whole development and blamed me, fairly enough, for the trouble it had caused. His people were better off, but increasing incomes and security only made them insubordinate. Alien traders did not salute him and spread revolutionary ideas. He did not want change. What I called progress, he saw, again quite fairly, as a backward step into social confusion, crime, political strife. And it was no good telling him that these things must come, in any case, from forces which neither of us could control, from world movements of power and opinion; he couldn't grasp such a notion. He had no idea of such forces. . . . And religion itself, the Mohammedan religion, meditative and world-rejecting, austere and puritanical, urged him to hate the new regime.
>
> Native traders, while making money, complained about foreign traders. As for the village chiefs, several of them complained that I was breaking up family life because the women could now sell food at good prices to the traders, and were neglecting their home duties. What is more, as my political agent informed me, the Emir and chiefs imputed the whole new economic policy simply to an anxiety on my part to get in the taxes. Nothing could persuade them that a large proportion of tax did not go into my own pocket.[20]

In such efforts to speed the processes of economic and social change and the industrial way of life, governments are operating in the interest of industrial urbanism. And they are effective, as observed by Forde:

> It is, however, important to bear in mind that part of the rural household production which is sold for cash—small as it may often be both in absolute

[19] Bert F. Hoselitz, "Cities in Advanced and Underdeveloped Countries," *Confluence*, Part II (The Relation of Advanced to Underdeveloped Countries), 1955, p. 329.

[20] Joyce Carey, "Policy for Aid," *Confluence*, Part II (The Relation of Advanced to Underdeveloped Countries), 1955, pp. 294-295.

> terms and even as a proportion of the rural economy—has in many areas great significance in connection with the urban developments with which we are here more closely concerned. Where cash crops are important, they provide the link between the rural household or village and the outside world from which comes the demand for payment of tax and the supply of manufactured goods, mostly of Western manufacture, that are highly valued for their quality or the prestige they confer. These goods are associated in the minds of the people as a whole with the urban centres from which, whether imported or locally made, they come. It is their ever-growing place in the lives of the rural population that endows the earning of money with such importance.[21]

Husain observed that urbanization might be less disturbing to rural people if they would begin to question their traditional values. He adds this thought, which applies to any region being reached by modern urbanism, "In this process there is, however, the danger that some of the finer elements of their culture may also be weakened or discarded at the same time, which would indeed be a pity."[22] When a culture is in process of rapid change, one cannot know which traits may be retained and which discarded; and even the traits that are retained may change to fit the new situation.

This new missionary-like role of the recently emerged modern city is essentially Western, almost entirely West European and American. Its influence in the more advanced countries, as it affects rural life, is usually identified as urbanization, but in the less advanced countries it may be called Europeanization, more commonly Westernization. In the countries of the Far East where industry is being promoted to relieve urban pressure, the efforts are frequently referred to as Westernization. Thus Japan, becoming more urban and industrial, is often referred to as a Westernized country.

## Specialization in Cities

While cities, especially in the West, were putting on modern dress and developing the modern way of life, they were also becoming specialized in their work. The trend toward specialization is a consequence of various influences, and the influences that determine the specialty of a place are as varied as the specialties themselves. Many cities, but probably more towns, find their

21 Daryll Forde, "Social Aspects of Urbanization and Industrialization in Africa, a General Review," *Social Implications of Industrialization and Urbanization in Africa South of the Sahara*, Paris, UNESCO, 1956, p. 22.

22 A. E. A. Husain, "Human and Social Impact of Technological Change in East Pakistan, Decca," *The Social Implication of Industrialization and Urbanization*, Calcutta, Unesco Research Center, 1956, p. 136.

special work in relation to the raw materials which may be easily accessible. A small city or town in a vegetable-growing valley may be a place where canneries are located. A coastal town may be occupied with catching, canning, and shipping fish. Another coastal town, if scenically situated, may become a vacation resort. A city in a ranching region may become a meat-packing center. Some towns or cities may be supported by their work in processing metals from ores brought from nearby mines. Cities strategically located with respect to land or sea routes or both are likely to become transportation centers.

Location with respect to large or very large areas of population density often influences the type of work performed by a city or town. Some may become market places. Every town and city is a market, but the size is dependent on the number of people in the conveniently near hinterland. The size and the nature of the market, in turn, may help determine the work an urban place will do.

One influence, then, in determining the work of a town or city is the supply of raw materials of one kind or other which is accessible. But the urban place is also a seller of goods, and the number of customers in its hinterland has an influence on what it sells or how much. This very simple double rule applied perhaps with more force to ancient cities than to modern ones. Today we find cities in the same general area, each of which may have a different specialization—cutlery, dishes, silk cloth, automobiles, chocolates—and each of which may have a world-wide market.

The cutlery plant might have started in any of several places with access to a wide market. Competitors for that market may be a cutlery town in the United States, another in England, and still another in Germany. As competitors, these specialized places, however distant, are as much aware of one another as if all three were in the same Connecticut valley. The advantages of transportation and communication which modern cities enjoy not only permit specialization, but put a premium on it.

## Classification of Towns and Cities

Since urban places are specialized, it becomes necessary to think of them in terms of categories. American students of urban communities have tried to classify cities in this manner. French scholars have also made their classifications, and two of these need to be mentioned. George, for example, sees cities as living through "generations," such as precapitalist, capitalist, colonial, and socialist. The great city is the metropolis. Colonization cities are those established in less advanced countries by other countries.[23]

Rosier finds that cities may need to be classified by different approaches. They can be grouped in terms of their founding, their evolution, their legal status, or their activity. Classified with respect to the conditions under which they were established, Rosier groups cities as "natural" or created. Versailles, built at the order of Louis XIV, would be a created city. The natural city is one that grows in an integrated relation to its site and market area. Classified in terms of evolution, a city may be fast growing, slow growing, or declining. But its evolution also concerns the form of its growth. It may be star shaped, extending along the roads of access, or it might take some other shape.

In legal terms, cities are classified by statute, the basis typically being the number of population. A smaller city might become a larger one merely by legally extending its borders and annexing its suburban communities, as in the United States.

Classed according to their activities, Rosier lists a diversity of categories. Some cities may fall into more than one of these.

1. Cities of consumption (residential or resort places)
2. Cities of production (mainly manufacturing)
3. Mixed activity cities (consumption and production)
4. Cities of storage and distribution (wholesaling)
5. River and seaport cities
6. Receiving and forwarding cities (free ports where goods are transferred, or cities like Le Havre that are ports for larger cities)
7. Cities of finance and credit
8. Cities of working men or artisans (residential)
9. Military cities (Verdun, Belfort, Malta, Kiel being examples)
10. Thermal or bath cities (health centers)
11. Climatic cities (Nice, Annecy, San Remo, Montreux, Miami)
12. Museum cities
13. University cities (function recognized by Rosier as being in cities like Paris that perform other functions)

Other special-function cities are religious centers, pilgrimage cities (Mecca, Lourdes), mining cities, and oil cities. Among the unclassified types are listed twin cities and declining or dead cities.[24] Among dead cities Rosier names Carcassonne in France and Bruges in Belgium. One point made clear

[23] Pierre George, *La Ville,* Paris, Presses Universitaires de France, 1952, pp. 28 ff.

[24] Camille Rosier, *L'Urbanisme, ou le Science de l'Agglomeration,* Paris, Dunod, 1953, Chapt. XII.

by Rosier is that classifying urban agglomerations, by whatever method, is arbitrary, but it is nonetheless informing.

Smith rated 92 American cities having 100,000 or more population in 1940. He then studied the composition of the labor force in each city, identifying the workers with certain categories of occupations. Thus he was able, for instance, to rate Gary, Indiana, with 68.9 percent of the workers in industry, as a manufacturing city. The 92 cities fell into the following categories:[25]

| | |
|---|---|
| Manufacture | Domestic and personal services |
| Government | Professions |
| Wholesale and retail | Finance, insurance, reality |
| Transportation | Amusement |
| Construction | Repair work, all kinds |

Gist and Halbert presented a classification of cities which is based upon the common knowledge people have about cities, mostly in the United States, sufficient to illustrate the fact of specialization. They found the following categories: Production centers, Centers of trade and commerce, Political capitals, Culture centers, Resort and vacation cities, and Diversified cities.[26] This group of categories could be used for classifying cities in different countries, or for comparison between countries when statistical data are not available.

Harris made a classification of 988 American cities of 10,000 population or more as of 1940. Since this classification shows a variety of categories, it is included below. Also included are the classification criteria used.

1. *Manufacturing (1st Group):* 74 percent of total employment engaged in manufacture, wholesaling and retailing, with about 45 percent of all workers in manufacture and mechanical work.
2. *Manufacturing (2nd Group):* 60 percent of total employment engaged in manufacture, wholesaling and retailing, with 30 to 45 percent of all in manufacture and mechanical work.
3. *Retail:* 50 percent of total employment in manufacture, wholesaling and retailing, but the number in retailing being 2.2 times greater than the number in wholesaling.
4. *Diversified:* 60 percent of total employment in manufacture and 20 percent in wholesaling, or 50 percent in retailing. Manufacture and mechanical work is usually 25 to 35 percent of the employment.
5. *Wholesale:* 20 percent of total employment in manufacture, wholesaling and retailing, and at least 45 percent in retailing alone.

[25] T. Lynn Smith and C. A. McMahan, *The Sociology of Urban Life,* New York, Dryden Press, 1951, pp. 97-101.

[26] Noel P. Gist and L. A. Halbert, *Urban Society,* New York, Thomas Y. Crowell, 1948, pp. 8-15.

6. *Transportation:* 11 percent of total employment in transportation, which must be one-third the number in manufacture and two-thirds the number in trade.
7. *Mining:* Cities over 25,000 having 15 percent or more of the employment engaged in the extraction of minerals.
8. *University:* Cities of 10,000 or more having 25 percent of the population occupied as teachers or students.
9. *Resort and retirement:* Cities having no special criteria for identification.[27]

Harris used his formula for making a classification of 1,041 American cities of 10,000 or more population for the International City Managers Association. For it he added a tenth category—residence, or "bedroom," cities. He discovered that specialization is found mainly in smaller cities, those in the 10,000-25,000 class, of which there were 643. Of the 643 cities, 157 were one-type (industry of one type) manufacturing cities, 124 were residence cities, and 95 were retail trade centers. Nearly 60 percent of all the one-type manufacturing cities were in the 10,000-25,000 class and 80 percent of the "bedroom," or purely residence, cities were also in the 10,000-25,000 class. Of 202 cities in the 25,000-50,000 class, 57 were one-type manufacturing, as were 30 of the 106 cities in the 50,000-100,000 class. Little evidence of specialization was found in cities of over 100,000. At this point, it seems, cities begin to acquire a diversified character. Of the 1,041 cities studied, none with 250,000 or more population was found in the categories of Education, Mining, Resort, Government, Wholesale or Transportation.[28]

## Cities and Their Hinterlands

If urban communities are specialized in their work it is partly because the situation permits it, and partly because the situation requires specialization. The work a city does is one phase of the two-way relationship between the city and its hinterland. This relationship is one that has developed with the evolution of the city, and it has figured in one way or another in all the changes through which the city has passed.

As a city grows and changes, so also changes its hinterland. In its early days as a trading center, one city may have its hinterland of customers in a relatively limited area. If an industry is established in the city it may draw raw

[27] Chauncy D. Harris, "A Functional Classification of Cities in the United States," *Geographical Review*, Vol. 33, 1945, pp. 86-99.

[28] *The Municipal Yearbook*, Chicago, International City Managers Association, 1950, p. 37.

materials from the limited local area, but its products may be sold over a wider area. Another industry established in the same place may bring its raw materials from another source, and its products may be sold in a market differently dispersed than those of the first industry. We may say that as a city evolves it acquires a variety of hinterlands, each involving a different city-hinterland relationship.

These relations with the different hinterlands, while not defined by political boundaries, may be limited by such boundaries, especially international borders where duties must be paid. Notwithstanding such barriers, the hinterlands of a city are as varied as are its different market relationships. This is because each type of industry in a city has its own special hinterland. The farm-machinery industry in and near Chicago sells agricultural equipment to any country in the world, but to rural, not to urban, people. The machine-tool industry in and near Chicago sells its wares to other metal-working industries, and mainly these are urban, not rural. The Connecticut hat makers sell hats to hat-wearing people everywhere, but silk hats (top hats) are sold principally in the larger cities and to upper-class people. Viewed in such terms, each of these hinterlands is also a special market.

The search for a market, as the search for supplies of raw materials, is an activity which usually starts out of the city. For each industry it is a special function and the cost must be included in the price of the product. Much of the effort in finding markets—by persuasion generally—is in inducing people to want and pay for the kind of goods offered. When we consider the many kinds of goods offered we come to the conclusion that city-hinterland relations are a very complex matter. A general picture might be obtained, however, if the market area for each of a hundred products distributed by a city is plotted on a map, and the data from the hundred maps superimposed. It will be found that the distribution of customers is scattered over a wide region, but the distribution of points is most dense in a zone ten miles from the urban center, more scattered in a zone ten to twenty miles out, and still more scattered in a zone fifty or a hundred miles out. On the basis of such a composite distribution of customers in a variety of markets, we are able to speak of a city as having its market area or hinterland.

Understandably, a city may in various ways try to exercise influence, even control, over its hinterland. Regardless of whether such control is by design or not, the city tends to retain a dominant position in its relation with the hinterland. Whether buying or selling, it has the advantage in price making. Not only is the city stronger financially than the hinterland, it is also better organized than the scattered hinterland can be, even though in the hinterland of the greater city there are smaller cities and towns as well as villages. Each smaller city or town, within its own orbit or hinterland, exercises a measure of dominance. Each competes with the other, and all compete with the large city. Thus, while the large city enjoys a favored position with respect to its hinterland, it must not overexploit its advantage. It must be aware not only of the smaller competitors, but also of other large cities that may encroach upon its hinterland.

## Inconveniences of Urban Growth

During the past thousand years of urban history there has probably never been a time when some observers did not view with concern the growth of cities. Congestion, for example, seems to be a continuing characteristic of cities. Yet they continue to grow—some rapidly, some slowly—and only an occasional one declines. We need to look at the phenomena of size, congestion, and growth in relative terms.

If urban people have sufficient work and are amply supplied with the necessaries, then congestion and urban size are not challenging problems. But if a city fortunate enough to be well supplied with work and food begins to grow to the point of discomfort, especially with respect to housing and streets, there may be other discomforts. That has been the experience not only of medieval and ancient cities, but of modern ones as well.

In 1953 the International Union of Local Authorities asked its member organizations in the different countries what measures, if any, should be taken "to prevent the further growth of large towns." In the main, the answers said, in effect, that nothing should be done. The Danish answer was that the size of a town was determined by its situation.[29] The Belgian answer mentioned the need of people to have employment, adding that towns grow because people find work there and so long as the town has employment size does not matter.[30]

If cities compete freely without political interference (and to some extent, in spite of such interference), each tends to influence the size of the others. But various types of influences, political and other, may hamper or enhance the growth of a city. It is sometimes pointed out, for example, that the national administration of France overcentralizes the economic, political, educational, and social life of all French communities in Paris, thus hampering free compe-

[29] *The Large Town and the Small Municipality,* The Hague, International Union of Local Authorities, 1953, p. 72.

[30] *Ibid.,* p. 51.

tition between hinterland places and Paris. This is apparently true of other countries as well—some South American countries, for example.

In the Orient are a number of coastal cities that were established by European colonial powers in places where only small towns had been previously. Calcutta is an example of a city that grew large in relation to foreign commerce, and at the expense of inland cities. Such primate cities later turned out to be overlarge in relation to the other cities in a developing country such as India.

> The primate or great city may have a paralytic effect on the development of other urban places and tend to be paralytic in relation to the remainder of the national economy. The fact that the great cities already exist creates the tendency further to centralize industrial, commercial and service development in them. Such a tendency may detract from the growth potential of other cities and may promote further concentration in the great city at the expense of the rest of the national economy.[31]

Such coastal cities grew in response to a colonial type of economy when these countries were much more agricultural than at present, but they are not oriented to a national economy of which they are now a part. As in the case of India, the economy previously was outward-looking, but it is now inward-looking, and it may take a few decades before the cities and towns of the interior strike a balanced relation with the coastal cities.

It must also be recognized that we have entered a world condition in which, more than ever before, urbanization is necessary if food-producing land is to be used to the greatest advantage. Such a condition was undreamed of two hundred or even a hundred years ago. There seems to be no other choice for many millions of people but to enter cities. The city, so long maligned as evil, is now assuming an opposite function—the refuge for the population surplus in the world. This subject will be further considered in the final chapter. At this point it is sufficient to observe that, after a long, painful, and indeed wasteful evolution, the city is arriving at a position of balance, inner strength, and leadership.

## Stages in Community Change

Our attention so far has been on the evolution of cities in general. The basis for this general examination is the idea that, whether in their structural

[31] *Conclusions* of the Joint UN/UNESCO Seminar on Urbanization in the Economic Commission for Asia and the Far East (ECAFE) region, Bangkok, August 8-18, 1956, *Urbanization in Asia and the Far East,* Calcutta, Unesco Research Center, 1957, p. 34.

aspects or in their way of life, cities have been changing, and in the course of these changes we can see evidence of progress. But the evolution of any town or city is also an individual matter, although even in this respect each can be classified. A city like Paris or London has moved through more stages of development than has such a later city as New York, while New York has lived through more stages of evolution than such a recent city as Los Angeles.

A modern city may move rapidly from one stage of development to another, an observation that applies not only to American cities, but to certain cities in the developing countries. Schmid, in his study of Minneapolis and St. Paul, the rival "Twin Cities," emphasizes the fact that a city in two or three decades may change to such an extent that it becomes a different type of community. While the Twin Cities are separate corporations, they are in fact a single metropolitan area. The following is a summary from Schmid:[32]

1. *The pioneer period:* Began with the founding of the cities in 1848 at the head of river transportation, ended with the close of the Civil War, 1865. A period of road making, land settlement, and covered wagons. Trade was limited. Population:

| | | |
|---|---|---|
| St. Paul | 1865 | 12,976 |
| Minneapolis | 1865 | 4,607 |

2. *First boom period:* Began with railroad building after the Civil War and ended with the financial depression of 1875. Further land settlement, town building, lumbering, construction. Population:

| | | |
|---|---|---|
| St. Paul | 1880 | 41,475 |
| Minneapolis | 1880 | 46,997 |

3. *Second boom period:* From about 1880 to the depression of 1893. Farming prospered. Lumber industry reached its peak. Flour mills got under way. Other industry started. Population:

| | | |
|---|---|---|
| St. Paul | 1890 | 140,292 |
| Minneapolis | 1890 | 192,822 |

4. *Third boom period:* From about 1895 to World War I. Lumbering declined. Flour milling declined somewhat, due to competition elsewhere. Some consumer-goods industries got under way. Relations with hinterland improved because of automobile transportation. Much rebuilding within the city. Population:

| | | |
|---|---|---|
| St. Paul | 1920 | 234,698 |
| Minneapolis | 1920 | 380,552 |

In the two volumes of his report, Schmid described the Twin Cities as of 1935. From the data in his report and from our general knowledge about

[32] Calvin F. Schmid, *Social Saga of Two Cities*, Minneapolis, Minneapolis Council of Social Agencies, Vol. I, 1937, pp. 4-7.

these and other American cities, we can with confidence add a fifth period to those summarized above:

5. *Period of early maturing:* From the end of World War I to the full employment era following World War II (which includes the Great Depression). A period of extensive internal readjustment with tall office buildings, more modern public buildings, buses replacing streetcars, automobiles replacing horses, appearance of chain stores. Increase of scientific farming in hinterland. Special farming (dairying, chickens) begins. Suburbs grow. Population:

| | | | | |
|---|---|---|---|---|
| St. Paul | 1940 | 287,698 | 1950 | 311,349 |
| Minneapolis | 1940 | 492,370 | 1950 | 521,718 |

Some might say that Minneapolis surged ahead of St. Paul because the leaders there were more energetic and resourceful, but such a reality-avoiding explanation cannot be used in comparing the Twin Cities. Apparently at the outset the topographical advantages favored Minneapolis: It was easier for the railroad to locate its lines there; good transportation encouraged the flour mills to locate there; it was the logical hotel center. Because of these advantages Minneapolis became the center for offices and shops. When two cities are so close together, there can be only one main central area. The human factor doubtless had a part in putting Minneapolis ahead, but that was secondary to the various natural or ecological factors.

Many American cities that grew from villages or small towns since 1850 have gone through the same developmental stages as the Twin Cities. Structurally and in their way of life, all are alike in hundreds of particulars, as in other particulars they are different. They are developing in much the same manner toward whatever the next stage of urbanism may be. To some extent this may be said of modern cities in other countries. But if we look back through the century or the centuries, we find great variety of development among the cities that were evolving through the Middle Ages. And, because cities with long-term backgrounds have their separate and unique histories, each tends to be somewhat unique today. While we can think of a "typical" American city of 100,000 up to 500,000, especially among cities that have emerged since 1850, for Britain, France, Germany, or Italy one cannot speak of any city as being typical of a population-size class.

As cities evolve from one stage to another they may also change their work.[33] This happens often to small places as they reach city stature. Another

[33] A. J. and J. J. McIntyre, *Country Towns in Victoria,* Melbourne, Agricultural College, 1944, p. 18. This is an account of villages established by gold miners which, when gold mining failed, became agricultural. The transition involved various structural changes as well as changes in the way of life.

example might be that of a steel plant locating in a university town, bringing in a different type of labor force. Rubber manufacture, especially automobile tires, after about 1910 changed Akron from a small trading city with plants that made cereals and clay products into a specialized industrial city. This "rubber city" grew from 69,057 in 1910 to 208,435 in 1920 to 275,605 in 1950.

## Boom and Depression in Communities

A period of boom with full employment may result in rapid urban growth and change, while depression with less than full employment may not only halt urban growth, but may lead to the decline or even the extinction of a community. Either prosperity or depression may affect a wide region—a nation or a group of countries—or only a limited area, perhaps a single city. Thus, a limited area may prosper if oil is discovered, and a village within that area may become a boom town. If the oil wells run dry, a depression results and the town may later be reduced again to a village. Technological change may affect an area either favorably or unfavorably. Seaports that flourished in the days of small wooden ships were in some cases isolated when the modern deep-draught vessels came into use. The ancient seaport of Bruges, Belgium, is an example.

Through the years a city may experience a succession of booms and depressions. New Bedford, Massachusetts, for example, once flourished as a whaling port, but its future looked uncertain when that industry declined. Then other industries entered and again the city boomed, but these industries also declined after a few decades, and once again the city seemed to be faced with "old age." Population fell from about 122,000 in 1920 to 109,000 in 1950.

Birkbeck, an English traveler, visited the United States in 1817. He was impressed with the vigor of the new country. Everywhere he saw building, working to get ahead, moving to get somewhere to take up land or start a business; and he also saw the concomitant aspects of the situation. He wrote this about Richmond, Virginia:

> Richmond contains 13,000 inhabitants, nearly half of which are Negroes. The hill on which stands the capital, a building of commanding aspect, is inhabited by the more opulent merchant and professional men, who have their offices in the lower town. Their houses are handsome and elegantly furnished, and their establishments and style of living display much of the refinement of polished society. The town is generally well built and increasing rapidly, whilst

> but little provision seems to be made in the country round for the inhabitants. The market is badly supplied. . . .
>
> The town is forced upward by the stimulus of commerce, while the surrounding country is groaning under the torbid influence of slavery; the cultivators are said to be jealous of the rising prosperity, instead of availing themselves, as they might, of the rising advantages it would afford as a market for their produce.[34]

Birkbeck saw in Richmond many English and Scottish immigrants, and in Pittsburgh, also a booming city but of later origin than Richmond, he met Irish immigrants in large number. They were all part of a town-building boom, and that boom atmosphere—as if people were working against time—prevailed in all the places he visited.

Los Angeles has experienced a series of booms and depressions. When this dusty little town was first reached by the railroad in 1876 it became a center for land speculators whose activities increased when the second railroad arrived in 1885. There was sharp competition between the two lines, and this brought thousands of passengers at less-than-cost rates from Kansas City and other places. Newcomers arrived by the trainload, and most of them came with speculation intentions.

> After Los Angeles city lots had been snapped up, promoters began to lay out new cities—in farming regions, on barren hillsides, in the desert, and even on mountain tops. These subdivisions were surveyed, marked with stakes and flags, and thrown on the market; almost overnight people flocked to the site, stood in line hours to buy. Some paid as high as $1,000 for a place at the head of the line. By the end of 1887 there were 27 mushroom towns along the Santa Fe railroad between Los Angeles and San Bernardino, and still others were springing up all over the Los Angeles area. One speculator sold about 4,000 lots in the Mojave Desert for as much as $250 apiece, the cost to him being about ten cents each.[35]

Most towns founded on the frontier, if they did not begin with a boom, at least experienced periods of boom and "bust."

Speculation and venturing, especially in modern times, are characteristic of the urban man. He trades in values—land values or the values of the market. The farmer grows the corn and wheat, but the market speculation in these commodities is urban. The country man may bet on a horse race or a dog fight, but he is not a gambler in the market. The urban man, with his urge to advance and to exceed today whatever was done yesterday, and with

[34] Morris Birkbeck, *Notes on a Journey to America*, London, 1818, p. 20.

[35] *Los Angeles, A Guide to the City and Its Environs*, American Guide Series, New York, Hastings House, 1941, p. 43.

his sporting concept of competition, is never quite sure what may be waiting around the corner. The values with which he deals are abstractions. As Mumford puts it, "The abstractions of time, energy and money are more real to him than any other system of abstractions, and from these he deduces a qualitative ideal of life. . . ."[36] The tempo and the tone of modern urbanism arise out of this climate of energy and venturing. This way of life assumes that there will be climbing up and falling down, not only for individuals, but for communities.

Depression may be a condition that develops quickly—and this seems equally true of the boom type of prosperity—but it may also develop gradually, not so much a fall as a sinking trend, chronic rather than acute. The chronic type of depression may be found in rural areas. Thus Dumazedier describes certain villages in France:

> But it is in the Brie district, with its mixed crops and cattle breeding, that the situation of the small farmers is the worst of all. Their buildings, tools and methods are old-fashioned and out of date; yields are very inadequate, and the produce is hard to sell, being greatly over-priced. In such circumstances, the small farmers tend to withdraw into themselves, cutting down their capital and leaving land untilled. They do not dare to borrow money for modernization purposes. They work themselves to death to achieve a standard of living which in many cases is very low. Even the wealthy among them have often grown avaricious, for fear of what the morrow may bring. The young men are heartily sick of the work, and the girls are ready to leave for the town at the first opportunity. Thus the villages are being drained of their population.[37]

This is chronic depression as it may affect a village, even though the city may be reasonably prosperous. It is the equivalent of industrial unemployment, but in this case both men and land are unemployed or underemployed. Villages, like towns and cities, may go out of existence for a variety of reasons; chronic depression is one of the conditions which can lead to such decline.

## The Rise and Fall of Cities

In this chapter we have considered some of the phases in the evolution of cities, an evolution which through the centuries has been gaining in momentum. It concerns on the one hand the artifacts which are sometimes called the material side of culture, and it includes on the other hand the ways of life

[36] Lewis Mumford, *The Condition of Man,* New York, Harcourt, Brace, 1944, p. 264.

[37] Joffre Dumazedier, *Television and Rural Adult Education,* Paris UNESCO, 1956, p. 118.

in communities. Many who have studied these phenomena in long-term perspective have been tempted to arrive at conclusions about the future of community life, especially in great cities. Spengler, who saw the basic values only in the rural man, regarded the city as something unnatural. Thus, he concluded, cities are headed for destruction and cannot be saved. Like so many cities of the past, modern cities may rise high but will eventually fall.[38]

Such "rise-and-fall" or "life-cycle" views have been expressed by others, or are at least implicit in the writings of those who hold to the idea that cities and towns live through periods of youth, maturity, and old age. Sert regards cities "as living organisms, as things that are born and which develop, disintegrate and die, for cities are no longer to be studied as immobile and finite units."[39] Often, in support of such conclusions, the fate of many ancient cities is brought to our attention—Memphis, Babylon, Troy, Ur, Shush, Carthage, and others. This is all convincing, up to a point.

To Brownell, the city is a parasite, essentially the view of Spengler, and it lives by aggressive relations with the "less aggressive" rural hinterland. It takes much and gives little in return, hence "it must conquer and continue to conquer, or die."[40] This thesis, also, can be supported, up to a point.

Although we recognize that many cities have died, we need to ask why particular cities died and were never revived, while others that were destroyed did not go out of existence. Before its demise, how did the dead city sustain itself? What was its hinterland? A city does not die alone. It is part of a complex, the center of a pattern which includes other communities of different size. When we think of a city rising and then dying we must think of it in relation to a total situation to which it was central.

Rosier, writing about cities and other agglomerations that have passed into history, recognizes that each city or town or village died for a particular reason and by a particular means. The contributing factors might have been changes in transportation routes, political changes, or natural conditions. A place may be destroyed by war, fire, earthquake, flood, and so on. What happens after the destruction is also a situational consideration.[41]

Even when a place is totally destroyed, another city is likely to rise in the same general area if the site is suitable economically and geographically,

[38] Oswald Spengler, *Decline of the West*, New York, Alfred A. Knopf, 1926.

[39] José Luis Sert, *Can Our Cities Survive?* Cambridge, Harvard University Press, 1942, p. 4.

[40] Baker Brownell, *The Human Community*, New York, Harper & Brothers, 1950, p. 134.

[41] Rosier, *op. cit.*, pp. 212-214.

and if political interference does not prevent it. Babylon is often mentioned as a city that died and did not rise again, but not so many miles away another city, Baghdad, later took its place. Cairo is the better-located successor of the ancient Egyptian cities.[42] Murphy mentions that a succession of seven cities have stood on the approximate site of Delhi, India. One of these, Tughlukabad, was totally destroyed in 1398 by Tamerlane. However, new cities grew up again because in this region a great city seems to have been needed.[43] This cannot be said of the ancient city of Carthage whose relation with its hinterland was artificially maintained by military power.

Some agglomerations, whether villages, towns, or cities, may live to a very old age and then decline or pass out of existence; others may decline or disappear a very few years after founding. In the development of the United States, many of the places that were founded and aspired to be towns did not reach this status. If such places did not disappear, they declined in importance. Most familiar of American ghost towns are those that were established in connection with the lumber industry or mining activities.[44] The same industry that enables a town to grow may, by moving away, cause the town to decline.

Inherent in the life-span idea is the assumption that a place is declining if new construction is not in process. The position of such a town or city, however, must be seen in a much wider relationship. Its decline, for instance, must be seen in relation to the changing status of other cities and towns, for this type of relative decline may be followed by a reverse trend. Whatever happens, in long-term perspective, depends of course in part on factors within the community, but it may depend more on outside and related factors. These are the realities which make it difficult to accept any general "life-cycle" hypothesis.

## Summary

The modern city has come through a long period of change and learning. In this chapter some of the salient elements in the natural history of cities

[42] This writer visited the excavations of the ancient city of Ur where the remains of several cities, one over another, may be seen. Each city had its "life span." There is hardly any need of a city in that locality today, either for land or water transportation.

[43] Rhodes Murphy, "New Capitals of Asia," *Industrial Development and Cultural Change,* Vol. V, No. 3, April 1957, pp. 225-226. Delhi grew from 192,578 inhabitants in 1891 to about 1,650,000 in 1956. Its functions have become more administrative and cultural than industrial and commercial.

[44] *Ghost Towns of Colorado,* American Guide Series, New York, Hastings House, 1947.

were considered. Technical in large part, this evolution has also been socio-psychological, and in the most objective use of the term, this evolution may be called progress. The experience of collective living has been cumulative and it has been transferred from region to region.

Ancient cities, seen in retrospect, were centers of experiment and preparation. People were learning to live in agglomerations of increasing size, learning to live and find sustenance detached from the land. The legacy of this accumulated experience moved westward, coming mainly out of the Middle East. Hardly any of this influence flowed from the ancient cities of the Far East.

The greatest progress in structures, the arts, and effective urban management was achieved in the less ancient cities around the Mediterranean. They also expanded commerce and improved on the art of war (an initial form of public work). Their influence expanded inland to the tribal peoples of Europe. Their ships found the way to Africa and still later along the Atlantic coast of Europe.

The new type of urban communities that began to emerge in Europe largely evolved out of the various European folk cultures. They also drew heavily on the heritage that came by way of the Romans, who had occupied parts of Europe for nearly four centuries. Beginning about the tenth century, this urban evolution was marked by a greater orientation to industry and trade and was more inventive than earlier urban civilizations had been. The evolution also moved faster and was more vigorous. It paved the way for the modern city.

The modern city must be associated with a variety of related developments, each of which had its own long history: transportation and communication, technology, public administration, social organization, and science. The greater emphasis on craftsmanship heralded the beginning of a trend toward a wider division of labor. Although the dates vary among countries, we can say in general that the beginning of the modern city was between 1750 and 1850.

The development of modern cities was also accompanied by a trend toward specialization, and on this basis cities and towns can be classified into different activity groups.

Industrial urbanism is only one of the identifying characteristics of the modern city, but it is perhaps the principal one. The city today is the center from which change-promoting stimulation radiates. Its way of life, although not carried on unaware of the state and the hinterland, tends to be independent of them; in fact, it tends to influence them.

Cities have always grown as circumstances would allow, but technological advance tends to encourage greater size today, as is necessary if surplus population is to be satisfactorily located and employed. Although urban size has frequently been a matter of concern, cities have continued to receive people "to the point of pain." The problem is serious only if urbanization takes place faster than work can be found.

Earlier cities, particularly the ancient ones, apparently were in a condition of continuous depression. Modern industrial cities, since they are more specialized, their industries more productive, and their markets wider, have been able to achieve higher living standards; but they have not been able to secure themselves against depression. Especially during the past century, most cities in the Western countries have experienced, some more than others, a series of booms and depressions. Whether these events are national or merely local, they may affect urban growth and urban morale, favorably or otherwise. Moreover, the effects may also extend into the hinterlands.

In a prolonged depression, smaller communities may go out of existence and larger places may decline. In the long history of urban communities, many cities—some of them famous—have passed out of existence. This fact has encouraged some to evolve theories about cities having "life cycles." It is more to the point to say that cities live so long as they have work to do, and, even if destroyed, they will rise again, or other cities in the general area will take their place. Cities, towns, or villages, if they decline or die, do so usually for technological, economic, or other reasons that can be identified—not because it is their destiny. In the course of intercommunity competition, some places are at a disadvantage. A prosperous, growing city may later decline, or a declining city may later revive. In either case, the process is dependent upon the work a city does and the amount of work it can secure. Moreover, the prospects of any city are closely linked with those of the towns and other communities in its hinterland.

CHAPTER 4

# The Village Confronted with Urbanism

THE VILLAGE IS THE MOST UNIVERSAL of communities and it is likely to remain so for a long time, adapting itself to urbanism as it has through its long history adapted to other influences. Hundreds of villages in a region may die, but *the village* remains, much as *the city* survives, although many cities have died. Nor is the present trend toward global urbanism likely to diminish the importance of the village in modern society. In spite of great numbers of people flocking to cities and their environs, it is better than an even guess that throughout the world more people live in villages today than ever before.

Expanding urbanism meets the resistance of ruralism mainly in the villages. In this chapter we need to ask if this resistance may not have a counterinfluence on urbanism. While no clear-cut answer will be found, we will do well to keep this question in mind.

As we considered the term in the opening chapter, urbanization is a two-part process: the movement toward urban places and the outward movement of urban influence. Thus as urbanization concerns villages, we can think of the process at different levels of completion.

(1) As people move into cities and as cities enlarge their areas many villages may be absorbed. Perhaps they lose their old village character but their names continue as the names of the areas where they once stood as rural communities.

(2) Villages within commuting distance of cities, although formerly occupied by rural people doing rural work and living the rural way of life, have come to be occupied by urban-oriented people doing urban types of work.

(3) Villages more remote from cities, although in frequent contact with urban centers, may become places of mixed rural and urban types of work, but

the people engaged in rural work may become increasingly urban in their way of life.

(4) Villages still more remote from cities and in less frequent contact with urban centers may not only continue to perform their rural work but, in spite of urbanization pressure, manage to retain in large measure the rural way of life.

(5) Villages, especially in underdeveloped regions, that are very remotely situated and have almost no contact with urban centers may still be little affected by the influences of urbanism. They remain wholly rural or primitive in their work and in their ways of living. They may be likened to a frontier, the outer line of extending urbanism.

Villages of the fifth level are numerous in some world regions, but they hardly exist in the United States, where even villages of the fourth level are few. The same would hold for such urbanized countries as the United Kingdom, Belgium, Holland, and the German Federal Republic.

Our task in this chapter is to consider a limited number of village types and to see if they can be located on an imaginary continuum between the most rural and the most urban. We may be able to come to some tentative conclusion regarding what urbanization means with respect to villages and the relation of villages to urban centers.

## The Nature of Ruralism

Let us begin again by asking what ruralism means. The idea is associated by some with peasantry, a way of rural life that no longer exists in the United States. In the traditional meaning of the term, in fact, it would be difficult to find peasantry in any advanced industrial urban country. Evans, the English geographer, observes, "We may find peasant values persisting among farmers who would resent the term. . . ." He does not find peasants any longer among the family farmers of England, but he recognizes a "modernized peasantry" in Denmark. "The word has always carried an implication of rustic inferiority, and we tend to apply it to countries other than our own. . . . Yet the word implies a permanent link with the soil—the *paysan* with his *pays*—so that we hesitate to apply it to the shifting cultivators of tropical Africa." The peasant is a family farmer who depends more on his labor and that of his family than on capital. His is a way of work that has not yet been encroached upon by specialization, mechanization, and the thinking of science.[1]

[1] E. Estyn Evans, "The Ecology of Peasant Life in Western Europe." See William I. Thomas, Jr., Ed., *Man's Role in Changing the Face of the Earth*, Chicago, University of Chicago Press, 1956, p. 220.

The French scholar André Siegfried in 1937 wrote a book about Canada in which he praised the French Canadian as a peasant. The Americans, he said, think of the peasant as a "serf attached to the glebe." He mentioned that there were no peasants in the United States, which he thought regrettable at that time of mass unemployment in American and Canadian cities. He observed that the French Canadian peasant was not greatly affected by the Great Depression because he could live with his tradition. In contrast to Americans, the French Canadian is rooted.

Siegfried sees the French Canadian as a "classic" rural man whose culture is oriented so much to his land and who is so stolidly contented with that way of life that opposition to Americanism is a natural outgrowth—Americanism seems the antithesis of his cherished attitudes. He lives for his crops, his children, and his church. Siegfried sees a "Biblical grandeur in such simplicity," and he seems convinced that it would be well for all countries if they had stolid peasantries as in the French part of Canada. He would, in short, have more of the old-fashioned rural people.[2]

Siegfried reported that the great worry of the priests in these French Canadian villages is that the people may come under the influence of Americanism, because the birth rate is high and the father is not able to provide all of his sons with farms. Increasing numbers migrate to the cities where "Americanization lies in wait," whether in Canadian or American cities. Once in contact with American urbanism, few are able to resist its influence.[3]

The French Canadians are almost the only first-settlement people on the North American continent who, after two centuries, can still be called peasants. They cling to a level of ruralism that has changed little since they brought it from France. Their economic condition might be very perilous were it not possible for many of their children to migrate into the more urbanized places.

Life among the first settlers in the New England colonies was also rural in a primitive way, as is suggested by the following item about New England in the early 1800's:

> The people in these times were a very plain people, dressed in homespun cloth. Every house had its loom and spinning wheel, and almost every woman was a weaver. Carding machines were just introduced at the beginning of the 19th century and clothiers had plenty of work. The first coat I had cost me a dollar and a half per yard, spun and woven by one of my best friends and I know not that I have ever had a better. For many years there was not a single wheeled

[2] André Siegfried, *Canada, An International Power,* London, Jonathan Cape, 1937 (revised 1949), p. 114.

[3] *Ibid.*, p. 72. For a study of such a local community, see Horace Miner, *St. Denis, A French-Canadian Parish,* Chicago, University of Chicago Press, 1939.

> vehicle in town. People who owned horses rode them, and those that had them not went on foot. Husbands carried their wives behind them on pillions. More than one half of the church-going folk went on foot. Sleighs or sleds were used in winter. I have seen ox sleds at the meeting house in Bath; and yet in the coldest weather the house was always full.[4]

Unlike the French Canadians to the north of them, these New Englanders were soon occupied with changing their way of life; in fact, before a half century had passed this section of the United States had become the dynamic center of American industrial urbanism.

If it were applied to France itself, Siegfried's estimation of the peasant way of life would not meet agreement among all his French colleagues. Some of them are concerned about what they regard as the backwardness of the French peasant. Dumazedier, for example, reported that of 21,500,000 rural people in France, 9,000,000 do not have running water in their homes; 95 percent live in villages without communal drainage and sewage systems; 70 percent live in communities without a village hall, and in 65 percent of the villages there are no sport fields. High percentages of the rural homes are without electric power.[5] Living standards are described as less favorable than the lowest levels for French industrial workers, and the cultural levels in these rural places are equally low.

Ruralism or peasantry involves a close and continuing bond between man and land, usually in the form of familial ownership of land and a familial work organization for extracting a livelihood from the land. These observations may apply equally to pastoral peoples. There is usually a sensitive relationship, often fear dominated, between man and nature. Attitudes about matters of life and work, the uncertainties of the future, and right conduct are usually entwined in a religious frame of reference overlaid with various folk beliefs that linger from generation to generation. Vidich and Bensman in their study of a modern rural community in New York found that work still holds a key place in the pattern of rural respect, along with religion, frugality, and self-improvement. The rural man is likely to hold in low esteem those who do not work hard many hours each day. "Industrial work with its regulated work day is viewed as offering a man little basis for respect. . . ."[6]

In rural life generally the ways of work are repeated with the seasons,

[4] *New Hampshire, A Guide to the Granite State,* Boston, Houghton Mifflin, 1938, p. 37.

[5] Joffre Dumazedier, *Television and Rural Adult Education,* Paris, UNESCO, 1956, pp. 118-119.

[6] Arthur J. Vidich and Joseph Bensman, *Small Town in Mass Society,* Princeton, Princeton University Press, 1958, p. 50.

as ways of living are repeated with the generations. Innovation is possible, but rare, and there is little to stimulate experimentation; in fact, the usual response to either is likely to be skeptical, if not condemnatory. Moral codes are not only strict and particularistic, but through primary group pressure they are severely enforced. To these generalizations one qualification must be added; namely, they tend to be most applicable to the most rural of people and to some degree less applicable to the more urbanized of rural people.

## The Village in Iran

In Iran are thousands of villages, most of them small and isolated. Many are reached only by trails and a good number of them are still encircled by walls. The Iranian peasant is rarely a landowner. As observed by Turner, in this country as in India may be found different types of absentee land ownership. While the types of ownership change through time, it has usually been a transition from one type of absent owner to another.[7]

Most of these villages must depend on irrigation, which means that their size is necessarily determined by the amount of water supply. For this reason many Iranian villages have not increased in population for generations, and with the generations their ways of work have changed very little. (This is less true today.) Nor have their ways of life changed appreciably, except that as they have gradually become identified with the money economy and therefore live less than formerly on a subsistence economy, they have been experiencing new hardships. Thus many, if not most, of the villagers must concern themselves with getting cash.

Neighboring villages may be from a mile to several miles apart, but there is a minimum of village-to-village contact, and the contacts that are made are often marked with caution. Strangers from far places may be well received by village families, but when they leave they are likely to be warned that they must be on guard against the people in the next village. The ways of work and life are the extreme of ruralism as described above. The social organization is familial; usually there are two or three family groups fitted neatly into a domination-subordination pattern, with each subfamily in each group having a definite status and each individual having his niche in his subfamily. Interfamily politics is a serious and continuous activity on the part of everyone and

[7] Ralph Turner, *The Great Cultural Traditions*, New York, McGraw-Hill, 1941, Vol. II, p. 1258. This writer spent two years in Iran and visited a number of villages of the type described here, villages owned by urban-dwelling families. Such an urban family may own several villages.

there is much interest in carefully arranged marriages which tend to keep the power arrangements in balance. Work arrangements are under direction of the *headman* who may be selected on agreement between family groups and the village owner. He is responsible to the owner, but he is also responsible to the villagers.

As already mentioned, water is the life giver for the village. Usually a tunnel is dug into the hillside from which the small stream comes. This *kanat* is designed to follow the stream to the source in order to increase the flow. Work on the *kanat* may continue for years.

The following is a description of an Iranian village about 1900, written by Sykes, a British Consul in Iran.

> The village of which I made a study is owned by a merchant. It consists of 32 domed houses built of sun-dried bricks around an enclosed square. It possesses no mosque, bath or caravansarai. The site occupies one acre of ground, and two walled gardens which adjoin the village and grow fruit trees, vines, willows, etc., have together over one and a half acres.
>
> The population is: men and youths, 20; women, 15; children (mainly unfit for agricultural labour), 15; total, 50. Following is the live stock owned by the village community: donkeys, 14; oxen and cows, 20; goats and sheep, 150; fowls, 50.
>
> The total area cultivated is 946 acres. Of this, 346 receive kanat irrigation, and the remainder is dependent upon rainfall. The quality of the soil is good. There is one kanat only, which is the property of the owner of the village, who keeps it in a proper state of repair. Should any work on it be necessary, the villagers are employed for the purpose and receive a small payment. Each villager receives water (for irrigation) every tenth day for about six or seven hours.
>
> The land is let to tenants-at-will. The owner has a representative in the village, whose duty it is to superintend the distribution of the water, to collect the master's rent and generally to act as steward of the property.
>
> The tenants are ten in number and the average area of their holdings is 94 acres. They own eight yoke of oxen. The chief crops grown are wheat, barley, oil seeds, opium and lucern. Cotton, millet and turnips constitute the autumn crops.
>
> The proprietor alone is responsible for the payment of the government taxes. They are all collected in two installments, at the time of the harvesting of the spring and autumn crops.[8]

If we knew this village and could visit it today, half a century after the description by Sykes, we would doubtless find that it is no less rural in its way of life. But some changes have taken place. The school comes in from

[8] Percy M. Sykes, *A History of Persia,* London, Macmillan, 1915, Vol. II, p. 495. (New York, St. Martin's Press.)

the outside and, although indifferently received, becomes established in the villages, as an urban influence.[9] Young people go to the towns and cities, and the villages come more under the money economy. More outside goods enter these villages and greater effort must be made to send produce out in order to get money. But many of these villages are still owned by urban families and the people have little direct contact with civil government; to them the government is a fearful thing and remote, and many hide when the census taker comes. We are here speaking of Iranian villages far from the roads. The villages on the main roads are more under the influence of the kind of urbanism that is found in Iran, and villages near the towns and cities are still more urban in thinking and behavior.

In the isolated villages of Iran people do not resist outside influences; rather, these influences are ignored. As Tumin has said of the Guatemalan Indian, he accepts his way of life and "positively affirms its goodness." He is "relatively untouched by the continuous and openly-expressed depreciation of his round of life."[10] The Iranian farmer continues to thresh his grain as in Bible times. We may think of him as passive and established, although between generations changes take place. Villages of this type can hardly be found in the more industrial countries, but we can find villages which seem to be at the fourth level of the classification offered at the opening of this chapter. The Irish village, mostly an open-country community, might be so identified.

## The Irish Village

The rocky, treeless countryside of the Irish Republic in 1800 was populated to the limit of its agricultural capacity. About 1830 the out-migration began. The total population in 1841 was about 6,500,000. It was down to 2,963,000 in 1926 and 2,958,000 in 1950. Some rural places since 1841 have declined in population by as much as 80 percent, while some urban places have gained in population.

Between 1820 and 1950 no less than 4,617,000 Irish migrated to the United States, and there were lesser migrations to other countries. Perhaps no other country in modern times has had such a high emigration rate, and no

[9] John Saville (*Rural Depopulation in England and Wales, 1851-1951*, London, Routledge and Kegan Paul, 1957) mentions that farmers in England complained about the Education Act of 1870, "Children, it was alleged, acquired an urban outlook from their schooling, and they became restless and dissatisfied with rural life." P. 198.

[10] Melvin M. Tumin, "The Dynamics of Cultural Discontinuity in a Peasant Society," *Social Forces*, Vol. 29, December 1950, p. 137.

country has benefited more than the United States. But Ireland has remained rural and rural Ireland has retained fairly intact its old way of life. Arensberg and Kimball describe the "Irish countryman" as a conservative person deeply loyal to his community and its traditions, and very devoted to the land; but his community is not so much a place as a kinship group.[11] The people live scattered on their stone-fenced farms. The farmer with five or six cows and 30 to 50 acres is said to be well-to-do. Perhaps he migrated to the United States in his youth and saved the money to buy the land, and perhaps his own children who migrated to the United States send him money. Later, some of these children may return to Ireland as adults, and the pattern is repeated.

Life is lived less in terms of the scattered community of individual farms than in terms of the individual farm house. The "farm family group spends its entire life sleeping, eating, giving birth and dying there, and sallying forth every day for work upon the fields."[12] And when the individual reaches his older years, especially a man of means, he will have his place in the informal group, the *cuaird,* which meets regularly in one house or another. The younger sit back while the older ones lead, whether it is "in conversation, singing or discussing the news" or telling tall tales.[13]

Ireland is a land of late marriage. In County Clare, according to Arensberg and Kimball, 88 percent of the men between 25 and 30 years and 73 percent of those between 30 and 35 are still bachelors. The marriage age for women is generally younger. The oldest son, who will inherit the farm, feels duty bound to postpone marriage until financial arrangements can be made for the younger sons. When he does marry, the parents move to the "west room" and the younger sons are expected to depart.[14]

Irish rural work is perhaps as advanced as the uncooperative land permits, but the rural way of life lingers on and is treasured. Evans remarked that the Irishman away from home is often considered lawless. He lacks the "disciplined tradition of self-government and respect for law and order" common to Englishmen:

> In the pastoral uplands loyalties are to kin rather than to community. The Irishman's inherent lawlessness and independence spring from this attachment.

[11] Conrad M. Arensberg and Solon T. Kimball, *The Family and Community in Ireland,* Cambridge, Harvard University Press, 1940. This study was carried out in County Clare, a middle-type county where the people are almost wholly identified with the rural economy.

[12] *Ibid.,* p. 31.

[13] *Ibid.,* p. 131.

[14] *Ibid.,* pp. 106-112.

> The countryman likes to have personal relations with those who govern him, and the outside world should be capable of being manipulated on a personal basis. It is all-important to "know the right people." "Friendship" keeps its original meaning of blood relationship, and the claims of friendship may be stronger than the claims of abstract justice. Civic virtues are poorly developed.[15]

Irish ruralism is not motivated by attitudes of world rejection, as in the isolated Iranian village, but is to some extent a conscious choice. The rural Irishman's chosen "traditional custom of life persists and continues to wield its power in essentially similar fashion decade after decade and generation after generation."[16]

## The Spanish-American Village

The Iranian type of rural life, not unique for the Middle East, is touched only lightly by expanding urbanism. The Irish type of ruralism lives satisfactorily in the midst of urban influence, but is able to conserve its old ways because of a favorable geographical isolation. For many Irishmen who know the world well, it is an ideal milieu to which one may one day retire. We come now to a third type of ruralism, that of the Spanish-American village, which is found both in the United States and in other countries. It is a type of ruralism which may be less world rejecting than indifferent, and although it may be surrounded by industrial urbanism, it maintains its own character.

Such a village is El Cerrito on the Pecos River in New Mexico, studied by Leonard and Loomis.[17] It is one of several villages on a Mexican land grant of 1822. Much of its land, good only for grazing, has been lost in tax sales but, at the time of this study (1939), some 85,000 acres remained of which 22,000 were leased to cattle companies at 8 cents per acre per year, hardly enough to pay the taxes on the entire domain. These "land-poor" people are also poor in that they do not have the resources for utilizing their grazing lands. They do not strive to build up herds and flocks. While they live in the "Americano" world, they do not feel any urge to adopt its ways, and if they try to their efforts are usually ineffective and may even be damaging to themselves.

In this village of 26 families, holdings range from one to four acres of

[15] Evans, *op. cit.*, p. 234.

[16] Arensberg and Kimball, *op. cit.*, p. 310.

[17] Originally from Olen Leonard and C. P. Loomis, *El Cerrito, New Mexico*, Rural Life Studies, No. 1, United States Department of Agriculture (now out of print). Citations are from a republication by Loomis, *Studies of Rural Social Organization in the United States, Latin America and Germany*, Lansing, Michigan State College, 1945.

irrigated land, and from 10 to 40 acres of dry farming land. Two families in the village were reported as owning old automobiles; the rest use wagons or saddle horses. The compact community is both economically and socially homogeneous. Children are as much at home in other households as in their own, although the community cannot be considered a "happy family." There are many rivalries. El Cerrito families can trace their lines back several generations to Spanish and Indian ancestry. Moreover:

> The Indian heritage was leaned upon heavily for local knowledge of terrain, the elements, and the means of combatting them. Agricultural techniques were of special significance, for many of the early Spanish colonists were not farmers. Evidence of knowledge gained from the indigenous people can be seen today in the houses, the crops and roads, the methods and tools of farming used by the natives. The process was one of borrowing rather than assimilation. Today the native still speaks in the language, enjoys the customs and is endowed with the superstitions of Andalusian Spain. Nor is there any conflict in the combination.[18]

Songs reminiscent of sixteenth-century Spain are still sung and institutionalized pastimes date back to the same source. Work ways, however, seem more to reflect Indian origins. Community attachment is strong, being a mixture of loyalty to family and to place. A person may speak, as do the Iranians, in derogatory terms of a neighboring community, but seldom does he make negative remarks about his own village. Of family loyalty, Leonard and Loomis wrote, "The authority of the father was never questioned. The oldest male in the group occupied a position that was never lost or relinquished until his death. His authority was felt in every function of village life."[19] This ancient formula for social order seems to hold for all rural groups.

The rural Iranian is a steady although not an ardent worker, and the Irish countryman is a hard worker. The Spanish-American villager, on the other hand, is casual about his work, not urgent in the American sense:

> These natives are able to see neither sin nor moral corruption in idleness and leisure time. They see neither virtue nor common sense in keeping busy for the sake of occupying the hands and the mind. Work is simply a means to an end—a means of accomplishing that which is valued or desired, and as such these people realize its importance. Children are taught at an early age the necessity of labor and are given ample opportunity to put it into practice. But the necessity of work lies in what it will bring in a material way—money, land and independence. Furthermore, the mere accumulation of material goods adds little to popular esteem for the individual. The prestige of the man depends

[18] *Ibid.*, p. 280.
[19] *Ibid.*, p. 283.

in no small measure upon his contribution to the fiestas or his activities in the political life and activities of the group.[20]

Although these attitudes of Spanish-American villagers are descriptive of a rural people, we must not make the mistake of assuming that they apply to rural Spanish-Americans only. To no small degree the thinking and behavior of people in the El Cerrito village may be very much like the thinking and behavior of other Spanish-Americans living in an industrial village. The difference might be that life in the rural village would be more compact and homogeneous, while that in the industrial village would tend to be disturbed by the distractions of urbanism.

One thing needs to be added regarding the El Cerrito villagers: Their way of life is basically, although not always piously, religious. Much of their leisure-time spending, not associated with family ceremonials, is linked with various religious and semireligious festivals. Again, this is not a strictly rural characteristic, although Tumin speaks of the "inter-connection of agriculture and religion" in describing the Indians of Guatemala, also under Spanish tradition:

> For this is true that the central core of religious belief and ritual is associated with the land. The primary religious celebrations are oriented to the spring planting and the fall harvest. The controlling deities ("disguised" as Catholic saints) are primarily agricultural in their concerns and controls. The most solemn occasions in which the greatest religious fervor and participation are evoked are periods of crises of the crops; prolonged drought, excessive rainfall, and the like.[21]

But these same rural festivals, perhaps with a different meaning, or no religious meaning at all, may be retained by people moving from the rural to the urban way of life. This is especially true in European countries. Many urban holidays such as May Day and Thanksgiving Day are of rural and religious origin.

## The English Village

American colonial life was largely centered in the village, and the traditions of the colonial village were largely English in origin and rural in char-

[20] *Ibid.*, p. 283. From Ralph L. Beals we get a similar report on work and wealth and one's obligations to give for fiestas. "No one can readily idȩntify the rich man. Partly, of course, this is because rich men, in order to avoid envy, are careful not to flaunt their wealth." *Cherán, A Sierra Tarascan Village*, Washington, Institute for Social Anthropology, 1946, p. 89.

[21] Tumin, *op. cit.*, p. 137.

acter. The concept of liberty for the individual and autonomy for the community, the village offices, rules for making roads, the town meeting, local responsibility for the needy—these are of English heritage. There were contributions from other Old World countries, but those from England were dominant. Our acceptance of the common law and the language of England are further examples. Trevelyan adds our "habits of self-government" and "Bible religion."[22] That English village, the rural type which was transplanted to the colonies, has all but vanished in the mother country.

The rural English village of the eighteenth century was indeed plain in its layout, structures, and life. While the people then were poverty burdened, community life had a wholeness that is quite absent from English rural places today. Each village had its lord of the manor around whose manse the village clustered, surrounded by the fields and pastures. Government was a mixture of civil and church administration, but these were already changing as was the position of the lord of the manor. As described by Bonham-Carter, the village was orderly, naturally but not severely so. Next to the justice of the peace, the important official was the constable, who busied himself with ordering, forbidding, chastising, officially concerning himself with every type of conduct.[23] It was quite normal and not at all an inconvenience for these villages to be isolated from one another:

> Only a small proportion of the villagers of either sex ever visited town. Most people remained all their lives under the influence of Pan and his magic . . . for the common people, untouched by the skepticism of the town, the fairies still danced in the woods, though when the wayfarer came around the bush they had already vanished. Books in the villages were few. . . . No city-made newspapers or magazines stamped a uniform mentality on the nation. In this isolation from the world at large, each shire had its own traditions, interests and character. . . . For gossip and sensation they were satisfied with the daily drama of their own village, with its poaching, affrays and smuggling adventures, its feuds and loves, its ghosts and suicides, its quarrels of miller and innkeeper, of parson and squire.[24]

The rural village in England's preindustrial period could live in isolation. This becomes increasingly difficult, however, with the spread of urbanism, and it was true even during the first half of the nineteenth century, according to Saville. He mentions that a village of 500 inhabitants at that time had a "well-balanced group of craftsmen; a thatcher, a carpenter, two tailors, a boot

[22] G. M. Trevelyan, *English Social History,* London, The Reprint Society, 1948, p. 215.
[23] Victor Bonham-Carter, *The English Village,* London, Penguin Books, 1952, p. 150.
[24] Trevelyan, *op. cit.,* p. 321.

and shoe maker, a baker, a blacksmith, a wheelwright, two masons, two dairymen and a marine store dealer,"[25] and usually one or two inns. By 1920 most of these craftsmen were gone; factories had taken over their work. So too, various small local industries, once at home in the English village, had to cease operating. This has been true to a lesser degree in American rural villages.

This changing of the work pattern in the rural village has been accompanied by a steady movement of population from English villages to towns and cities and into industrial work. Saville notes that in 1851 approximately 49.8 percent of the population in England and Wales lived in rural districts, compared with approximately 19.3 percent in 1951. But he also notes that, while rural people over the past century have migrated in large numbers to industry and to cities, many who still live in the villages no longer have rural occupations. The problem is further complicated by the spreading influence of towns and cities into rural areas:

> The difficulties involved in the definition and measurement of urban and rural populations are not easily resolved. The nearer we approach our own day, the greater is the mixing of urban and rural and the greater the impact, at every level of economic and social influence, of the urban upon the rural; and an analysis which seeks to disentangle rural and urban populations becomes involved in complexities which increase rather than diminish as we move into the twentieth century.[26]

Saville's findings seem to confirm the conclusion that the more rural and isolated the village, the more likely it is to lose population and status in the face of spreading urbanism; if it is flexible and able to adapt, however, it may find a new status.

The little English village is no longer self-contained economically, and with this change the same becomes true culturally and socially. One must go to a town for his fun—to see a movie, to dance, or witness a sport spectacle. Morris described the lot of Ickham, a somewhat isolated village in Kent. He mentioned the signs of decay. Some villages, he noted, have adapted while others stand dazed. "It is extraordinary how drastically the pattern of life in this place has been shattered and how swiftly the customs of its 200 inhabitants are losing their old flavour." The old flavor included "the gentry," all of whom have gone except some widows; no more families use servants. Farming continues but "half the young men of Ickham prefer to go to the

[25] Saville, *op. cit.*, p. 23.
[26] *Ibid.*, p. 64. See especially pp. 59-69

towns for their jobs, so that it is a village of elderly people. . . ." Present interests of the people are outer-directed; for example, "the retired clergyman, handsomely bearded, writes his film scripts with dignity and panache. The well-dressed gardener studies astronomy. . . . What sense of dynamism there is to Ickham comes from the outside, and the village declines gently among its meadows."[27]

## Evolution of a Danish Village

The rural village in Denmark has had a long and stable history, and this is true of the other Scandinavian countries. However, within the past half-century, as these countries have become more industrial and hence more urban-oriented, the rural communities have been confronted with the necessity of adjusting—and most of them have adjusted. A report on this development in Denmark was made by Skrubbeltrang.[28] This intriguing document gives a general summary of developments since about 1750 and presents the story of the village of Vejen and the changes that took place therein.

Danish villages for generations have been governed by their local councils whose decisions were final. "Peasants had to fence their corn land at the proper season and take great care that fences and gates were intact and that no loose animals did damage to the crops. No person must move boundary stones, plow too close to another man's land, or inconvenience a neighbor by neglect of any kind."[29] There were rules for the harvesting and the exchange of labor. Fines imposed for breaking rules were held by the *oldermand* and each year the council decided to what community need the money should be assigned. Neighbors were expected to aid one another in times of distress. The clergy had a leadership position in practical as well as spiritual matters, although it was more advisory than authoritarian. The clergy kept the records of military service, taxes, and other matters, which were sent to the government. Sometimes the parson or rector was suspected of favoring the landlord to whom the peasants were beholden for work and the use of fields. The village by-laws "required regular attendance at church and godliness, but belief and superstition were often strongly mingled."[30]

[27] James Morris, "The Changing Village," *The Manchester Guardian Weekly,* April 25, 1957.

[28] F. Skrubbeltrang, *Agricultural Development and Rural Reform in Denmark,* Rome, Food and Agricultural Organization, 1953.

[29] *Ibid.,* p. 9.

[30] *Ibid.,* p. 12.

The village of Vejen in 1784 had a population of 114 peasants, some of them farm tenants and the rest laborers, but all responsible to the landlord. Work ways were fairly primitive. Wooden harrows and plows were used. Wheels lacked iron tires. Chairs were beginning to appear in the homes, as well as pewter plates and copper kettles. The clock arrived in Vejen in 1779. A few peasants owned hymn books, but few could read and fewer could write their names. Poverty was general.[31]

Vejen in 1801 included a population of 296. The peasants had now acquired the land. In the village were such occupations as blacksmith, locksmith, weaver, clog-maker, wheelwright, thatcher. There was one small store and an inn.[32] By 1834 population had increased to 476 and there were about 60 landowners. The farmers owned 127 horses, 338 cows, 443 sheep, 140 beehives, and 53 pigs. They were better equipped and the community was more prosperous than it had been under landlord control. There were 15 craftsmen and in addition the village had a tannery, a cloth-weaving plant for wool, a brickmaking plant. There was a growing interest in education. The village engaged a school teacher but did not achieve a school house until 1863.[33]

Vejen in 1860 included 751 persons. Farms had increased to 89. There were now 166 horses, about 750 cattle, 718 sheep, and 149 pigs. The increase of land holdings was largely due to cottagers acquiring small plots of two acres or so. The village council had been replaced in 1841 by a form of local civil government. Officials had to learn to keep records, to be responsible for education, for the care of the poor and for making roads, to collect taxes and expend public moneys, to regulate the sale of liquor and grant licenses for stores. Social life took another turn—there were dancing parties, rifle clubs, and singing groups. There was by now a growing interest in the folk high school movement and the village was becoming equally interested in politics.[34]

From the 1860's to the present the story of Vejen is one of continuous change. The railroad came in 1874. Population grew from 865 in 1870 to 1,431 in 1890 and to 2,110 in 1901. In 1901 only 833 persons were in agriculture, while 663 were in handicrafts and industry and 527 in other urban occupations. In 1906 there were 272 urban properties and 158 rural, of which 50 were farms and the rest were cottages with adjoining plots of land. Already Vejen had become more urban than rural in its work. A citizens' association had already been formed in 1898 to promote the well-being of the

[31] *Ibid.*, pp. 33-36.
[32] *Ibid.*, pp. 95-97.
[33] *Ibid.*, pp. 149-154.
[34] *Ibid.*, p. 150.

community. Now the village had four brickmaking plants, a margarine factory, and other small industries.[35]

In 1945, Vejen with its population of 3,664 was already becoming a small town. Over a considerable period the agricultural group had averaged about 900. Town people had by now taken the initiative for civic improvements such as sewers, a piped water supply, electric power (The rural landscape was cleared of windmills, the old source of power.), street paving, and public buildings. Although the farmer segment has opposed most of these improvements on grounds of tax costs, these rural people have not isolated themselves from the evolving urbanism taking root in Vejen. They have identified themselves with a large variety of rural organizations in Denmark, mostly cooperatives for buying and selling. They are willing to adopt any available means to improve their land and to increase its productivity. In the most enlightened use of the expression, they have taken on urbanism as a way of life while still working their land.

## An Urbanized Rural Village

The Danish village, so briefly described, was for centuries a poverty-burdened feudal community until partly through its own efforts and partly because of situational changes it became an independent village on its own responsibility. Out of tradition-bound peasantry, it was able to follow a course of development leading from a condition of complete ruralism to a condition of advanced agricultural development, an evolution which was accompanied by a high level of social and cultural development.

Sublette, Kansas, is a rural village that began without tradition, occupying a 24-mile-square area which in 1887 was organized as Haskell County. From the start it occupied itself with wheat growing. In 1887 the population of the county was 2,666. It was down to 2,088 in 1940, of which 582 lived in Sublette, the rest on scattered farmsteads. Today an increasing number are finding it convenient to move from the farms to Sublette. Horses are few and becoming fewer; tractors are used for working the land and automobiles for transport and hauling. Sublette is a frontier village that began in a situation of land speculation and it has little sentimental interest in land as such. As reported by Bell, it has ever lived either in a boom or in a depression atmosphere, prospering in seasons of rainfall and declining in seasons of drought.[36]

[35] *Ibid.*, pp. 226-241.

[36] Earl H. Bell, *Sublette, Kansas,* Rural Life Series No. 2, Washington, United States Department of Agriculture, 1942, p. 25.

When the railroad arrived in 1912, nature was smiling on Sublette. There were bumper crops and farmers could pay their lean-year debts and back taxes. After repeated experience with lean and fat years, however, many farmers found it unprofitable to own land; they permitted the bank to keep the mortgage and possess the land, on which they now pay rent.

Wheat growers who live in Sublette visit their farms only when work needs to be done. Annual use of the family car ranges from 25,000 to 75,000 miles per year; mileage depends on whether the farmer has his land five miles or 15 miles from the village. It is not uncommon for families to drive weekly to Fort Dodge 40 miles away.

Sublette is an integrated place where little of the rural atmosphere can be detected. It has a high school, church groups and social clubs, a tennis court, sports arena, and swimming pool—all very much in use. The village supports a movie. Some complain that Sublette has too much delinquency, hard drinking, and gambling, but the statistics belie the charges. Village families average 2.19 persons compared with 3.5 persons for scattered farm families.[37]

Agricultural experts feel that Sublette, dependent on the one crop, wheat, is in an economically hazardous position, and they are trying to encourage other agricultural activities. It is understandable that if the one crop fails the loss must be shared by others in the community. Overurbanized, overmechanized one-crop farming places the farmer in a position no less insecure than that of industrial workers in the face of unemployment.[38]

## Depressed Rural Communities

Changes wrought in rural areas by advancing urbanism, whether beneficial or not, are rarely distributed evenly. This fact was brought dramatically to attention in the United States during the Great Depression. Most rural areas suffered economically, but some were especially inconvenienced. Areas of fertile land and mixed farming were better situated than either poor-land areas or areas of single-crop farming. There were several kinds of marginal areas. There were, for example, the areas where forests once stood but where the trees had been cut and the stumped acres sold to people hoping to clear the land for farming. There were many such settlement villages along the northern border of the United States, cut-over areas in Michigan, Wis-

[37] *Ibid.*

[38] Walter C. McKain, Jr., "The Western Specialty Crop Areas." See Carl C. Taylor and others, *Rural Life in the United States,* New York, Alfred A. Knopf, 1950, p. 438.

consin, and some in Minnesota. These farmers who lived under great difficulty in good times were hardly able to subsist when the depression came.

Another such area included the worked-out cotton lands in certain states of the South. Two other areas were in mountain regions; one was in the Ozarks and the other included part of the region being reclaimed by the Tennessee Valley Authority. Finally, there was a north and south strip of country lying between the wheat belt and the Rocky Mountains. This "short-grass" country is a region of occasional drought which is much more suitable for grazing than farming. A great many people from these distress areas of few towns but many small villages had migrated to the cities for employment. When unemployment caught them in the cities many retreated to their home villages.[39]

It was realized during the Great Depression that there were many hundreds of villages which even in normal times held a marginal position in the industrial urban economy. Their security, like that of other places, depends on their ability to produce and to sell and buy and on being linked in a firm bond with the wider industrial urban economy. If they cannot do so, they become victim villages dependent on outside support. One of the several functions of the Tennessee Valley Authority is to help the communities of this marginal area to find a new economic base and a higher self-sufficiency.[40]

With respect to "victim villages," the United States is better situated than some other countries for Americans seem more able to relinquish or abandon a place. Many of the communities established on the frontier had to be later abandoned. Hamlets and villages went out of existence in the United States with the coming of hard-surface roads and the use of the automobile by rural people. Crossroad hamlets lost their function. Victim villages can also be found in other industrial countries, but there is a greater reluctance to abandon them. Some of the people living there will fight for the right to remain, even at great inconvenience.[41]

Mitchell studied three neighboring villages in England, all of which had

[39] For a report on these depressed regions, see P. G. Beck and M. C. Forster, *Six Rural Problem Areas,* Washington, Federal Emergency Relief Administration, Monograph No. 1, 1935.

[40] On the relation of the Tennessee Valley Authority to rural rehabilitation, see Philip Selznik, *TVA and the Grass Roots,* Berkeley, University of California Press, 1949.

[41] A study was made of such a victim village in the forested regions of Sweden. Industrial changes had undermined the economic base. Some of the people migrated, others would not or could not. Morale in the village declined. The sociological report on Forestville by Bengt G. Rundblad has not been published, but it was read by this writer. Another member of the research team, Joachim Israel, did make a report on the psychological problems, interpreting the situation in terms of a failure of traditional authoritarian patterns: "Personality Design in a Socially Disturbed Rural Community," *International Social Science Bulletin,* Vol. VII, No. 1, 1955, pp. 15-22.

been left behind in the march of industrial urban change. All three in 1870 and earlier were healthy and self-sufficient, but road changes and lines of transport had favored a fourth place, Tradelet. One of the three victim villages with 650 inhabitants in 1950 has gained 45 inhabitants since 1871, but it has lost its flour mill and brewery although a small textile mill still operates. The village has two stores, two church groups, a garage, two inns, and a football club. The second village dropped in population from 750 in 1871 to about 200 in 1950. It has no sewer or piped water and only half of the households have electricity. There are no leisure-time facilities and people must go elsewhere for entertainment.

The third village, Southam, had 1,448 inhabitants in 1871 but only 250 in 1950. There are no sewers, piped water or electric power, and no recreational facilities. Attitudes of Southamites are urban rejecting and there is criticism of the urban-oriented teaching in the schools, a view which was found in the other two villages to a lesser degree. The houses here are in poor repair and the land is poorly cared for. Mitchell found the rejection attitudes of the people to be unconscious:

> When we speak of an unconscious element in the reaction to social change, we do not necessarily imply a group-mind theory. We can assume that in any community where relationships are familiar and relatively static, any change is regarded as a threat to its very existence, even where the change promises to bring indulgences. Indeed, the urban influences promise better sanitation, electricity and other social benefits, but these provoke the same reactions as the less desirable ones.[42]

Industrial urbanism, whether in the United States or another industrialized country, develops a competitive exchange network in which all places, large and small, nearby or remote, are included. Every village "belongs" in this urban-centered network to the extent that it is able to participate in a way that is mutually meaningful to itself and other places. If the village is unable to find a way of living, and perhaps prospering, within the industrial urban economy, then it is unlikely that it will live well and gain status outside. The least urbanized community must have its links with the more urbanized, and the more urbanized places depend for security on the least urbanized.

## Urbanization and the Village

Let us now consider briefly some of the ways in which the urban type of life and work operate to influence the rural village. Some of these have already

[42] G. Duncan Mitchell, "Social Disintegration in a Rural Community," *Human Relations*, Volume 3, No. 3, 1950, p. 303.

been mentioned in previous sections of this chapter. We have noted that the influence relationship is mainly one of urban-to-rural; the rural community occupies its area and goes about its affairs, but it also is in the path of the radiating urban influence.

## *Outpost Industries and Company Villages*

It is not necessary here to review the early stages of American industry. It is well known that many of the early "infant industries" started in villages and many of these villages, especially in New England, have become towns. But American industry also has had a history of expansion. Textile mills and small shoe factories migrated to rural areas in the South and Midwest. Sawmills were established in many forested areas; mines were opened and ore-processing plants were located in sparsely inhabited areas. Most of these enterprises drew their initial labor supply from surrounding villages, and found it necessary to provide housing and facilities for them near the work place. Thus evolved the "mill village," a property of the plant owner and employer, who often assumed the right to direct the workers' lives.

We need mention only in passing that the company-owned mill village, coming as it did to "backward" areas, was looked upon by many as "a promise of a fuller life."[43] On the other hand, it was considered by others as a form of labor exploitation. Workers could remain in the houses only while employed. The only store available was owned by the company. Company officials named the school teacher and the officers in the village. Workers who complained were dismissed and were obliged to leave the village at once. Unwelcome strangers, such as labor organizers, if not arrested, were asked to move on. But our immediate interest in these industrial villages concerns their role as industrial outposts. To some degree or other, they were—and still are, where they exist—small communities of radiating influence. The labor force drawn from the rural area became industrial workers, a class apart from the rural workers.

In many cases these company villages have vanished or have become absorbed by other communities that grew beside them. Some of these places, in turn, have become small industrial centers fully urban in their life.

## *Factories in the Fields*

A second type of urban industrial extension into rural areas is concerned with the processing and packaging of various products. The activity may in-

[43] For a report on such villages in the South, see Harriet L. Harring, *Passing of the Mill Village*, Chapel Hill, University of North Carolina Press, 1949.

volve processing and canning in a fishing village, or canning, packaging, and shipping fruit, vegetables, or other agricultural products. A sugar-making plant may be erected near an area of sugar-beet or sugar-cane farming. The types of plants are as varied as the types of products, and they are now being established in various underdeveloped countries. They serve the useful function of getting food supply to cities with a minimum of waste.

In different ways the factory that goes to the field has a dynamic influence on the rural way of life, and it may change rural work patterns. The sugar factory could not exist if the farmers in the vicinity did not continue growing the sugar-producing crops. Not only is the farmer encouraged to specialize in one kind of crop, but he usually produces under contract with the factory, and thus is indirectly a company worker. A plant for canning vegetables may serve to encourage farmers to turn from other types of farming to growing vegetables. Or the specialization might be in the growing of fruit: In Hawaii specialization concerns the growing of pineapples; in parts of Africa it may be cocoa or peanuts.[44]

Once a rural village has become a place of specialized production, it has become completely identified with the industrial urban economy. If it had been an area of mixed agriculture previously, it had been subject to the hazards of drought, excessive rain, frosts, etc. These may affect the specialized crops also, but now the security features of mixed agriculture are absent, and there is the new form of insecurity, the demand for the product in the world markets. Moreover, with specialized crops, the agricultural village loses much of its former local economic autonomy and becomes attached in a subsidiary relationship to the great corporations. There is little reason for assuming that this type of specialization may soon diminish.

## *Agricultural Change in Response to Other Change*

In the United States many examples may be found of cattle ranchers who have become sheep ranchers, although previously they may have had a deep bias against sheep raising. The reason is in part ecological: The grazing areas would no longer support cattle, while the nature of the grass supply was found to be more suitable for sheep. There are also examples of ranchers who, under pressure of increasing population, divided their land into farms, selling it to farmers interested in producing field crops. Land values had been increasing

[44] For a presentation of the comparative advantages of family and corporation farming, see Carey McWilliams, *Small Farm and Big Farm*, New York, Public Affairs Committee, Pamphlet No. 100, 1945.

and it was no longer advisable to use land merely for grazing. The growth of population had rendered it necessary to change from a less profitable to a more profitable use of land. Again, this has an ecological basis.

As we approach urban communities we find that the amount of land devoted to open-field crops tends to diminish and the amount used for intensive types of farming tends to increase. As land values rise the farmer is under pressure to produce the crops that will yield the greatest return. Also, as one approaches urban centers the size of farms tends to be smaller. As the demand for land increases, the larger units tend to be divided into smaller ones. When we come very near the urbanized area we may find that the price of land has become so high that it can hardly be used profitably for the most efficient and intensive farming. At this point it is likely to be divided into still smaller units and used for urban occupancy. Thus from the urban center outward the use of all land tends in large measure to be determined by the money value of land, although other factors are also involved.

The influence of land values and population increase on the uses made of rural land is not difficult to observe in the United States where sometimes much of this land was first occupied only two or three generations ago. Furthermore, the effects of these changes on rural village life are easier to trace in the United States than in some other countries. The major stimulations for this change we must recognize as industrial and urban.

### *Part-time Farming*

About one fifth or more of all farm operators in the United States may be described as part-time farmers, which by usual definition is a farmer who for hire or profit works 100 days or more per year outside of his farm. But members of the farm family may work in nonfarm jobs. As we come nearer the cities, according to Nelson, "The occupational heterogeneity of the country is being enhanced not only by the fact that farm people are choosing nonfarm occupations, but also by large numbers of former urbanites who have interpenetrated the open spaces adjacent to the cities."[45]

Part-time farming is a type of substitute for migration to the urban labor market. It enables the farmer and his grown children to live with "one foot on the land" and at the same time to enjoy the benefits of nonfarm incomes. This type of rural-urban relationship may be expected to increase in the United States as transportation facilities permit, and as it has increased in

[45] Lowry Nelson, *American Farm Life*, Cambridge, Harvard University Press, 1954, p. 64.

European countries, where farms and towns are not far apart. The German coal-mining industry began with using the spare labor of farmers. Hence came the tradition among German coal miners of wanting to live in separate cottages with gardens and space for pigs and chickens. This rural tradition, however, is passing as mining areas become more congested and urbàn.

In certain underdeveloped countries may be found types of part-time farming which verge on an abandonment of the land, or which leave the farming to women and old men. Thompson and Woodruff found examples of such land desertion by natives in Rhodesia and Nyasaland:

> This system does perhaps form a vague kind of social insurance for the African; if he becomes unemployed he can return to the village and live off the land. It may also enable him to acquire consumer goods in the towns, but an area denuded for much of the time of most of its able-bodied males is not likely to progress agriculturally, while the contribution of the African who engages only temporarily in paid employment is likely to be limited. The magnitude of the problem can be gauged from the fact that 70 percent of the able-bodied indigenous natives [of South Rhodesia] spend some part of every year working in the European area.[46]

This is not exactly part-time farming in the Western sense. Normally the part-time farmer is interested first in his land, but he manages to get other income as opportunity permits. However, it is possible that the farmer may earn so much more from outside sources that he is tempted to neglect his farm. Our interest for the moment is not in part-time farming as such, but in part-time farming as one of the consequences of urban expansion and the extension of urban influence to the rural village.

## *Rural and Urban Service Relations*

In the days of walled cities the town man seldom went into the country. It required too much effort, roads were impassable, and it was dangerous. The very opposite is true today. Urbanites of all classes enjoy ready and frequent access to the rural hinterland. As roads improve and transport facilities become cheaper and more available, more people travel from city to country, especially on the holidays, and this going to the country tends to increase as more and more urban workers are allowed annual vacations. They range in their automobiles in all directions, sometimes seeking out the most rural places. It has become quite the vogue for middle- and upper-class groups

[46] C. H. Thompson and H. W. Woodruff, *Economic Development in Rhodesia and Nyasaland*, London, Dennis Dobson, 1953, p. 131.

in the American cities to move to the country in summer. City families may "board" their children with farmers.

Such contacts of city people with the rural population are both numerous and varied. They constitute another source of income for rural people, but they are also another type of urban influence on the rural village. They add up to a vast number of rural-urban personal contacts which supplement the impersonal contacts rural communities have with urbanism through the movies, television, radio, and the press.

### *Urbanization through Mandate*

We may assume that these different methods by which the influences of industrial urban life are extended to the rural village are common to most industrial countries. They radiate from all urban centers, even in the Communist countries. However, one feature of Communist policy with respect to rural life needs to be mentioned. Communism must be regarded as essentially an urban industrial ideology. Accordingly, in most Communist countries systematic efforts have been made to change the ways of life and work of rural people. The farmer must be separated from his traditions.[47] Thus efforts have been made, apparently with much resistance, but still with a measure of success, to force farmers into various types of modernized collective groups.[48] Whatever the merit of these urban-conceived schemes, their objective is to hasten a process which might in the long run take place anyway, and which seems to be taking place in other countries, especially in the United States and other land-frontier countries.

## Rural Influences on Urbanism

In 1951 at a meeting of French scholars convened to discuss the "civilisation urbaine et civilisation rurale en France" Georges Davy in his opening remarks observed that very much was being said about the influence of town on country. He suggested that some study might also be made of the influence of the country on the town.[49] Sociologists in Poland since 1957 have been using the expression "ruralization of the city" to denote the excessive migra-

[47] "All eastern Communists stand for the elimination of the peasant," wrote D. Mitrany, *Marx Against the Peasant*, London, Weidenfeld and Nicolson, 1952, p. 223.

[48] See for example, John W. Riley, Jr. and William Schramm, *The Reds Take a City*, New Brunswick, Rutgers University Press, 1951, pp. 165 ff.

[49] Georges Friedmann, Ed., *Villes et Campagnes*, Paris, Libraire Armand Colin, 1953, p. 4.

tion of rural people into the cities after World War II. Evidences of rural thinking and conduct were noted in the urban communities.[50]

It would be a mistake to assume that, however much urban influence radiates outward, there would not be a return seepage of some rural influence. The problem is to identify these evidences of ruralization. A prominent city businessman may like to visit occasionally a restaurant where he can eat pickled pigs' feet because they remind him of his youth on the farm; people of rural origin in some southern cities in the United States like to eat grits for breakfast; many European urbanites like to sleep in feather beds—perhaps these examples may be called a type of rural influence.

Earlier in this chapter it was mentioned that certain traditional holidays enjoyed by city people really had their origin in rural places, especially holidays that originally were concerned with the planting or the harvest. Among the Mormons the 24th of July is celebrated as the day of arrival of the first settlers in Salt Lake Valley. Associations with this day for many years were rural, but the event in the cities comes to be celebrated as any other holiday. The old rural celebrations continue, but with new meanings. This seems to be true of the carnival in European countries, a festive event originally of rural origin. Originally these events were escape devices from the dullness and toil of everyday life and, although modern work is less dull and severe, people still like to escape from it.

Rural influences are perhaps most evident in some of the institutions of urban life and in some of our deep-rooted attitudes. Some have argued that the American devotion to work is of rural origin, as are our traditional views regarding individual and family relationships. Americans—among them many urbanites—found it very difficult to accept the idea of a social security system. Even in the big cities at the beginning of the Great Depression the idea was generally accepted that no person should be given help before he had passed a work test. Attitudes of suspicion toward the jobless existed in the cities but they were most in evidence in rural places.

American attitudes regarding the use of leisure—that it should be used for service or self-improvement—certainly hark back to ruralism with its precepts about "the little busy bee" and the idea that "the devil finds some

[50] The "ruralization of the city" idea is expressed by Antonina Kloskowska who reports a study of "Changing Family Models in the Popular Magazines in Poland," see Nels Anderson, Ed., *Recherches sur la Famille*, Göttingen, Vandenhoeck und Ruprecht, 1958, pp. 159-182. Kloskowska finds that rural attitudes expressed in letters from urbanites to editors in 1950-51 were much less evident in letters from urbanites to editors in 1956-57. Her interest was in letters concerning family problems.

mischief still for idle hands to do." In spite of the spending habits of many urbanites, the idea of frugality and saving is still firmly entrenched, and this too is out of our rural heritage. These ideas of saving and service are strongly fixed in the urban conscience, as witness the fact that in no other country has so much private wealth gone into foundations and trusts.

Further evidences of ruralism can be found in examining the charters of the many private associations found in our cities—those for the care of children and the aged, for protection of animals, or for safeguarding morals. This heritage, which we identify largely by inference, gets established also in the law, and there as in the institutions and rooted attitudes it functions in a sort of stabilizing role in the changing scene. That generations of rural people, without too much difficulty, have been assimilated by the city may be good evidence that the city has also assimilated some of these basic traits of the rural culture.

## Summary

For generations villages have been the most universal of communities, and the most useful. In all societies the village is the nursery of culture, antedating cities; yet wide differences exist today between the more rural villages and the ways of life and work in urban communities. In general, all villages are in process of change and the direction of change is toward urbanism, although some villages may be at a lesser level of urbanization than others. Rural villages may become increasingly urban in their ways of life while still remaining agricultural in their work. Some may become urban in both their life and work.

Examples of villages were presented, ranging from the isolated semi-primitive Iranian village to the most urbanized wheat-growing village in Kansas in which a highly urban way of life is found. The Irish rural community is a type in which the people are well acquainted with urbanism but where they choose to preserve their traditional ways of life, which is possible because of geographical isolation and the natural situation.

In the Spanish-American rural village is found a way of life which tends to be indifferent to urban expansion, showing no interest in the work and success values of Americans. These attitudes of indifference may be found in rural villages elsewhere, even in Spanish-American villages of urban character.

The old English rural village is seen as the principal source of American rural patterns for work, group life, institutions, and social values. In England this village has lost most of its old flavor and character, although it still persists in the United States where the change processes began later.

A brief review was given of the evolution of a rural Danish village from feudalism to democracy, from the most primitive to the most advanced farming methods, from primary group control by a traditional council to a fully shared civil government, and from depressing poverty to an urbanlike level of living and culture. The village has become increasingly urban in its life while becoming more successfully agricultural in its work.

With respect to depressed rural regions and "victim" villages, the examples presented indicate how necessary it is for all places within reach of the zone of influence of the industrial urban economy to participate in the general system of competitive exchange. Within this system they gain status and sustenance, but outside of this order they fare badly. The more effectively they can participate, the more secure becomes their position.

Industrial urban influences extend outwards in other ways than those usually identified with trade, the press, the cinema, radio, and the like. Industry may establish in remote places and start industrial villages. Factories may be established in rural areas and thereby encourage specialized agriculture. Population pressure and rising land values may stimulate changes from one type of agriculture to another. The nearness of industry may encourage rural people to become part-time farmers. Easier access of urban people to rural places, because of improved roads and transportation, may increase rural-urban personal contacts. All these are change-stimulating influences. While the processes of urbanization move naturally in most countries, there have been systematic efforts in countries under Communism to speed the process by mandate.

Little can be said, except by inference, regarding the counterinfluence of rural ways of life upon urban populations. However, especially in American cities, traces of rural influence can be detected in attitudes toward work and leisure, social values regarding the family, and ideas regarding welfare and service, particularly in the institutions and the law. It may be assumed that had such transfers not taken place, the generations of rural people migrating to the cities could not have been so readily assimilated.

CHAPTER 5

# Community Location and Space Occupancy

MANY QUESTIONS ARE ASKED about communities—questions regarding size and location, shape and appearance, and differences in the work or life of communities in different world regions. To some of these questions the geographers try to find the answers. Perhaps the anthropologists and the economists are also interested. In this chapter we will be primarily interested in the efforts of a relatively new group of scholars, the human ecologists, to understand community life. We must begin, however, with a brief examination of the efforts of geographers to understand the nature of human communities.

## Geography and Human Settlements

One does not need to be a professional geographer to appreciate the importance of various phenomena which are essentially geographic. It is generally recognized that certain parts of the earth are more favorable for human habitation than others, and population density is likely to be greater on areas of fertile land than on land that is less fertile. It is also recognized that, whether land be fertile or not, population density will depend on other limiting or favoring geographic facts, such as climate and rainfall. But there are still other geographic factors that may need to be considered—topography, the presence or absence of water bodies and streams, and the distances between areas or regions. The layman's approach to the complex phenomena of geography is soon exhausted.

Among the earlier geographers to undertake research into the relationships between man and his environment were Edmund Demolins of France and Friedrich Ratzel of Germany. Their approach, which concerned itself with the influence of environment upon man, is called by some the doctrine of environmental determinism,[1] or environmentalism. This theory, which regarded man as the creature of his physical surroundings, is much less emphasized by geographers today. There is, in contrast, a growing interest in man's influence on the environment—his building of transportation lines, damming of streams and irrigation projects, the influence of deep wells and so on.[2] The more moderate view of the relationship between man and his environment recognizes that geographic facts are affected, for example, by a Panama Canal or a tunnel through a mountain. The French geographers, especially Paul de la Blache and Jean Bruhnes, were among the first to take this more moderate view. Bruhnes states:

> They [humans] live *on* the earth. They are subject to the atmospheric and terrestrial conditions. They belong to certain climates, to certain altitudes, to certain zones. Besides, they live *from* the earth; it is by coordinating themselves to natural phenomena that they ensure to their bodies the necessary conditions of life; and to their faculties, development and expansion.[3]

For man to coordinate himself with natural phenomena does not mean being at war with nature; rather it means that he is operating within limits established by nature until he finds a way to extend himself beyond these limits. For example, man may pipe water to a desert location and establish a mining operation. If the heat is unbearable, he may provide himself with air-conditioned habitations. This, says Tatham, is not necessarily to be regarded as a relationship of antagonism between man and nature; man goes forward as he is able to pay the price.[4] But geographers as well as other students of human phenomena become aware of the fact that certain sections of the world's population are more active than others in efforts to break through the limits set by nature. This is well recognized by Westerners who try to establish industrial urbanism in the various underdeveloped regions of the world.

Erickson, a sociologist, makes the observation that geographical limitations tend to lose force as man gains knowledge of them. "Succinctly stated,

[1] Ellen Churchill Semple, an American student of Ratzel, faithfully reflected his *Anthropogeographie* in her book, *Influences of Geographic Environment*, New York, Henry Holt, 1911.

[2] Thomas H. Buckle placed much emphasis on environmentalism in his efforts to understand history and he thus tried to explain the creativity of some peoples compared with the backwardness of others. *History of Civilization in England*, New York, Longmans, 1903.

[3] Jean Bruhnes, *Human Geography*, Chicago, Rand, McNally, 1920, p. 46.

[4] George Tatham, "Environmentalism and Possibilism." See Griffith Taylor, Ed., *Geography in the Twentieth Century*, New York, Philosophical Library, 1951, p. 162.

as the social heritage grows, the immediate geographic factors assume a less important role in the interpretation of life in modern cities."[5] This observation, although pertinent, is only relative. It does not apply with the same force to the farmer as to the urban factory worker, for example. However, man may be the more stimulated by his environment if he understands it and can visualize possibilities. Watson holds that "a difficult environment, in challenging man's idealism and his ingenuity, may stimulate far greater advances in human thought than an easy one."[6]

Watson prefers the term "social geography" because it is "interactional":

> The geographer focuses attention on the synthesis in space—on the region; the sociologist on the synthesis in life—on the community. Since the community is a vital entity, it is dynamic, and has certain functions. Social geography studies the community in the region, and finds *function* and *form* more or less fused together in what he has to call the functional region.[7]

Thus geography begins to include social phenomena in its province, while "cultural geography" concerns itself with cultural phenomena.

## Geography and Human Ecology

Today, geographers in increasing numbers are turning their attention to human ecology. Evans, for example, is interested in the ecology of peasant life, with particular reference to the evolutionary aspects:

> An ecological approach which considers the interrelations of cultures and environments seems to be necessary, but it must allow for the inheritance of traditions from earlier phases of adjustment. We need intensive studies before it is too late, of the dialects and traditional dresses, the crafts, oral traditions, folklore, and folk music of our peasantries—work in which Swedish scholars have pioneered—but we also need systematic, anthropological surveys of peasant communities as functioning entities.[8]

Thus when the geographer turns from the study of the physical environment and approaches his subject through man it is difficult for him not to be intrigued by ecological, social, and cultural data. A geography by White and Renner is called "an introduction to human ecology." The discussion on

[5] E. Gordon Erickson, *Urban Behavior*, New York, Macmillan, 1954, p. 119.

[6] James W. Watson, "The Sociological Aspects of Geography." See Griffith Taylor, Ed., *Geography in the Twentieth Century*, New York, Philosophical Library, 1951, p. 471.

[7] *Ibid.*, p. 472.

[8] E. Estyn Evans, "The Ecology of Peasant Life in Western Europe." See William I. Thomas, Jr., Ed., *Man's Role in Changing the Face of the Earth*, Chicago, University of Chicago Press, 1956, p. 234.

"types of social adjustment to natural environment" deals with such subjects as population distribution and social habits of living; agglomerative forms such as villages, towns, and cities; types of land use such as hunting, pasture, fishing; special arrangements of the people such as tribes or other groupings; methods of using labor such as slavery, caste, or the modern system; and cultural traits such as clothing, housing, religion, etc.[9] Another part of the book deals with "political adjustment to the environment," and still another with economic adjustments. The ecological elements are very general.

Geography is also turning its attention to the various human agglomerations, and within this area is urban geography, which typically includes ecological implications. Among the basic interests of urban geography, according to Dickinson, are situation, site, and nodality. Situation concerns the location of the community with reference to the general surroundings, including other communities, but it may also concern the origin and development of the place. Site is more specific and concerns the place where the community rests, its geographical advantages and disadvantages and the pattern of occupancy.

Nodality, according to Dickinson, concerns the relationships, nature of contacts, and accessibility of the community in terms of other places. What lines of transport, for instance, radiate from the place or pass through? "In other words, nodality should be measured on the basis of the functions of the town as a focus."[10]

We cannot concern ourselves with the debate over whether or not geographers may properly venture into the realm of human ecology. It is difficult to understand why they should not if their efforts prove fruitful, but a word is in order about the approaches of geography and human ecology. Hawley observed:

> Geography treats men and their activities in their visible aspects and so far as they may be regarded as distributed phenomena. It does not concern, except incidentally, the interrelations among men. Human ecology, which is also interested in the relations of man to his geographic environment, fastens its attention upon the human interdependencies that develop in the action and reaction of population in its habitat. In other words, while geography views the adjustment of man from the standpoint of modifications of the earth's surface, human ecology makes a detailed analysis of the process and organization of relations involved in adjustment to environment.[11]

[9] C. L. White and G. T. Renner, *Geography, an Introduction to Human Ecology*, New York, Appleton-Century, 1936, pp. 21-27.

[10] Robert E. Dickinson, *The Western European City*, London, Routledge and Kegan Paul, 1954, p. 5.

[11] Amos H. Hawley, *Human Ecology*, New York, Ronald Press, 1950, p. 72.

According to this distinction, geography is more macroscopic, while human ecology is more microscopic. Hawley adds the thought that geography approaches the problem by examination of the environment, while human ecology starts with the organism. Geography "takes a one-time picture," describing things as of a moment, while human ecology is concerned with development and becoming. The human ecologist can hardly help being historical in the study of his materials. We must bear in mind, however, that all these distinctions between geography and human ecology are relative and general. All would not be agreed to by all geographers nor even by all human ecologists.

## Access and Urbanism

"The basis of any city, or indeed of any town, with the possible exception of purely religious centers, is access."[12] A city, or a smaller urban place which we may call a town, functions in behalf of and in relation to a certain area, and that area functions in relation to the town or city. The effectiveness of this relationship depends very much upon the ease and cheapness of contact, or access. The outside area of a great city must include other cities as well as towns, all of which form a network. While all the urban places may be within a major network with lines radiating from the major city, each has its own network. The relationship is competitive, but different places in this network enjoy relative degrees of dominance, with a type of overall dominance being enjoyed by the major city, the metropolis.

This dominance, to whatever measure it exists in any urban place in a regional network, depends on a variety of interrelated factors: type and volume of natural resources, types of populations in areas, industrial development, and, not the least important, access to places in the network.

Obviously, access between cities or between cities and their hinterlands depends on various geographic conditions: harbors, water courses, and the absence of barriers such as mountains. But distance is also a factor. It may also be an advantage if a place is accessible by different types of transportation such as sea-going vessels, river craft, railroads, and highways.

Cooley, then writing as an economist, published a treatise in 1892 advancing the thesis that "population and wealth tend to gather where there is

[12] Rhodes Murphy, "New Capitals of Asia," *Economic Development and Cultural Change*, Vol. V, No. 3, April 1957, p. 216. Murphy adds, "Thus cities grow up at points from which they can command the largest tributary area, in competition with other cities, or in other words at points where access is maximized."

a break in transportation."[13] By "break" is meant an interruption in the flow of goods at transportation meeting points; for example,

(1) Where water and land transportation meet or where types of water or types of land transportation meet (sea transport, river transport, railroad, or road). Today air transportation would be added.

(2) Where goods must be handled, transferred, or packaged; or where they "break bulk" going from a large to a small carrier (ship to boat, ship to railroad, railroad to motor truck); or where goods "change form," as when cotton is made into cloth, leather into shoes, corn into starch, and so on.

Perhaps at a break in transportation very little work is invested in the handling of goods, but a considerable investment of work may be involved if goods change form. The more work involved, the greater the aggregate of population at the break in transportation. The greater the number of people collected, the more services are needed for them. At the point where an oil pipe line meets a harbor, the oil may be pumped into ships or there may be a refinery; thus comes the city of Abadan in Iran. Where coal and iron ore meet (water and land transport), as in Gary, Indiana, there develops a city of vast plants identified with the steel industry.

The Cooley thesis in substance had been proposed by J. G. Kohl, a German, in 1841[14] (certainly unknown to Cooley), and it has been stated by others who certainly did not read either Kohl or Cooley.[15] For the present chapter we need to examine the ecology of towns and cities. Here we find that access is a dominant factor in determining how space is used. Factories must be most conveniently situated to the kinds of transportation needed for getting raw materials in and finished goods out. Access to transportation, which saves labor in handling and rehandling goods, determines the location of warehouses and storage plants. The center of the city must be so situated that it is easily reached by the maximum number of people, a fact that determines what uses are made of the space at the urban center. When housing congestion at the urban center or near the urban center becomes excessive,

[13] Charles H. Cooley, "A Theory of Transportation," republished in book by R. C. Angell, Ed., *Sociological Theory and Social Research*, New York, Henry Holt, 1930, p. 75.

[14] J. G. Kohl, *Der Verkehr und die Ansiedlungen der Menschen in ihrer Abhangigkeit von der Gestaltung der Erdoberfläche* (2d ed.), Leipzig, 1850.

[15] Douglas Ridgley ("Geographical Influences in the Study of Cities, *Journal of Geography*, Vol. 25, February 1925, pp. 66-78) presents similar views of routes and breaks in transportation. See also Chauncy D. Harris and Edward L. Ullman, "The Nature of Cities," *Annals of the American Academy of Political and Social Science*, Vol. 242, November 1945, pp. 7-17.

the pressure may stimulate the building of new lines of transport, which may provide outlet. Thus suburban areas become more accessible, and around the great city grows up a network of suburbs, each to some extent conditioned in its occupancies (and life) by its access to the central city. In one way or another the occupation of all space within a metropolis depends on access to other areas.

Just as access tends to determine how space is used, both inside and outside an urban community, so it determines the demand for space. Where the demand for space is great, people are encouraged to concentrate their activities into narrower limits, but they also resort to multiple uses of the same space. Thus they erect on the same site buildings of several floors or many floors. The demand for space is objectified in terms of rents and land values.[16]

## Human Ecology Defined

With respect to the occupation and use of space, human ecology is a special approach to the study of man's relations with his physical environment. Man's environment includes other space-occupying and food-consuming organisms, people and animals whose needs may be very much in competition. On most parts of the earth, particularly in modern countries and especially in areas more accessible to urban places, animals share space as man permits them to. A region occupied by man and different types of animal life, domestic and wild, exemplifies what is sometimes called a "web of life." Here are people (human ecology), various animal forms (animal ecology), and plants of many kinds (plant ecology). Plants provide the basis for life of all the animal forms. But the plants are in continuous competition for the food supply in the ground. Man tries to improve on nature by tilling the land, protecting the plants most useful to him and eliminating others (weeds). He may eliminate some, if not all, of the wild animals, to the advantage of himself and the domestic animals. He may then eliminate some of the domestic animals, as when he replaces horses and mules with tractors and motor vehicles. This saves space for feeding horses and makes more land available for food.

But our interest is primarily in ecology as it applies to the urban com-

[16] Darell Garwood (*Crossroads of America, the Story of Kansas City*, New York, W. W. Norton, 1948, p. 201) tells of Thomas H. Swope who owned much land and sold it as the city expanded. He would lament that land he sold had gained in value many times in a few years. A clay hill and a bog near the city center which he could not sell came to have high space value later and yielded him $50,000 a year in rentals.

munity. We need to recognize that the ecology of an urban community does not stop at the political boundaries, although as it extends outward the implications change. According to Hawley, "In at least one of its aspects the human community is an organization of organisms adjusted or in process of adjustment to a given unit of territory."[17] By "organisms" in this sentence he doubtless is referring to people. The humans are in competition for sustenance and space much as the plants are in a natural habitat, but man is a social animal.

This raises a question about which human ecologists are not agreed. Are we to regard human beings distantly, as they might be observed by a man from outer space, or should we think of them also as social beings? Actually, in the community we are confronted with the social environment as well as the physical. Chombart de Lauwe made the observation in connection with his ecological study of Paris that the same land area on which a community stands may be defined in terms of a variety of interests in it—geographic space, economic space, demographic space, or time space, and also social space. The community is a "collectivity of social beings in relation to one another and to their environment." These other space concepts—geographic, economic, etc. —have no meaning "except for people in their social relations."[18] We will consider this subject further.

In the urban community, space therefore acquires more than a mere areal meaning, for man builds both upward into the air and downward into the earth. Thus space is horizontal or vertical. The utilization of vertical space hinges on man's skill and technology; he must overcome the problem of cost and other limitations. Furthermore, horizontal limitations may have a restrictive use on vertical space. For example, as buildings in a city rise higher there is a corresponding pressure for space in the streets and greater loads are imposed on the transportation facilities. Thus urban man tends always to be confronted with space-use difficulties to which he must make one type of adjustment or another.

As a definition of human ecology as it seems to apply to urban and near-urban situations, we might state the following:

> Human ecology is the study of the spatial relations of man in his habitat as well as inter-habitat relations, and is concerned with his efforts to utilize his

[17] Hawley, *op. cit.*, p. 68.

[18] Paul Chombart de Lauwe, Ed., *Paris, et l'Agglomeration Parisienne,* Paris, Presses Universitaires de France, 1946, Vol. I, p. 24.

habitat to his advantage, as well as his efforts to cope with the limitations of the habitat. These man-habitat relationships are seen also in a time perspective, as changing in time as conditions in the habitat change. It is recognized that man is himself a dynamic factor in initiating such changes, although major changes may not be due to purposeful or conscious efforts on the part of man, but incidental to them.

We need to emphasize that the human ecologist is intent on observing phenomena in process of development. He must see the present distribution of uses of a habitat in relation to previous ones and still earlier ones. Thus he is able within limits to project trends, and it is in this area of effort that sociology is served by ecology.

## Ecology and Social Environment

The social environment must be regarded as real. It is the contemporary content of the social heritage. If we speak of social organization we have in mind the structural aspects of the social environment. However, we find that ecologists have different views about the inclusion of data which tend to be sociopsychological in nature. Man uses the area of his habitat for different essential purposes. Some of these are clearly economic uses: space for work, for storing raw materials or finished goods, stores where goods are sold; and space where he can live—apartments, hotels, single houses and so on; space for collective use, including roads and streets which facilitate access to his work, bridges, parks, and play areas, and for public buildings. He needs space for private collective uses such as club houses, golf courses, yachting associations.

But in various ways, as Firey reminds us, these uses are affected by social sentiment.[19] A monument must stand at a certain place; an historic building whose location is economically not justified may stand in the path of planned ecological change but for sentimental reasons is not removed. But the intrusion of sociopsychological influence is seen in other ways. The idea that land values are economically determined holds with respect to space used for work purposes, but often social values intrude with respect to space used for residential purposes. Some people move away if people of another race, nationality, religion, or social class occupy houses near them. Or they may form groups for preventing such unwanted people from locating near them.

[19] Walter Firey, "Sentiment and Symbolism as Ecological Variables," *American Sociological Review,* Vol. 10, April 1945, p. 148.

Here we are confronted with social values which may, temporarily at least, affectively influence economic values and ecological change.[20]

While the human ecologist is justified in avoiding elusive sociopsychological materials involving sentiment or bias, there is nothing to prevent him from working with objective materials which may in some way reflect the attitudes and associations of people. He can make maps showing the distribution of suicides, mixed marriages, or residential change as easily as he can map economic information, information about transportation problems or change, or information about the buildings that are erected. None of this information is of any final value unless it is interpreted in terms of phenomena in the social environment. It is true that in the long-term perspective these social influences and the force of sentiment tend to fade and are forgotten. But people live not so much in terms of the long-run perspective as with present facts, and the human ecologist must live with them, too. He may be a better ecologist if, while being "scientific," he is also able to square his data with the flow and foment of life in the area being studied.

## Space-occupancy Patterns

There tends to be order in the way in which people, whether in a village, town, or great city, occupy their land. As there is a semblance of order in the position, size, and arrangement of communities with respect to one another, so within any community there is a functional distribution in the uses of space. We may assume that all uses of space are competitive and that each use involves certain minimum conditions. All uses, however, tend to be based on one common characteristic: that they be as near as possible to the center of the community.

Uses of space differ widely in the minimum amount of space needed for adequate functioning. A furniture dealer needs more space than a tobacco seller; a manufacturer of dresses needs more space than a furrier, and a manufacturer of men's clothing needs even more. A hotel for transients is more compact for residence than a building of one-room apartments, and the latter in turn is more compact than the apartment buildings for housing families. Thus cost of land and rentals are selective factors. Occupiers who handle goods

[20] We may find near the center of a changing city little islands of old houses occupied by a few "first families" who refuse to move in spite of all pressures for change. Sometimes with court action and at financial loss they hold their place. Although change may come in the next generation, the ecological process was held up.

that are both compact and high priced (if turnover of goods is high) have an advantage in competition for the most-desired space.

Some occupiers must be near the railroads or ships, or if they use auto transport, they must have more room. They cannot locate near the urban center as can a lawyer or doctor who needs but two or three rooms. Thus access is a selective factor as well as land values and rents. But there are still other selective factors. Some types of spatial uses tend to cluster. Dress manufacturers must have their show rooms within a limited area. They must be near one another, and the area must be near the hotels. For reasons of competition they cannot scatter.

When these various factors that influence location have had their effect, the result is that the total occupancy of a habitat tends to have a related unity. The distribution of space users tends to form a pattern, the result of hundreds or thousands of individual choices. Occupiers tend to fall into groups, and the arrangement is not without logic. We may say that the pattern centers at the point of most desired access at the community center and extends outward beyond the limits of the community.

In a competitive situation, especially if population in the community is increasing and economic activity expanding, the relationships between areas within the occupation tend to be in continuous change. This dynamic character of space occupancy is clearly visible in American cities, and it is somewhat the same in cities and towns of other new countries. Although the processes of change are not so evident in the older cities, some clustering of activities can be seen. Cressey, however, found that some of the characteristics of the ecological process, so clearly seen in American cities, were either absent or scarcely evident in Rangoon, Burma.[21]

As competition for space in the urban community determines the location of occupiers and the relationships of areas into patterns, so does it influence the use of space inside the buildings. We are all familiar with the space-saving methods used in apartment buildings, but if we study the use of space in a great office building, say of an insurance company, or the arrangement of sections and counters in a department store, we will find in miniature space-use patterns not unlike those in the community itself. It is the work of a specialist, a "space-control" man, to arrange the offices in the insurance company building so there is a "flow" of the work. The arrangement must also

[21] Paul Cressey, "The Ecological Organization of Rangoon, Burma," *Sociology and Social Research*, Vol. 40, No. 3, January-February 1956, pp. 166-169. But if Rangoon is seen in long-term perspective the verdict might be different.

take into account the authority structure of the company. In the department store, space-control experts must arrange the display of goods with the convenience of the customers in mind. Some goods "must" be on the ground floor near the front entrance. Other goods are more appropriately placed in the basement. House furnishings are usually on one of the upper floors. Areas of the floor space least in demand for merchandising are used for packaging or storing goods. The entire operation forms a pattern.

## The "Lay of the Land"

Ecological patterns in any community may or may not be clearly seen. A specialized activity, unique for a particular city, may give rise to specialized areas not found in other cities. However, the shape of the city, the flow of the traffic, the character of intersecting streets, and other physical conditions may give one city a pattern of space occupancy quite different from another. Not infrequently, the pattern of occupancy is due to a mixture of influences, two of which need special mention: the historical and the geographic. Aström of Finland thinks that Americans have neglected the historical factor in their ecological studies.[22] It is perhaps true that American ecologists have had little to say on that subject, but it is also true that they have studied cities that are relatively new. Such history as these cities have has been taken into consideration under the concept of "succession," of which more will be said later.

Among other things, a study of the historical development of a city must deal with past efforts of the people to adapt to an inconvenient or threatening geographic site—an effort which for some cities has been heroic. New Orleans, for example, rests on an area of flat land which is lower than the level of the Mississippi River and which must continually be guarded against floods. Amsterdam, Holland, is a city built on land most of which was originally under water. Much of the occupied space in this city was dug out of the canals or dredged out of the harbor. There are several rings of canals which for modern traffic needs offer serious circulation problems,[23] and it is difficult to establish solid foundations for the erection of tall buildings. Yet in spite of an unfavorable site a great city grew here. Alexandria, Egypt, is built on a long, low sand

[22] Sven Erik Aström, "Kaupunkiekologiset Teordat ja Historiallinan Todellismus," (Urban Ecological Theories and Historical Realities), *Suomal Soumi,* Vol. 24, No. 2, 1956, pp. 77-82.

[23] This writer made the suggestion to an Amsterdam citizen that certain of these canals might be filled with earth and the space used as wide streets. He answered, "If they took away the canals it would no longer be Amsterdam." This is an example of sentiment as it affects the use of space.

bar and its shape has been so determined. San Francisco has been hampered in its growth by a rugged shore line and high hills. New York is stretched over three islands with water bodies limiting its shape and growth. Stockholm, Sweden, is also built on irregular bodies of land intersected by areas of water. Other examples could be cited; in fact, some of the great world cities are most inconveniently located as far as site is concerned.

It must be said, however, if the site on which a city rests is one that provides maximum access to the various hinterlands, and if there are good economic and other reasons for population and work to be centered at that point, then the city will adapt and thrive. We may say that it will fit itself into the "lay of the land." All the functions pertinent to any area of human habitation will find their place in whatever ecological pattern that develops, and special functions peculiar to the particular city will be more favorably situated than others in the pattern.

This task of adapting to the lay of the land is not a matter that concerns the city alone. In one way or another it concerns nearby areas which may be suburban to the city, since the pattern of occupancy extends outward with various networks of communication and access. A factory nearer to the urban center may need to locate its warehouses in a suburb; the in-town publisher may locate his printing plant in a distant suburb. The outer areas that are used and the way in which they are used also depends on what is possible and practicable.

## Distance and the Use of Space

Lösch, a German economist, wrote a book entitled *Die raumliche Ordnung der Wirtschaft*, which is a theoretical study of space as it relates to the location of industrial enterprise. While he does not use the term, it is indeed an ecological approach. One of the central ideas focuses on the role that distance plays in the location, size, and activities of any human agglomeration. The man who makes bread to sell to his neighbors is limited in the area of his market by, among other factors, the distances over which he must deliver his product, or over which people will come to buy it. The barrier of distance takes form in time and travel cost.

According to the highly abstract presentation of Lösch, distance is a factor which figures prominently in determining the nature and amount of economic activity found in any location. It determines the size and reach of the local market and it affects relations with markets elsewhere. He recognizes, of course, that distance between places may be difficult to negotiate because of barriers, as he also recognizes that an improved technology can overcome some of the obstacles, making it possible for man to move over wider areas at less cost in time and money.[24]

The uses of land are in many ways conditioned by its location with respect to other places, a thought that has already been expressed under the term "access." In a society where everyone walks and where four or five miles is the radius of convenient contact, the implications of distance are not so involved as in a community where different types of transportation are available and the time-money costs of moving goods or going places tend increasingly to decline. Thus access takes on a new meaning with each improvement in transportation, and changes resulting with respect to distance are reflected in the ecology of space occupancy.

It needs to be recognized that such changes in access which affect distances between places also affect adversely or favorably different areas and communities. The small town, the larger town, the small city, or the great city tend generally, although not always, to become more accessible to more people, ". . . hence," says Hawley, "the territorial scope of the community and, to a larger extent, the number of individuals who may live in close mutual dependence. Similarly, the distribution of units within the community varies with the time used in movement. A temporal pattern is implicit in each and every spatial pattern."[25]

The objective of facilities for reducing the time-money cost of distance is to make access to the community center easier for more people and more goods from a wider area, and to make these wider areas more accessible for the town or city. As Hawley indicates, this greater accessibility may affect space uses in the community itself. Thus it is to be expected that certain functions will expand and need more space. Other and weaker space occupiers will be squeezed and will need to find a new location. The direction of this expansion within the community is likely to be along the newly created lines of access or very much in relation to them. If a new superhighway enters a city, outward extension of the city will usually follow that highway, more so than along less well-improved roads to the city. Especially it is to be expected

[24] August Lösch, *The Economics of Location*, New Haven, Yale University Press, 1954. Translated by William H. Woglon assisted by Wolfgang F. Stolper. Lösch finds that very small communities tend to have their restricted simple market areas, which he shows in a hexagonal pattern. Groups of these form a larger hexagonal network around a larger market (subregional), and these larger markets form a hexagonal pattern around a major or metropolitan market, which is linked with world markets.

[25] Hawley, *op. cit.*, p. 288.

that new residential areas will locate along the new highway, relieving pressure on residential areas in the city.

Hawley mentions the temporal pattern as being implicit in the space pattern. The temporal pattern concerns distance in terms of cost in time and money, especially in time. If one were to draw a series of irregular circles from the urban center outward, called the 20-minute zone, the 30-minute zone, and so on up to an hour, such a temporal map of the city would show that some areas that are geographically near the urban center, say within a mile or so, may be more distant in terms of time than are other areas up to five miles away. Differences in time-distance, as they influence space occupancy, may result in areas within the city being isolated. In the same way, as Dickinson points out, areas and places outside the city may become isolated from one another, for, while each may be on a rapid transit line with the metropolis, local access networks tend to lose importance. Each village or town now acquires a niche in the city-hinterland relationship.[26]

## Zones of Occupancy

From time to time students of community life have offered hypotheses describing the processes of urban growth and space occupancy. One very much debated formula was that offered by Burgess in 1925. Based on his studies in Chicago, Burgess concluded that areas of occupation, which he called "natural" areas, tend to be grouped into types in terms of distance (geographic) from the urban center. That the urban center has a character of its own and could be identified was not debated; difficulty arose in pinpointing areas of occupation just outside the urban center, or those two miles or five miles distant. Burgess undertook to group these areas into "zones." He concluded that a number of zones or "concentric circles" could be identified. These zones tend to be circular except as interrupted by topographical features, water bodies, or works of man such as railroads, parks, or major highways.

Outside the central area Burgess identified a zone of light manufacture and slum residence into which manufacturing and other nonresidential uses of space were spreading. The next ring included a better class of homes for workingmen. The third ring was occupied largely by heavy industry, although it also contained residential areas. Next came the suburban areas, which are likely to be mixed in character since suburbs tend to have their individual

[26] Robert E. Dickinson, *City, Region and Regionalism,* London, Kegan Paul, Trench, Trubner, 1947, Chapt. 6.

identities to some extent. Within the frame of the zonal arrangement Burgess noted the presence of special districts: "hobohemia" where homeless men gather, the "gold coast" where the rich concentrate, certain areas of vice, and finally, ethnic centers like "Chinatown."[27]

## The Urban Pattern

Watson, a geographer, already quoted, found a zonal pattern in the city of Hamilton, Canada.[28] Taylor in his geographic study of Charlottsville, Prince Edward Island, Canada, found the following zones or areas: (1) public administration, (2) professional offices, (3) better shops, (4) smaller shops (all at the center); then (5) church zone, (6) school zone, (7) third-class houses, (8) second-class houses, (9) first-class houses, and (10) institutions (jail, hospital, race track).[29]

Hoyt made a study of different American cities to learn something about the nature of urban growth. He concluded that growth tends to follow the main arteries of travel, that new sectors develop along these arteries. Each of these sectors begins to expand until intervening space is occupied. Although this well-recognized pattern of growth has been regarded as contradictory to the Burgess hypothesis, it must be remembered that Hoyt's study is concerned with only a part of the whole. Hoyt tells how one type of growth takes place and Burgess tells what the entire urban pattern looks like when occupied.[30]

Although not a social scientist, Hurd more than half a century ago was looking for the answer to the same questions: How do cities grow and what governs the building up process? He noted that there were outward "pulls" as well as the pulls toward the center, and these pulls were reflected in rents and land values.[31] Hawley is of the opinion that Hurd's explanation of urban growth is a clear application of the idea of concentric circles as new outside areas are added to the city.[32]

Chombart de Lauwe in his ecological study of Paris found, as did Burgess,

[27] Ernest W. Burgess, "The Growth of the City." See Robert E. Park, Ed., *The City*, Chicago, University of Chicago Press, 1925, p. 55.

[28] Watson, *op. cit.*, p. 484.

[29] Griffith Taylor, *Our Evolving Civilization*, London, Oxford University Press, 1946, p. 220.

[30] Homer Hoyt, *The Structure and Growth of Residential Suburbs in American Cities*, Washington, Federal Housing Administration, 1939, Chapt. 6.

[31] Richard M. Hurd, *Principles of Urban Land Values*, New York, The Record and Guide, 1903, p. 39.

[32] Hawley, *op. cit.*, p. 384.

the many types of areas, called *quartiers*, and these areas tend to fall into groups or *secteurs*. He found that the *secteurs* tended to be bounded by public buildings, railroad lines, wide avenues, parks, factories, and the like.[33] But he also found that the areas and *secteurs* tended to be arranged in a sort of pattern around the urban center, a sort of zonal distribution as follows:

1. The *noyau de Paris*, the urban center.
2. A cosmopolitan cultural zone of strangers and tourists, from rich to poor, refugees, students, artists, some racial groups, railroad stations. Area laps over the central district.
3. Near residential zone, small and medium enterprises, some rich areas of residence, residences of different classes.
4. Residence and industry, larger industries surrounded by workers' housing, small rural or semirural areas.
5. Mixed residential and small suburbs. Zone line becomes very irregular. Average distance 10 kilometers from urban center.
6. Zone of *lotissements* or building developments, some more self-sufficient suburbs. Zonal line extends outward on lines of traffic like fingers.
7. Zone for distant commuters where city merges into the hinterland.[34]

With respect to terrain, Paris is spread over a somewhat flat region with rolling hills, not unlike the site of Chicago. There are no difficult barriers to development, but Paris is not spread over such a wide area as Chicago. It is a city with hundreds of years of history, while Chicago is little more than a century old. The processes of growth and formation in Chicago have been rapid by comparison, but the distribution of space occupancy in zones around the urban center is fairly comparable.

## Ecology and Competition

Enough has been said about the spatial relations of people in their use of land in the urban community to emphasize the role of competition between the occupants of land in the community. Although we will not examine more closely the characteristics of such competition, we do need to consider some of its implications. Property rights of many kinds are involved, and these often involve very expensive court cases. These contests relate not only to the rights over ownership, but also rights to a reasonably free use of land. There are rights regarding renting space as well as rights of access. These contests over rights in land are not peculiar to great cities alone but are common to all

[33] Chombart de Lauwe, *op. cit.*, Vol. I, p. 55.
[34] *Ibid.*, Vol I, p. 45.

communities, even primitive villages, where the issues are judged on the basis of custom, and the rules of custom may be as complex as those of the law.

Contests about rights may affect individual occupancies here and there, but they do not prevent a type of collective competition between a cluster of occupants and different clusters surrounding them, in which a particular area occupied by a cluster of space users may be in process of contracting or expanding. If it is contracting, space may be yielded to adjacent clusters that may be expanding. If the cluster is expanding, it encroaches upon one, or another, or different areas around it. If it is not possible to expand in this manner, the entire cluster may move to a new location and there become stronger than before.

While adjustment to space problems involves individual choices, different individuals tend to make similar choices, since the realities that must be considered are similar or the same. An important automobile dealer may rent space for show rooms at an intersection not too distant from the urban center, a vicinity formerly occupied by old tenements. Owners of the new buildings conclude that high rentals can be had from the automobile dealers, already crowded by the building of a new hotel in the area they presently occupy. Other automobile dealers move to the new location. Auto-supply dealers follow. A used-car show room locates nearby. The result very soon is a cluster of like-interest occupants, and this area is identified as a specialized market.

## Specialization of Space Users

Specialization in the production of goods generally develops as an adaptation to a competitive situation. It enables the maker of goods to operate on a larger scale, producing more goods at a lower price. Doctors, if they are specialists, need to be located near the urban center. A large number may concentrate in one or more office buildings, where the different specialists necessary to one another will be found. The X-ray laboratory will generally be here in the midst of its market where it is most needed rather than in an office building occupied by lawyers. Such a group of specialists cluster not alone for economic reasons; they constitute a service whose value to the community is greater if they are clustered and hence accessible to more people in a more effective way.

Not all urban space occupiers are so dependent on clustering. A mail-order firm may need to be near the urban center, perhaps in the first zone outside the center. It does not need to be accessible to customers, but to transport facilities—railroads, postal and express services. A junk yard, a specialty very necessary in the urban community, needs to be accessible to all areas, but it does not need to be clustered with other junk yards. It must, on the contrary, locate in an area where its presence is not resisted. However, the junk market (old metal, old clothing, paper, bottles, etc.) may cluster in areas that have become isolated in the space competition race.

The different specialties found in the urban community may be classified according to a number of major groups, somewhat as follows:

(1) *Financial group,* which may include banking, insurance, the stock market, export and import and shipping offices. These tend to cluster, although establishments of each type may be scattered.

(2) *Marketing group,* includes a variety of markets, some of which need to cluster, particularly wholesalers catering to wide markets or specialty shops catering to exclusive customers. The less exclusive and expensive the merchandise, the more the market is spread out. While specialty shops may cluster, others such as fur or jewelry shops need to be near the urban center but do not need to cluster. Stores selling commonly used goods may be widely scattered.

(3) *Professional group,* includes a variety of specialists, some of whom cluster while others need not, although all need to be centralized. Besides lawyers and doctors, this group would include tax experts, accountants, advertisers, editors and publishers, fortune tellers, faith healers, and artists—the list is long.

(4) *Catering and social-life group,* found in every urban community and often centrally located, includes hotels and special restaurants, theaters and night clubs, exclusive- or not-so-exclusive-membership social clubs or university clubs.

There are, of course, other groups, but this list is enough to indicate the relationship between urban space occupancy and specialized activities.

## Succession and Ecological Equilibrium

In almost every city bronze plates may be found on the sides of buildings which report that "on this site stood the building" in which some man famous a hundred years or more ago was born. The old house was removed and perhaps another building used for another purpose took its place, and perhaps that building also was replaced by still another, the present one, which is used for still another purpose. In the course of a century this same bit of land may have been occupied by three different structures, each more modern and more economically productive than the previous. The different occupants, coming and going, have put this space to different uses. Such a parade of occupations is common to American urban life, but is not always so evident in Old World cities where space occupancies change much more slowly.

This process of occupancy change is generally known as "succession." It may refer either to a succession of occupants in a structure or to a succession of structures on the same plot of land, although both types of succession are evident when a particular land unit or area is seen in a long-term perspective. In American cities succession is conspicuously seen in the occupation of residences. A new apartment building intended for, and occupied by middle-class families may within a decade be regarded as "getting old" and therefore not a first choice of middle-class families; before another decade has passed it is receiving occupants of lower middle-class status at comparatively lower rentals. Within another decade the building is no longer modern and the new occupants are of the upper lower class. Most of the people moving away are in search of better housing. The social and economic status of each new wave of occupants tends to be lower than the previous one. Ultimately the structure reaches the point at which it can no longer be profitably used for residence. It will be replaced by a more modern apartment or by a structure used for another purpose.

As this process of succession is evident in single buildings, so it can be seen in groups of buildings, not only in residences, but in stores, office buildings, hotels, theaters, and restaurants—any structure that may be outmoded with passing time and technological advance. These compelling forces for change are present even in buildings and areas where space occupancy remains relatively constant. This is illustrated in the case of a hotel in Darmstadt, Germany. In the 1300's it stood near the city gate, a wine-making place on a corner, where customers could drink wine. It later expanded into a hostel. In the 1400's it was an inn and remained an inn from then on, although several times during 500 years a new building replaced an old one. It was completely destroyed by the 1944 bombing but has been rebuilt as a modern hotel. Here succession found expression in several renewals of structure resulting in improvement and rising status with each change.[35]

The name and the site of the Darmstadt hotel remained the same but there were changes from outmoded to more modern buildings. In Brussels, for example, the modern center is quite removed from the center of centuries ago.

[35] Darmstadters bombed out during the war tried hard to return, and many tried to return to the same streets where their families had lived for many years. Business enterprises tried to relocate in the old places. This was true even of poor people. On this subject, see Fred C. Ikle, "The Effect of War Destruction upon the Ecology of Cities," *Social Forces,* Vol. 29, May 1951, pp. 383-391. Also see Leo Grebler, "Continuity in the Rebuilding of Bombed Cities in Western Europe," *American Journal of Sociology,* Vol. 61, No. 5, March 1956, pp. 463-469.

The monumental old center is retained, jealously guarded against change, while the rest of the city behaves quite ecologically. But, without altering their appearance, the old structures, still used, tend to be modernized. Local pride and sentiment behave nonecologically. In Vienna, as in Brussels, Paris, or other cities, are certain very old restaurants which *must not* be changed regarding site or appearance, although they strive to be ultra-modern in their service (and prices). They stand firm although ecological change goes on all about them. Thus sentiment may be a barrier to ecological succession—a delay but not, however, a final block. Such sentiment may apply to some degree to different buildings, streets, or places, especially in older cities.

Succession, ecologically speaking, is concerned with the change of use of a unit of land or the change of occupants within a use group, or it may be expressed in structural change to meet changing needs of the same use group. It is to be found, in one form or another, in all urban communities, although it will be most evident in rapidly growing communities. Succession can scarcely be absent in communities that are being stimulated by technological change. Even if technological change is at a minimum, succession may result from competition between space users. In such cases the displacements resulting from competition may be apparent only in occasional shifts and, seen at any one time, the local situation may appear to be in equilibrium. Equilibrium, however, is a highly relative term.

As Sears points out with respect to plant or animal communities, equilibrium involves a balance between dominant and subordinate forms, and the dominant forms "are often quite dependent upon the activity of inconspicuous and seemingly subordinate forms."[36] For different reasons, dominant forms may lose their virility while the "seemingly subordinate forms" are gaining advantage. We cannot say that this does not happen in human communities where, states Riemer:

> . . . instinctive adaptation is replaced by folkways and culture. The human ecologist starts with the assumption that equilibrium is achieved in the urban environment by a process of automatic adjustment. A relatively stable equilibrium is considered to result from the struggle for survival on the part of individuals. Unless distorted in its delicate internal balance, the equilibrium is expected to perpetuate itself. Like the plant or animal ecologists, the human ecologist focuses attention upon conditions existing in the community as a whole rather than upon its component parts, the individual city dwellers.[37]

[36] Paul B. Sears, "The Ecology of City and Country." See Elmer T. Peterson, Ed., *Cities Are Abnormal,* Norman, University of Oklahoma Press, 1946, p. 51.

[37] Svend Riemer, *The Modern City,* Englewood Cliffs, N. J., Prentice-Hall, 1952, pp. 120-121.

An approximate equilibrium is found between the existing order and pressures for change. These pressures may result from influences which intrude from the outside or from changes within, as when a group in some area increases in size and in activity. Such an equilibrium, continuing in its tendencies, also exists for regions as well as for whole communities, although each community has its own processes of change. Ecological changes are never total (except perhaps in disaster), but are distributed to different sectors of community life. Thus, each area is affected either by internal change or change in neighboring areas; only gradually are groups of areas affected. The wider community is not thrown out of equilibrium. Normally, life in the modern community is flexible and tolerant of change, and maintains balance.

## Utility of Human Ecology

One important function of any social science is to afford understanding of society, its structure and processes. Our purpose in this book is to gain a better understanding of community life, especially of the ways of life and work in the urban community. Since such an understanding is of great value to those living under urbanism, we will do well to make use of any approach that will increase that understanding. Human ecology is one of many methods for gaining knowledge about community life. One would not expect to use ecology to obtain data that can be secured better by a psychological approach; but if we want to study social problems in their spatial relationships, the techniques of ecology can prove effective. This approach involves certain assumptions which have been stated by Dunham in these terms:

1. That human communities have a certain organic character in that they expand, change and decline, with the probability that the process will be repeated. This cycle constitutes a dynamic equilibrium.
2. That in this expansion a process of distribution takes place which sorts and relocates individuals and groups by residence and occupation over a given land area. In ecological theory the expansion was a function of competition, and it has been demonstrated that certain conscious motives often operate in relocation of persons.
3. That this selective process creates "natural areas" which develop their own characteristics and can be delimited.
4. That each area with its particular characteristics leaves its cultural "stamp" upon the people who reside there, and affects them in numerous and diverse ways.
5. That this cultural "stamp" will be registered in each area of frequencies

> of numerous types of both acceptable and unacceptable behavior which will differ according to the character of the area.[38]

It should be kept in mind that, while human ecology to date is largely an interest of American sociologists, it is also being used by other disciplines. Although sociologists in other countries may not use the term, they may carry on similar research under some other name. In Germany it is *Raumforschung* (space research).[39] In England the approach is one of social planning, a good example being a study of Middlesbrough by Glass[40] which resembles in many respects the American ecological approach but she makes no reference either to ecology or space.

Quoist made a study of Rouen, France, which in some respects is not unlike the study of Paris made by Chombart de Lauwe and associates mentioned above. While Chombart de Lauwe called his work ecological, the term is not used by Quoist, who speaks of space relations as geographic. He speaks not of natural areas but of *groupements naturels* and he uses maps of different kinds to locate these groups and to locate spatially other phenomena reported in his study.[41]

It should matter little to the sociologist who may be interested in this type of research whether the approach is called human ecology or something else. What is important is that urban sociologists in other countries are making what Americans call the ecological approach to the study of community life. That is much to be desired, since the results of ecological research elsewhere should widen the perspective of American sociologists in their own ecological studies.

## Summary

Human ecology is an approach which permits intimate acquaintance and understanding of the community as a place on the ground. It has come to be an interest of geographers, engineers, town planners, and others. Geographers,

[38] H. Warren Dunham in a comment following an article by John A. Clauson and Mervin L. Kohn, "The Ecological Approach to Social Psychiatry," *American Journal of Sociology*, Vol. 60, No. 2, September 1954, pp. 150-151.

[39] Renate Mayntz (*Sociale Schichtung und sozialer Wandel in einer Industriegemeinde*, Stuttgart, Ferdinand Enke, 1958) includes a chapter on *räumliche Gestalt der Gemeinde* (spatial pattern of the community).

[40] Ruth Glass, Ed., *The Social Background of a Plan*, London, Routledge and Kegan Paul, 1948.

[41] Michel Quoist, *La Ville et L'Homme*, Paris, Les Editions Ouvrières, Economie et Humanisme, 1952. See especially Chapter XII, "La Dimension Geographique."

interested in relations between the earth and man, now study the location, the resources, and the life of communities, and they may call their approach human geography, social geography, or ecology. Sociologists became interested in ecology through the work of plant ecologists and animal ecologists, from whom they borrowed terminology for their own work.

Human ecology is the study of man in his community life, adjusting to his habitat. It concerns his spatial relations within the community, how the competing interests in the community manage to share the space available for such uses as residence, work, leisure, as well as for collective public or private purposes. All of these uses tend to be competitive and concern the use of land or space. Space has economic value and can be rented or sold, a source of profit. It is also limited, and hence multiple use, such as building into the air or digging into the earth, is encouraged. Space that is very accessible acquires high value, as in the urban center where it is compactly used; its value is low when it is least accessible.

In the use of the urban habitat man tends toward specialization in the occupation of space. Some areas are used for residence, others for industry, others for commerce, and so on. But there are different types of residence and industrial areas and many types of commercial areas, which may in the great city be called special markets. In different ways these areas are occupied by users who tend to cluster. In some cases these areas are expanding, in others contracting; but whether growing or declining, neighboring areas are affected. Changes in one area effect changes in other areas.

Seen in wider perspective, these different related uses of space tend to be interlocked from the urban center to the outer suburbs, and even beyond. Relations between points within this extended pattern are affected by distances between them and by the ease with which they may be made accessible to one another. It is being recognized that the extended region of occupancy from urban center to the remotest suburbs has a unity which can be graphically presented in terms of a series of occupation zones from the urban center outward.

One ever-present result of competition between occupiers is that any particular bit of land or area may witness a succession in users and uses as well as in structures. These changes tend to be continuous, although they may move faster in some urban communities (and in some countries) than in others. However, such change is never complete. The effects tend to be scattered, which permits the existing order to remain in equilibrium, although it is a moving equilibrium in communities of rapid change.

CHAPTER 6

# Urbanism and Population Phenomena

COMMUNITY CHANGE when measured in terms of population change is a subject of intriguing interest, whether it concerns population increase or decrease, population mobility, or changes in the characteristics of a population. By statistical reporting through the use of charts and tables and with different mathematical devices we are able to measure with increasing accuracy the various phenomena relative to population. This is the important field of demographic research to which the sociologist in his study of community life, urban life in particular, must frequently turn. Much important information about urbanism and population phenomena is not reached by the approach of demography, but much of such nondemographical material, once assembled, may be enriched by the findings of demography. For example, from economics we can learn much about the global trend of urbanism, but such information is greatly clarified when matched with demographical information about urbanization trends.

We must see urbanism not in terms of cities alone, nor in terms of counting cities of different size. It must be seen as clusters or networks of communities. This is being recognized by some American sociologists in their population studies for metropolitan areas. The objective of this chapter is to assemble materials which serve to describe urbanism in terms of population phenomena.

## The Stimulating Influence of Urbanism

We have already recognized that the industrial city cannot grow, indeed can hardly live, if it does not have outside markets and sources of supply.

This has always been true of cities, but the rapidly growing industrial city or the commercial city linked with industry must have wider and more frequent contacts. A network of competing industrial cities engages in a race to find new markets anywhere in the world and to draw supplies from any source. As the cities themselves grow and change they also stimulate change in outlying areas, even areas remotely situated. When nonindustrial regions where the population has been in equilibrium for generations come into contact with industrial urbanism, they suddenly experience population increase, even in the face of very limited possibilities for increasing food supply.

Davis cites the case of Egypt with a population increase of 14 to 24 million since 1927, with a birth rate of about 44 per 1,000 for 1950-1953, and with a declining death rate. Population increase has been three times faster than the increase in area of cultivated land, while manufacturing since 1927 has increased by only about 8 percent. Population now averages 1,250 persons per square mile of cultivated land (35 in the United States). The cities are filled with refugees from rural places in which they cannot gain a living.[1]

Egypt is not so far, economically, from cities that use Egyptian cotton. In other parts of Africa there are population changes stimulated by industrial urbanism. Industry enters to create new labor markets, and people move from their native villages to join a new kind of economy and to experience a new kind of life. They take with them certain of their old ways, but quickly adopt the new ways in the town. In one way or another the populations are thrown out of balance. It may be that changes occur in the ratios of males to females or of younger people to older ones, or it may be that the birth rate suddenly increases.

Population balances are disturbed by *any* changes that take place with respect to industrial development or urbanization. At the beginning of the Industrial Revolution these effects were fairly local and on a smaller scale. Today we find that many industries are world-wide in their operations and cities are joined in a world-wide network of competition and exchange. Those stimulating influences which radiate from the industrial urban centers of the more advanced countries will be found in operation wherever new supplies of raw materials can be found and wherever new markets can be developed.

Although today these developments assume the character of a global urbanism, this was less true in 1850, for example, when wide regions of the

[1] Kingsley Davis, "The Political Impact of New Population Trends," *Foreign Affairs*, Vol. 39, No. 2, January 1958, p. 297.

earth were beyond the outer fringes of such influences. These outer regions were in a state of equilibrium with respect to their populations. The problems due later to arrive in the outer regions were then confined to a limited number of countries which are today designated the advanced countries of the West. In this chapter we can only touch on the global implications of industrialization and urbanization, but we do need to be aware of these wider implications as we consider population phenomena in a more intimate relation to the more advanced urban communities.

## Industry in the City

The beginnings of industry in the United States after the Revolutionary War, particularly in the New England states, must be recognized as one of the most romantic phases of American history. But in England industry had already been emerging for nearly a half-century and was beginning to develop at a faster rate. New labor markets were being evolved in the towns, towns soon to become cities, since by 1800 the new factories were already attracting people from the villages. This movement of villagers in large number to the towns and cities did not begin in the United States until after the Civil War, really not until after 1870.

Industrial development on the European continent was under way in Belgium, France, Holland, and Germany during the first half of the nineteenth century, but it was not until after 1850 that factories began to multiply in the cities. The development in Germany was faster after 1870, after the many small states and free cities joined in a nation.

The influx of rural people in the European countries caused somewhat of a "ruralization" of the industrial centers. These workers retained the rural family pattern. Birth rates were high. Always the city had more people than could be employed. Wages were low, hours of work long, and job security hardly existed. Houses were crowded and poorly provided with the common amenities for comfort and health. Death rates, especially of children, were high. To eke out an existence, women and children in large numbers had to enter factory employment. Factory owners came to be regarded as exploiters, and to a large extent they were, but it is also true that industry was itself in a state of trial and error. Factories were poorly equipped, often insecurely financed, and the idea of industrial management had not yet emerged.

However, in the course of a century—that is, by about 1870—English industry was well established, and during the next fifty years manufacturing and industry-connected commerce were well rooted in all the Western countries. In the cities living standards were beginning to rise, labor conditions were improving, the earlier low-level types of housing were disappearing, working hours in the factory were shortened; in fact, the life of industrial workers had greatly improved over earlier decades. Excessive urban birth rates had declined, but the trend of change had moved from one extreme to another, from rates recognized as too high to those that seemed too low.

By 1900 industry had also taken root in the United States, and each passing decade has seen its growth. Urbanization was then moving at a rapid rate, and continues to now. The American situation was favorable to such development for several reasons: Raw materials were available, initiative and experiment were least hampered by rooted work ways and traditions, and there was the additional stimulus of labor shortages. The American industrialist had to compete with the land frontier for labor supply, a problem that was met by importing workers from Old World countries, many of them people of rural origins. During this period of industrial emerging, American cities also had the problems of high birth rates, congestion, and slums, but the magnitude of the misery was never so great as it had been in European cities. Moreover, the population problems of American cities when they were being industrialized, serious though they were, were not the same as the present population problems in these cities.

In general, we can say that the expanding industrial urbanism has everywhere been, and still is, a disturber of population balances. Initially these disturbances were local, within the immediate hinterlands of cities, but the hinterlands tend to be wider today. Furthermore, the problems of population not only tend to be continually present even within the cities, but they change in nature in response to other changes. They differ at any one time between urban community, rural community, and underdeveloped region, and in each type of community these problems are changing.

## The Rapid Growth of Cities

While the bulk of the world's population is still not urban, there is an increasing urban trend. We are reminded by Davis that in 1800 only 2.4 percent of the world's population lived in cities of 20,000 or more inhabitants, in 1900 the cities of this class contained 9.2 percent of the total, and in 1950 20.9 percent of the world's population were living in cities of 20,000 or more. Correspondingly, cities of 100,000 or more inhabitants held 1.7 percent of the world's population in 1800, 5.5 percent in 1900, and 13.1 percent by 1950.[2]

Table 1 presents such figures as could be assembled by the United Nations showing recent urban growth in a few selected countries. The percentages for each country show growth of the urban section of the population over a period of about 50 years. It must be recognized that the definition of what is an urban community differs from country to country. In the intervening period, overall populations in these countries have been growing. Finland's population increased from 2,712,000 in 1900 to 4,286,000 in 1950; Japan's from 55,157,000 in 1920 to 83,199,000 in 1950, and that of the United States from 75,994,000 in 1900 to 150,697,000 in 1950. In the meanwhile the Union of South Africa grew from 5,175,000 in 1904 to

Table 1. Percentage urban in population of selected countries*

| *Country* | *Percent* | *Year* | *Percent* | *Year* |
|---|---|---|---|---|
| England and Wales | 77.0 | 1901 | 80.7 | 1951 |
| New Zealand | 43.1 | 1901 | 61.3 | 1951 |
| United States | 39.7 | 1900 | 59.0 | 1950 |
| Canada | 37.5 | 1901 | 57.4 | 1951 |
| Union of South Africa | 23.2 | 1904 | 42.4 | 1951 |
| Japan | 18.1 | 1920 | 37.5 | 1950 |
| Finland | 10.9 | 1900 | 32.3 | 1950 |
| Algeria | 16.6 | 1906 | 23.6 | 1948 |

* *Demographic Yearbook,* United Nations, 1952, p. 14.

12,646,000 in 1951. As for the world, its estimated population grew from 1,813 million in 1920 to 2,438 million in 1951.

We tend to think of urbanization in terms of big cities. In Table 2 we have selected out of the 64 countries reported in the *Demographic Yearbook* ten countries that are most urban and ten that are least urban. Here we see the percentage of the population that is urban and the percentage living in cities of 100,000 or more inhabitants. The figures are revealing, but they may also be misleading. For example, some countries include in the larger cities the people in "fringe" communities. Others do not. In the Federal Republic of Germany, the table shows that only 27.1 percent of the population lives in cities of 100,000 or more, but seven out of ten Germans are urban. Many live in cities from 50,000 to 100,000, and many of the smaller cities are side

[2] Kingsley Davis, "The Origin and Growth of Urbanism in the World," *American Journal of Sociology*, Vol. 60, No. 5, March 1955, p. 433.

by side. Of the least urban countries, India is the most striking. With an increasing population (356,829,000 in 1951) nearly equal to that of all Europe, excluding Russia, only 17.7 percent of this population is urban and only 6.8 percent live in cities of 100,000 or more. India has 76 of these larger cities, of which 19 are of 300,000 population or over, and four (Bombay, Calcutta, Madras, and Hyderabad) were in the million class in 1951.

Table 2. Percentage urban (by national definition) and percentage of the total population in cities of 100,000 or more inhabitants*

| *Most and least urban countries* | *Percentage urban* | *Cities of 100,000 or more* |
|---|---|---|
| | Ten most urban countries | |
| England and Wales, 1951 | 80.7 | 51.9 |
| Israel, 1951 | 77.5 | 39.9 |
| Fed. Republic of Germany, 1950 | 71.1 | 27.1 |
| Australia, 1947 | 68.9 | 51.4 |
| Denmark, 1950 | 67.3 | 33.5 |
| United States, 1950 | 63.7 | 43.7 |
| Belgium, 1947 | 62.7 | 25.8 |
| Argentina, 1947 | 62.5 | 40.6 |
| Canada, 1951 | 62.1 | 36.7 |
| New Zealand, 1951 | 61.3 | 32.8 |
| | Ten least urban countries | |
| Turkey, 1950 | 25.2 | 8.3 |
| Bulgaria, 1946 | 24.6 | 8.0 |
| Philippines, 1948 | 24.1 | 9.3 |
| Romania, 1948 | 23.4 | 8.0 |
| Iran, 1950 | 20.0 | 9.3 |
| Korea, 1949 | 19.6 | 14.7 |
| India, 1951 | 17.7 | 6.8 |
| Yugoslavia, 1948 | 16.2 | 6.3 |
| Ceylon, 1946 | 15.4 | 5.4 |
| Burma, 1931 | 10.4 | 3.9 |

* *Demographic Yearbook*, United Nations, 1952, p. 11.

It is hardly necessary to parade the figures in order to underline the thought that cities in all countries are growing, although in some countries the rate of growth is faster than in others. We may add, too, that urban growth everywhere is at or above the capacity of cities to absorb people—an observation that would apply to cities in any age.

## Rural-urban Differences

Whether a man is urban or rural, more urban or less urban, can hardly be learned from statistics. Is he rural because he lives, as defined in the United States, in a place under 2,500 inhabitants, or is he identified as rural because of his occupation? He may be a very modern farmer, but he thinks and behaves much as his neighbor who lives in the small town but works in a factory. The demographer, however, is expected to distinguish between rural and urban people. He does it by the practical device of identifying rural and urban with the sizes of places where people live, knowing full well that urban people (by occupation) will be living here and there in rural areas, while some rural people (by occupation) will be living in areas designated as urban. "As a result," to quote the *Demographic Yearbook:*

> the characteristics of the urban population, no matter how it is defined, are heavily weighted with the characteristics of the more highly urban population. Conversely, the characteristics of the more rural portion of the population make a strong contribution to the characteristics of the rural population as a whole, no matter how defined.[3]

Elsewhere in this report it is noted that "Because there is no point in the continuum from large agglomerations to small clusters or scattered dwellings where urbanity disappears and rurality begins, the division between urban and rural areas is necessarily arbitrary."[4]

An urban area in the United States, as defined by the Census Bureau, includes all places of 2,500 population or more; all densely settled fringe areas around cities of 50,000 or more (a fringe area is one having 2,000 or more population per square mile outside a major city), and unincorporated areas of 2,500 or more outside the fringes of any urban place. By this definition, according to Bogue, 64 percent of the population of the United States in 1950 was urban and 36 percent rural, most of the rural being urban-oriented.[5]

What is called urban and rural in other places? In Canada as of 1951 any community having 1,000 or more inhabitants, whether incorporated or unincorporated, including suburban parts, is called urban. Places of less than 1,000 are rural. In Mexico any place of less than 2,500 is rural. In Argentina

[3] *Demographic Yearbook,* United Nations, 1952, p. 12.

[4] *Ibid.,* p. 24.

[5] Donald J. Bogue, "Urbanism in the United States, 1950," *American Journal of Sociology,* Volume 60, No. 5, March 1955, p. 471.

any place of less than 2,000 is called rural. Only places of 5,000 or more in India are called urban, and the same is true for Belgium, while in Denmark any place is urban if it has 250 or more inhabitants. In Spain a community is called urban if it has 10,000 or more inhabitants, while the number for urban status in Holland is 20,000. These differences remind us of the difficulty that must be faced in making international comparisons of urban or rural people.

For our limited review of urbanism as it relates to population phenomena, however, rural-urban comparisons within countries are quite sufficient.

## Migration

As most frequently used, urbanization relates to migration, generally from less urban to more urban places and from smaller to larger places, or from one very urban place to another. Cities have always received migrants. Fulford says of migrants in London in the last century, "From the harsh countries of eastern Europe, driven out by terrors, purges and privations, they drifted to London in the certainty that they would be free from persecution and in the hope that they might be able to scratch together a livelihood."[6] That could be said about many great cities; they are places of refuge as well as givers of work. They have always assimilated people and have needed to do so.

The subject of migration will be considered more fully in Chapter 7 on "The Moving About of the Population." We need mention here only the importance of migration. For example, American cities for decades had to look to other countries for labor supply. In large part, the arriving immigrants settled in the cities, mainly in the cities of the northeastern and northern states, and in these cities they congregated in their own quarters where they could speak their own languages, eat their own kinds of food, and continue certain Old World culture traits. Bercovici in the 1920's found that in New York City one could see in miniature most of the major cultures of the world.[7] Here is a brief description of the ethnic groups found in the city of Cleveland:

> In the anonymous masses that make up the living city, Cleveland is almost a Midwest anomaly. The white stock of native parentage comprises only a quarter of the population; eight percent are Negroes and the remaining 67 percent are either foreign-born or the offspring of foreign or mixed parentage.

[6] Roger Fulford, "Jubilee London." See Ernest Barker and others, *Golden Ages of the Great Cities*, London, Thames and Hudson, 1952, p. 274.

[7] Konrad Bercovici, *Around the World in New York*, New York, Appleton, 1924.

> Once almost entirely Nordic or Celtic in makeup, Cleveland was transformed by the expanding steel industry into one of the most racially diversified communities in the United States. Forty-eight nationalities have representatives here; more than forty languages are spoken in the city. First in number are the Czechoslovaks followed by the Poles, Italians, Germans, Yugoslavs, Irish and Hungarians. Where they concentrate in nationality groups, their native tongues are spoken almost as commonly as English.[8]

Any cosmopolitan city will have its special areas occupied by in-migrants of people of particular background if they are present in sufficient number to concentrate. Americans in foreign countries tend to gather in specific areas also, and the British when abroad may deliberately form their own areas of residence, as do other Europeans when they go to the less advanced countries.[9] Southall described the situation in Kampala, Uganda, where in 1952 there were 4,250 Europeans (mostly British), 17,000 Asians (mostly Indians), and 16,800 Africans, with another 30,000 Africans living in slum villages around the fringe of Kampala. The Africans were segmented by tribal groups.

> The stereotype is that, while Europeans are politically dominant through their control of the formal political institutions, Indians dominate the economic scene, and Africans hold the lowest status, supplying all the unskilled labour. Although there are many exceptions to this stereotype, and there is no statutory color bar, but rather considerable administrative enforcement of non-discrimination, nonetheless racial discriminations are actually institutionalized to a considerable extent. Europeans, Asians and Africans have to attend different schools, and unofficial members of the municipal council are selected on the basis of their racial origin. The very attempt to secure equality between the races in many situations emphasizes racial divisions.[10]

Three different racial groups, migrating to Kampala, are occupied with joining their labor to make it into a city. An ecological division of work and space occupancy has resulted, but each racial group presents a different kind of population problem. The African labor supply is the most inconvenienced by this urbanward migration. For the present at least, racial differences stand as a bar against the assimilation of these different groups of migrants. This factor is a vanishing problem for European migrants to America.

[8] *The Ohio Guide,* American Guide Series, New York, Oxford University Press, 1940, p. 217.

[9] Cyril Sofer and Rhona Ross, "Some Characteristics of an East African European Population," *British Journal of Sociology,* Vol. II, 1951, pp. 315-337. This is a report on the clannishness of 549 Europeans, 73 percent of whom were British, in an African town. Seventy percent of the Europeans are males, mostly between 20 and 45 years of age. They systematically avoid social contact with Asians and Africans.

[10] A. W. Southall, "Determinants of the Social Structure of African Urban Populations, with Special Reference to Kampala, Uganda," in *Social Implications of Industrialization and Urbanization of Africa South of the Sahara,* Paris, UNESCO, 1956, p. 560.

## Rural and Urban Sex Ratios

In all three racial groups in Kampala the males outnumber the females, particularly among the Africans. This imbalance is found generally in new cities in underdeveloped countries, and they therefore assume somewhat the character of former American frontier settlements where males predominated. A somewhat reverse condition seems to develop as cities become more modern, more varied in their industry, and equally varied in their commercial life. With these changes, cities also become both more accessible for female migrants and more receptive to their coming; more receptive because the modern urban labor market offers more work opportunities for women.

But countries differ widely in the extent to which males and females are included in both urban and rural populations. This observation is amply supported by the comparative figures in Table 3 which gives both the urban and rural sex ratios for a number of selected countries. At the one extreme in this table is the Republic of Ireland with 85.5 males per 100 females in the

Table 3. Sex ratios of urban and rural populations in selected countries*

| *Country* | *Year* | *Males per 100 females* | | *Excess of rural over urban* |
|---|---|---|---|---|
| | | *Urban* | *Rural* | |
| Republic of Ireland | 1946 | 85.5 | 114.2 | 28.7 |
| Poland | 1949 | 85.8 | 92.5 | 6.7 |
| France | 1946 | 86.3 | 94.4 | 8.1 |
| Fed. Republic of Germany | 1950 | 87.6 | 89.6 | 2.0 |
| England and Wales | 1951 | 90.5 | 101.8 | 11.3 |
| Denmark | 1945 | 91.2 | 112.0 | 20.8 |
| Sweden | 1950 | 93.0 | 107.7 | 14.7 |
| United States | 1950 | 94.1 | 105.7 | 11.6 |
| Canada | 1951 | 95.8 | 114.1 | 18.3 |
| Japan | 1950 | 96.9 | 95.7 | – 1.2 |
| Argentina | 1947 | 97.2 | 119.7 | 22.5 |
| Greece | 1940 | 100.9 | 97.8 | – 3.1 |
| Israel | 1951 | 101.5 | 111.9 | 10.4 |
| Egypt | 1947 | 101.8 | 96.5 | – 5.3 |
| Korea | 1945 | 104.5 | 101.6 | – 2.9 |
| Federation of Malaya | 1947 | 117.8 | 110.3 | – 7.5 |
| India | 1941 | 122.8 | 104.8 | – 18.0 |
| Ceylon | 1946 | 139.0 | 108.9 | – 30.1 |

* *Demographic Yearbook*, United Nations, 1952, p. 15.

cities compared with 114 per 100 females in rural areas; at the other extreme is Ceylon where the respective figures are 139 males per 100 females in cities and 108.9 in rural areas. That some of these urban-rural ratios seem out of balance is only a reflection of the ratios for the total population. Ceylon, for example, with under seven million inhabitants in 1946, had a ratio of 112.6 males per 100 females, while for India in 1941 there were 106.9 males per 100 females. The Republic of Ireland, with less than three million people, had 102.4 males per 100 females. The ratio in England and Wales was 92.6 males per 100 females, while in France it was 89.9, and in the United States, 98.1.

The ratio of males to females is rarely in balance, and it is subject to variation from one decade to another. In some industrial countries the excess of women is in part due to their longer life expectancy. But there can be extremes which may give rise to social problems. In Southall's report on Kampala he noted that the town-dwelling males of the African Ganda tribes far outnumber the town-dwelling Ganda females. Thus these women of marriage age are in a uniquely dominant position which finds expression in types of indifference to family and marriage tradition—an example of attitude change in response to a numerical imbalance between the sexes living in a new situation.

> The Ganda women undoubtedly have a very great influence on the type of marriage and family life which is prevalent in the urban area. Many of them prefer to have a series of lovers and retain their freedom than to tie themselves to any lasting union. This meets the needs of the men temporarily in town, including, of course, those who are stably married but have left their wives behind. This group of women has a very low fertility, and, when children are born, they are often sent after weaning to the country with relatives of either the mother or the father.[11]

This is perhaps a primitive approach to female emancipation, but it has its counterpart at a more sophisticated level in the modern city where women are able to enter occupations and become economically self-sufficient to the degree that they can afford to have their own separate apartments. Many such emancipated modern women prefer their careers to marriage.

## Birth Rates

It is generally true that countries with half or more of their population living in urban communities have lower birth rates than those reported for

[11] *Ibid.*, p. 557.

countries that are 40 percent or less urban. In general it is also true that, whether countries are more urban or less, urban birth rates are lower than rural birth rates. The figures in Table 4 are based on birth-registration records, but since both rural and urban records are so fallible in most countries, only a few countries are included in the comparable group. The figures show the number of children under five years of age born to each 1,000 women 15-49 years of age for the specified countries, including urban and rural communities. For each country shown, the number of children is higher for rural than for urban women; the greatest difference among the countries shown is in Canada, where the ratio is 422 children for urban women to 637 for rural women.

Table 4. Children under 5 years of age per 1,000 women 15-49 years old in selected countries*

| *Country* | *Year* | *Ratios of children to women* | |
|---|---|---|---|
| | | *Urban* | *Rural* |
| Australia | 1947 | 353 | 506 |
| Canada | 1951 | 422 | 637 |
| Denmark | 1948 | 372 | 457 |
| Norway | 1930 | 194 | 386 |
| Sweden | 1945 | 301 | 368 |
| United Kingdom | 1951 | 332 | 358 |
| United States | 1950 | 383 | 505 |

* *Demographic Yearbook,* United Nations, 1952, p. 17.

The least urban-rural difference for the countries shown is in the United Kingdom, whose ratio is 332 children for urban women to 358 for rural women. In that country there are almost no rural places that are not embraced by the urban way of life. It should be noted that the figures in Table 4 are almost identical for Australia and the United States, both new countries and both highly urban.

This method measuring fertility trends is one of three that are coming to be widely used. The three methods have been compared by Stolnits as they apply to the figures of the 1950 United States Census.[12] The results are as follows:

1. Number of children under 5 years ever born to 1,000 native white women 15-49 years of age . . . . . . . . . . . . . . . . . . . . . . . . . . . . . . 404

[12] George Stolnitz, "Population Composition and Fertility Trends," *American Sociological Review*, Vol. 21, No. 6, December 1956, p. 740.

2. Number of children under 5 years ever born to 1,000 native white women 15-49 years of age *who have ever been married* ........... 527
3. Number of children under 5 years ever born to 1,000 native white women 15-49 years of age *who were married once and live with their husbands* ........................................ 571

Each method selects a certain category out of the total number of women who have had children, and each serves a purpose. The first method is the one used in the demographic yearbooks of the United Nations.

In the 1920's Whelpton made a study of the number of children born to native white women in certain agricultural states, semi-agricultural states, and industrial states. This was later enlarged to include the census figures for 1940. He found that for each 1,000 women 16 to 45 years of age in the United States 1,000 children were reported for 1800 and 336 for 1940. For the most agricultural states the figures were 1,043 for 1800 and 441 for 1940. In the heavily industrialized states he found that the number of children per 1,000 native white females had fallen from 786 in 1800 to 306 as of 1940.[13]

In 1935-36 the United States conducted a health survey of the American people. The national sample included 700,000 families of all social classes. Of this sample, 632,000 families were in urban communities and included 596,474 females 15-44 years of age, of whom 336,226 were married. Karpinos and Kiser made a study of these married women in urban communities, which they designated a study of nuptial fertility. Their attention was on the number of children born to women of different economic levels and of different levels of education.

They found that wives in families with incomes of $3,000 or more had 84.6 children per 1,000, while wives in families with incomes of less than $3,000 had 132.9 children per 1,000. They also found that wives with college education had 96.9 children per 1,000, those with high school education had 102.5 children per 1,000, those with seventh- or eighth-grade education had 117.5 children per 1,000, while wives with less than a seventh-grade education had 130 7 children per 1,000.[14]

Huber, Bunle, and Boverat made a comparative study of births per 1,000 females 15-49 years of age in France for the periods 1846-55 and 1935-39.

[13] P. K. Whelpton, "Industrial Development and Population Growth," *Social Forces*, Vol. VI, 1928, p. 462. The 1940 figures were added by Warren S. Thompson and are included in his *Population Problems*, New York, McGraw-Hill, 1942, p. 167.

[14] Bernard D. Karpinos and Clyde V. Kiser, "The Differential Fertility and Potential Rate of Growth of Various Income and Educational Classes of Urban Population in the United States," *Milbank Memorial Fund Quarterly*, Vol. 17, October 1939, pp. 367-392.

Fertility was related to the number of births per 1,000 females, the number per 10,000 population, and the number per 100 marriages. The figures for all France, irrespective of urban or rural residence, were:

1. *Births per 10,000 population:* 264 average for 1846-1855 and 148 average for 1935-1939, a decline of 44 percent.
2. *Births per 100 marriages:* 338 average for 1846-1855 and 227 for 1935-1939, a decline of 33 percent.
3. *Births per 1,000 females 15-49 years of age:* 101 average for 1846-1855 and 60 for 1935-1939, decline of 40 percent.[15]

These scattered bits of evidence tend to confirm in quantitative terms the generally accepted idea that urban populations are less fertile than rural populations, and, furthermore, that rural populations are tending to follow a downward trend.

## Death Rates

A community increases its population either by migration or birth—usually by both methods—and the community loses population as people move away or die. This is, obviously, an oversimplified statement of a very complex subject, particularly with respect to urban communities. Questions must be concerned not just with the number of people entering a community by birth or migration, but with the kind of people. This also holds for exits from the community. Death rates in many ways have a sociological importance. The death rate may be higher for males than for females, or higher for children than for adults; or it may be that in the course of a few decades the life expectancy for all age groups has been extended by several years, which will give rise to various problems.

Death rates in all the more advanced countries have been declining during the past several decades. In 1953 in the United States the rate was a little under 9.6 per 1,000 population. For Canada the same year it was slightly above 8.6 per 1,000, while in Iceland it was 7.4. In the United Kingdom, where the median age level of the population is high, the death rate in 1953 was 11.4 per 1,000 population. Such rates are low compared with those of some less developed countries. For example, the 1950 death rate in India was 16.6, that of Guatemala was 19.6, and that of Chile was 20.6 per 1,000 population.

[15] Michel Huber, Henri Bunle, and Fernand Boverat, *La Population de la France*, Paris, Libraire Hachette, 1954, p. 141.

Such figures for death rates are for countries as wholes. Comparative rural and urban rates are difficult to obtain, and we can do no better than to quote the *Demographic Yearbook* of the United Nations:

A compilation of national crude death rates and similar rates for major cities in 1949 gives no clear pattern, either for industrialized or for less urban countries. In one-third of the countries, crude death rates for the principal cities were lower than the national averages, in a slightly larger number of countries the city rates were higher than the national rates, and in the remainder the city and national rates were virtually the same. On the basis of these data no conclusion can be drawn as to whether mortality is higher in urban or in rural areas.[16]

Table 5. Deaths in Great Britain by age and sex per 1,000 population for 1880-1882 and 1951*

| | *Males* | | *Females* | |
|---|---|---|---|---|
| *Ages* | *1880-82* | *1951* | *1880-82* | *1951* |
| All ages | 20.8 | 13.4 | 18.6 | 11.8 |
| Under 4 | 61.3 | 7.7 | 52.5 | 6.0 |
| 5 to 9 | 6.4 | 0.7 | 6.1 | 0.5 |
| 10 to 14 | 3.4 | 0.6 | 3.6 | 0.4 |
| 15 to 19 | 4.9 | 0.9 | 6.1 | 0.7 |
| 20 to 24 | 6.4 | 1.4 | 6.3 | 1.0 |
| 25 to 34 | 8.2 | 1.7 | 8.0 | 1.4 |
| 35 to 44 | 12.6 | 3.0 | 10.9 | 2.4 |
| 45 to 54 | 19.1 | 8.8 | 14.9 | 5.5 |
| 55 to 64 | 33.5 | 24.3 | 27.6 | 13.4 |
| 65 to 74 | 66.6 | 58.8 | 57.4 | 37.8 |
| 75 to 84 | 141.3 | 138.2 | 125.9 | 107.0 |
| 85 and over | 290.3 | 314.6 | 264.8 | 268.3 |
| Infant death rate† | 150.0 | 35.0 | 124.0 | 27.0 |

* *British Annual Abstract of Statistics*, 1953, p. 26.
† Infant mortality, deaths per 1,000 live births.

The difficulty generally is that rural deaths are not always registered. Information from scattered places is difficult to assemble, and apparently in some countries no effort is made to get complete reports. The same observation holds for the rural reporting of births.

As the age level of a population rises one of the results is that the percentage of older people in the population increases. This means that over a

[16] *Demographic Yearbook*, United Nations, 1952, p. 18.

period of time the age distribution of deaths will change. This is clearly seen in Table 5, which gives the death rates by age in Great Britain for 1880-82 and 1951. The table is included because in 1880 Great Britain had already become the most urbanized country in the world—more urban, even, than the United States in 1950. For each age group, except for persons 85 and over, the death rates declined over this 70-year period, although the decline in rates appears to have been more favorable to females. The decline in infant mortality is striking: the death rate among male infants went from 150 in 1880-82 to 35 in 1951, and among female infants from 124 to 27. In the urban slums, both in the United States and in Great Britain, infant death rates were high during the decades before 1900; this was regarded by many of the urban poor as something to be expected. Ferguson wrote thus of Scottish cities in connection with the plagues which took many children:

> The reaction of the people to them was a curious mixture of panic and vicious zeal to combat infection, and of a fatalistic belief that epidemic prevalence was to be accepted as a visitation of Providence, best countered by religious fasts; that somewhere behind the plagues, especially those that affected children mostly, there was probably an element of benevolence—the kind of outlook that regarded small pox as a poor man's friend because it saved him the trouble of bringing up a family, and thought little was to be gained by removing one major cause of death of children, because some other disease would probably come along and take its place—the so-called Doctrine of Replacement.[17]

While poor people in that day of miserable and dangerous urban living were finding comfort in such reasoning, many intellectuals at that time, in the United States not excepted, were rationalizing such infant death rates in terms of "the survival of the fittest" doctrine. It was even argued that medical science in lowering the death rate was performing a social disservice, for thus the weak were given the chance to survive to produce weak offspring.

In a study by Febvay and Croze of infant deaths in France it was found that death rates differ according to the occupations of fathers. They were found to be 1.68 per 100 live births for fathers in the medical profession. Among other rates according to father's occupation were: professor, 2.28; farmer, 4.80; sailor, 5.10, and miner, 8.60. Predominantly big city occupations had the lowest infant death rates. The point was made in the conclusion of

[17] Thomas Ferguson, *The Dawn of Scottish Social Welfare*, London, Thomas Nelson, 1948, p. 2.

this study that low income is not a sufficient explanation for higher infant death rates in some occupational groups. To a large extent ignorance and inattentiveness may also be responsible.[18]

While we do not have much information about urban as compared with rural death rates in general, there have been some specific studies in this area. Kemp and Smith assembled information regarding causes of death for Negroes and whites in urban and rural communities. They found that the death rate from all causes for 1940 in urban communities was 337.5 per 100,000 population for whites and 314.8 for Negroes. In rural communities white people had a death rate of 243.6 per 100,000 population compared with 187.3 for Negroes. Deaths from cancer and from diseases of the heart are higher for whites than for Negroes in both urban and rural communities, and for both races diseases of the heart head the list of causes of death. In other leading causes of death Negroes had higher death rates than whites, and in most cases urban rates were higher than rural rates for Negroes. With respect to tuberculosis, for example, the rates for Negroes were 116.5 in urban communities and 93.7 in rural communities, while for whites they were 38.5 in urban communities and 34 per 100,000 population in rural areas.[19]

## Life Expectancy

A decline in the death rate—and this is taking place in most countries—means that more people in each age group from infancy to late adulthood will enjoy a longer life expectancy. This is not an isolated phenomenon in the development of the industrial urban society; rather, it is an accompaniment of other changes, most importantly the rising standard of living and the increase of leisure, especially in urban communities. Sauvy sees some connection between the lengthening life expectancy and the income level of people in different countries. He compared 14 countries, each having life expectancy at birth of 64 years or more for 1949. The United States (life expectancy of 68 years) ranked first in per capita income but eleventh in life expectancy. Norway (72.3 years) ranked ninth in per capita income but first in life expectancy. Canada (69 years) ranked second in per capita income and seventh for life expectancy. France (65.5 years) was twelfth in both per capita

[18] Maurice Febvay and Marcel Croze, "Nouvelles Données sur la Mortalité Infantile," *Population*, July-September 1954, p. 390.

[19] Summary included in T. Lynn Smith and C. A. McMahan, *The Sociology of Urban Life*, New York, Dryden Press, 1951, p. 274.

F

income and life expectancy. Sweden (72 years) was fourth in income and second in expectation of life.[20]

How the aging trend, which includes the lengthening life expectancy, operates with respect to a particular city is shown for Paris by Landry. Persons under 20 years of age declined in Paris from 271 per 1,000 in 1901 to 204 per 1,000 in 1936, while persons 60 years of age and older increased from 80 per 1,000 in 1901 to 115 per 1,000 in 1936. This was a decline of nearly 25 percent for persons under 20 years and an increase of nearly 33 percent for those 60 years of age and older. During this period, too, the number of persons 20 to 59 years of age increased from 649 per 1,000 in 1901 to 681 in 1936.[21]

By contrast, we must mention a study by Chombart de Lauwe of the rural region of Pays de la Garonne where he found in 1940 that of each 1,000 persons there were 272 under 20 years, 538 between 20 and 59 years, and 190 who were 60 years of age or older. This also suggests an aging population, thrown out of balance by urbanward migration. There are relatively fewer in the working age group (20 to 59 years) than in Paris.[22]

It is generally recognized that a country with a younger population is likely to be more venturesome and imaginative than a country with an older population. It is also assumed, and it may be true, that the younger population is more venturesome, while the older population is more security minded. But within a country of older population there may be significant differences, as shown for France, between the age distribution of population in cities and in rural areas. Cities will attract unduly large proportions of the active age groups from the country, leaving behind large proportions of older people and of the very young. It may be asked if such a development may take place in a young country such as the United States where the life span is also getting longer and where cities are still growing. Such a development is under way in the United States and it may move faster in the future.

An aging population means an increasing number of persons living to quite an advanced age, and it usually means an increasing number of old couples living separately from their families. They need less housing space, and single old people need still less. Families with children become relatively fewer and the number of children fewer, which may create another type of

20 Alfred Sauvy, *Théorie General de la Population*, Paris, Presses Universitaires de France, 1952, Vol. II, p. 92.

21 Adolphe Landry and others, *Traité de Démographie*, Paris, Payot, 1949, pp. 126-127.

22 Jean Chombart de Lauwe, *Bretagne et Pays de la Garonne*, Paris, Presses Universitaires de France, 1946, p. 74.

housing problem. An older population presents different political problems than a younger population. Its consuming habits and health problems are also different from those in a community with many children. Institutional demands are different; fewer schools may be needed but more adult educational facilities may be required. Whatever these differences between an older and a younger population, they are likely to be more conspicuously present in an urban than in a rural community.

## Worries about Population Trends

Population changes resulting from rising or lowering birth rates or death rates and from the natural entries and exits of people are likely to be unnoticed by the man on the street. But he may become very much interested if there is a sudden moving of people into or out of the community. This applies especially to in-migrations, which may cause uneasiness to local residents. Whether such moving-in causes inconvenience or not depends on the community situation at the time. The community may be a village where a single stranger may be a disturbing element, or a great city that receives thousands of strangers daily. It may be a community highly homogeneous with respect to the race, religion, and occupation of its people, or it may be very heterogeneous in all these respects. At a particular time the community may be prosperous and expanding, or it may be depressed in its economy and declining in population.

A very cosmopolitan urban community may be disturbed if it receives large numbers of newcomers from a particular nationality or racial group who agglomerate in certain areas, or who crowd unduly into certain types of employment. Perhaps in prosperous years such newcomers crowd into employment where their presence is least objected to; but in times of unemployment their presence may be noticed and perhaps strongly resisted. Thus in the Great Depression there was agitation in different American cities to deport the foreign-born who had not gained their citizenship. Organizations took form, largely for the purpose of resisting certain in-migrations or for expelling certain types of unwanted people and harassing them.

People of other than the dominant racial group are perhaps the most inconvenienced. In the United States, for example, in Pacific Coast states during the unemployment period of the early 1890's Chinese workers were mobbed and driven away. Many who joined in the persecution were themselves immigrants, but white immigrants from Europe. The problems of Negroes, trying to move from rural villages to urban centers, are also well known. They have been confronted with opposition both in the labor market and in their efforts to find residential space. The issue is overwhelmingly racial.

Worries about population changes may be based on the idea that certain types of people in the urban community have too high a birth rate, and that these people may one day move into the occupations and the residential areas of the now dominant groups. On the basis of such fears there may be demands for restrictive immigration laws, and in particular communities various restrictive ordinances may be enacted.

## Population Planning and Control

Almost every country has laws which in some way are designed to cope with population problems. These may be laws forbidding practices calculated to limit population, or those which permit and encourage some kind of limitation of the birth rate. More often such laws relate to immigration. A country of higher living standards may wish to restrict immigration from countries of lower living standards. Perhaps such legislation establishes a policy of selective immigation. Such restrictive measures are fairly recent; prior to the Industrial Revolution in Europe there were almost no barriers to cross-border migration.

From the extreme of open international borders is the extreme of closed borders now maintained by the "Iron Curtain" countries. According to this policy emigration is not permitted; such a policy is little concerned with immigration.

Immigration laws, being essentially selective, usually give preference to persons of certain age groups or of certain occupations, or they may exclude persons because of political or other views. Or, the laws may give immigration preference to persons of certain national origin and limit immigrants from other countries. The United States, for example, has such a national preference policy. Some countries, while they permit their nationals to migrate to other countries, do not permit them to surrender their citizenship. In theory at least, they can be recalled by their home country for military service or to perform other citizenship obligations.

Urban communities within countries do not have selective powers, but because they are more articulate than rural areas, they have considerable influence in the formulation of national policy regarding both immigration and emigration. The urban industries of certain kinds in European countries may demand certain types of foreign labor, and national policy is framed accordingly. While some foreigners are admitted as immigrants, others are accepted on labor contracts. They are permitted to remain so long as they continue in the specified industries, and only so long as they are needed. In seasons of unemployment they may be deported. Such controls, which operate in the industrial urban sector of industrial life, are essentially motivated by economic self-protection.

## Controls and the Natural Increase

It is not part of our task to weigh the many arguments that have flared around the doctrines of Malthus. While similar statements had been made before Malthus, his essay caught attention at a time when interest in population problems was growing. England was becoming rapidly urban, and the high birth rate of the urban people, especially the poor, added more burdens to their poverty. This is a condition to be found in the cities of underindustrialized countries today, particularly in the Far East and Middle East. Many leaders in England were worried lest the country become overpopulated and be confronted with food shortages. Malthus gave no comfort. He declared that population increases naturally faster than the food supply, at least it can and it had. He admitted that there are restrictive forces also.

In large measure, the debate concerns what *may* happen and what *can* happen. There are those who devoutly declare that, whatever the birth rate, man will find a way to meet the requirements of increasing population. Landis, for example, holds that the "limits of nature seem at present to be beyond exhaustion."[23] On the other side, Hertzler reminds us that there is always the possibility, under given conditions, for an increasing population to "press upon the supporting power of the land."[24] This is not a new danger by any means; it is, rather, one of the most ancient of realities.

Davis, after mentioning that Communist China had adopted a policy of birth control, following the example of Japan, India, and Puerto Rico, con-

[23] Paul H. Landis, *Population Problems*, New York, American Book Company, 1943, p. 38.

[24] J. O. Hertzler, *The Crisis in World Population*, Lincoln, University of Nebraska Press, 1956, p. 88. After the present work had gone to press, the writer saw the volume of the 1954 Harris Foundation Lectures at the University of Chicago. One paper, that by Woytinsky, is especially pertinent to population and supply. It deals with foods, fuels, metals, and other forms of supply, and their prospects and limitations as resources. The problem of supply is seen in its global dimensions. It takes into account the trends in technology and the finding of new food sources. W. O. Woytinsky, "World Resources in Relation to Population," see Philip M. Hauser, Ed., *Population and World Politics*, Glencoe (Illinois), Free Press, 1958, pp. 46-78.

cluded that different countries are beginning to recognize that their strength does not lie in sheer numbers alone. In fact, the have-not nations may be handicapped by an overly high birth rate. He remarked that population changes are taking place all over the globe with different effects that are differently disturbing:

> The most salient demographic change is the astonishing rise in the rate of growth of the world's total population. Another fact of the greatest importance is that the poorer countries are now contributing far more than their share to the inflated growth of the world's population. This means, for one thing, that the areas from which the industrial countries draw many of their raw materials are becoming glutted with people. It means that the greatest advances in science and technology are being made in those countries which have the least need of it in terms of population expansion. Above all, it means that the gap in wealth and power as between the rich and poor nations is becoming wider.[25]

In support of his thought about the widening gap between rich and poor nations, Davis mentions that in 1938 the 15 richest industrial countries had an average per capita income about ten times greater than that for the 20 nonindustrial nations included in his study. But in 1952-54 the same 15 industrial nations had an average per capita income *eleven* times greater than that of the same 20 nonindustrial nations. Moreover, these nonindustrial nations are underconsumers, compared with the rich nations, although they produce more than a fair share of the world's raw materials.[26]

England has passed through a cycle of population problems, from a high birth rate in the industrial cities, great congestion, and deep poverty to the present situation with a low birth rate, a fairly high standard of living, a universal system of social security, and an aging population. During the first half of the nineteenth century various reform movements were getting under way and certain English reformers were getting wide attention. One of these was Annie Besant who advocated birth control as the solution for most of the country's social and economic ills. She reprinted in England an American book, *Fruits of Philosophy,* for which she figured in a famous court trial which did much to broadcast her cause. While this propaganda had little influence on family size at the time, its influence was felt later.

> There can be no doubt from the records of the last quarter of the 19th century that birth control propaganda had a powerful influence on public opinion at that time. The eminence in science and literature of those who drew attention to the pressure of population and the known disinterestedness of Mrs. Besant

[25] Kingsley Davis, "The Political Impact of New Population Trends," *Foreign Affairs,* Vol. 19, No. 2, January 1958, p. 294.

[26] *Ibid.*, p. 295.

and Charles Bradlaugh and other open advocates of birth control helped to loosen the taboos that had previously surrounded the subject and so to diffuse the idea of family limitation. But Mrs. Besant and Charles Bradlaugh had been openly advocating birth control for many years previously, and there had been other eminent persons similarly engaged at different times throughout the preceding 50 years. A burst of publicity similar to that of 1877 onwards, whether through court proceedings or other means, might have flared up at any time, but did not. It is therefore still necessary to explain why this propaganda should have continued for over 50 years (*circa* 1820-1877) before it produced any traceable effect on the birth rate.[27]

Thus in England, even after the taboos began fading away from the idea of family limitation, more than a generation passed before family limitation became fairly general and subsequently gave rise to the declining population problem.[28] The Royal Commission on Population, just quoted, had been assigned the task in 1944 "to examine the facts relating to present population trends in Great Britain; to investigate the causes of these trends and to consider their possible consequences." If Great Britain needed population growth, the Commission asked what the disadvantages might be. They found these did not concern land room alone, but "growing numbers have to be supplied with capital equipment and productive resources have to be devoted to this purpose which might otherwise be used to raise standards." On the advantage side, increasing numbers would:

1. Facilitate an increase in the scale of production and supply a stimulus to technical improvement;
2. If due to high birth rate, [be] associated with a low average age of population;
3. Make the economic system more flexible and . . . thus make it easier to avoid the waste of productive resources through obstinate mass unemployment; and
4. Tend to increase the nation's international influence and so in various ways to strengthen its economic position.[29]

It appears that the industrial urban countries have population problems that are very different from those of the less industrialized and less urbanized

[27] *Report of the Royal Commission on Population,* London, Stationary Office, 1949, p. 36.

[28] Kurt Wicksell, a university student in Sweden in the 1880's, began a one-man crusade advocating birth control. Although it was a shocking idea to most people, he did gain some support. But the birth rate did not begin to drop until a generation later, when Sweden was becoming industrial and living standards were rising. In summarizing the efforts of Wicksell, Alva Myrdal observed that his facts were true at the time but they did not apply later when his preaching had become practice. Sweden was an immigration country then but it is now receiving immigrants. Alva Myrdal, *Nation and Family,* London, Kegan Paul, Trench, Trubner, 1945, p. 20.

[29] Royal Comm. on Pop., *op. cit.,* p. 102.

countries. Some of the underdeveloped or less industrial countries are inconvenienced by congestion on the land, high birth rates, and far too little industrial employment. Some of the industrial countries are confronted with low birth rates, high age levels, and a lack of food-producing capacity. Highly industrial and urban Great Britain is able to eat only if less industrial food-producing countries are willing and able to trade food for her goods and services.

The United States, also highly industrial and urban, is in a more favorable position. Americans are able to produce the food they need and have a surplus for export, and they are not, as of now, confronted with a problem either of overpopulation or of underpopulation.

## Trends and Prospects

While demography affords many clues to future trends in population phenomena, it is not a predicting science. There are always the unexpected developments, such as those emerging from technology and scientific discovery. One cannot say, for example, how life span may be changed by developments in medicine or how certain basic inventions may affect the work day or the movement of population or the standard of living. It is also impossible to predict with accuracy just how changes in one area of human activity may affect other areas.

A few observations about the future can be made with some assurance. For example, it is likely that the world's population will continue to increase, perhaps for some decades. It is also likely that increasing numbers of people will move from rural to urban living and from land occupations to industry. This means a continuation of the global extension of urbanism and an equally global spread of industrialization. These related networks, urbanism and industrialism, can do no other than extend. To describe them as networks is to take account of the many types of interdependence by which industrial urbanism is linked from continent to continent.

Although we cannot make an overall prediction as to what this global linkage of industrial urbanism holds in store for the different populations of the world, we can mention certain possibilities. A legion of towns will become small cities and many more villages will become small towns. Many small cities will become greater both in area and population. There will probably be less of urban population congestion and, as transportation improves, cities will spread outward. This may mean an increase in part-time farming, with more farmers becoming part-time industrial workers and more industrial workers becoming part-time farmers. In this way the urbanization process will increasingly embrace people who have either one foot or both feet on the land. Here we can ask some very pertinent questions about the possible influence of this urbanization process on such population phenomena as, for example, birth rates, life expectancy, and mobility. And pertinent questions can be asked about the effects on population if, as a result of industrial urbanism, living standards are raised in such underindustrial countries as India where at present only 6 percent of the population lives in cities of 100,000 inhabitants or more. Were India as urbanized as the United States 170 million people instead of 16 million would be living in cities of 100,000 or more.

Some of the most challenging population questions for the future concern the underdeveloped countries. Western nations, especially the United States, are engaged in systematic programs for advancing industry, mining, and agriculture in these frontierlike regions, and these will call for the building of cities.[30] Whatever the motivations for these efforts, we may expect that they will give rise to a variety of social problems, problems involving citizenship and community organization; there will be population problems as well, some of which are already acute.

These are but a few of the more apparent possible developments of the future as they concern urbanism and population phenomena. They are not problems that will be easily resolved; rather, they foreshadow a great amount of social disorganization and perhaps social conflict. For the urban sociologist who has some appreciation of the dynamic character of this modern industrial urbanism they should be an intriguing challenge—a challenge because during the next few decades large segments of the world's population will be uprooted in this transition from one way of life to another; in fact, the uprooting and transition process is already under way.

## Summary

Before the emergence of great cities, the world population was fairly well settled according to various geographical isolations, although populations near

[30] See *The Development of Africa,* Strasbourg, Council of Europe, 1957. This report was made by a group of European and African experts on the possibilities of closer cooperation between the countries of Western Europe and their African colonies and independent countries in Africa. The importance of the document is that it states frankly that Europe needs the products of Africa, but Europe must establish new cooperative relationships with these countries and must expend money not only to develop the resources of Africa, but to raise the living standards of the people there so that the Africans will be able to buy the products of Europe. It is tacitly recognized in this report that the old colonial methods can only bring negative results.

cities and towns may have been influenced by urbanism. Cities assumed a different and more dynamic character as they became increasingly industrial. Decade by decade their influence has spread to all regions, and population has been affected in various ways. Among the consequences has been a rapid rise in birth rates, especially in the least industrial countries. This has meant increasing pressure on the productive land and migration from rural to urban places, often to nonindustrial cities or less industrial cities.

As they became industrialized, cities in Western countries went through a cycle of population change from high birth rates and deep poverty to low birth rates, high living standards, and a rising age level. In the United States the evolution of industrial urbanism has moved under more favorable conditions, since it is a new country with many natural advantages. We may ask whether newly industrial countries will have to pass through a like pattern of experience, similar, for example, to that through which industrial urbanism in England has evolved.

The study of population phenomena, with differences from country to country, shows differences also between rural and urban peoples. In some countries the sex ratios, family size, or age distribution may differ between urban and rural people. There is a strong and persistent tendency for urbanward migration to include the younger and more able of rural people. In general, particularly in the more advanced countries, life expectancy is becoming longer, and this gives rise to a variety of social and economic problems. Death rates tend to decline faster in the more advanced countries, and it appears that they are declining faster in urban than in rural communities.

The nearness of cities and their influence on far places tend to stimulate migration. Cities attract people of different racial, national, and cultural backgrounds. The presence of certain kinds of people in a community may disturb other people who may be in a dominant position. Various ideas and movements may develop and certain of these viewpoints may figure in forming national policy regarding immigration control. There may also be concern over the birth rate of certain classes. Normally, in the city, people of like racial stocks tend to become assimilated, but there may be continuing problems in relation to people of other races; for example, Negroes and Asiatics in the United States.

It seems fairly clear that countries in which there is rural population pressure on the land will turn more and more to urban life, provided work can be found in the cities. This means an increasing demand for industrialization as people by the millions come to be identified with urbanism.

CHAPTER 7

# The Moving About of the Population

ENOUGH HAS BEEN SAID in previous chapters to underline the importance of mobility in the evolution of communities. People must move and goods must be moved, else there could be no city, and many cities could not have achieved greatness had they not been able to draw abundantly and continuously on the surplus population of less urban places.

Towns and cities bank their future on the assumption that the roads will be kept open so that people can come or go and goods can be carried. Efforts of communities to grow and become more secure are usually linked with transportation improvement and better-access schemes. Cities in the past have waged war to keep transportation routes open, or to maintain a monopoly of such routes, and nations have waged war for the control of the sea lanes.

We begin this chapter with a general overlook which will consider the importance of roads and routes in the evolution of industrial urbanism. In this connection also we need to consider the relation of the frontier to mobility, as well as the stimulus of the Industrial Revolution. The later sections of the chapter will deal with mobility in more specific respects, as it is involved in the changing life of the modern community.

## Are People "Naturally" Mobile?

This question is often asked, although it is not likely ever to be finally answered. Mobility statistics in the United States might encourage an answer very opposite to that derived from statistics for a country like Spain, for example. The view is not uncommon among Americans that "every man has in him a streak of wanderlust." Walt Whitman, among others, helped to glorify this idea. For him the road inspired fantasy: It came always out of the unseeable past, from all places on the earth, and carried everywhere the experience of the ages.

> You road I enter upon and look around, I believe you are not all that is here,
> I believe that much unseen is also here.

Whitman warned that only the strong and venturesome should enter the road, an idea that was carried on by such scholars as Ellsworth Huntington who regarded migration as a selective force: The strongest and most venturesome migrate and the strongest go farther. But, whether one moved or stayed home, the road was always there beckoning to him, and once on the road one may never rest, except briefly:

> You but arrive at the city to which you were destin'd, you hardly settle yourself
> to satisfaction before you are call'd by an irresistible call to depart.[1]

But the primitive man did not depart; he remained for generations in the same habitat, unless driven from it by enemies or natural change. If he was curious about what might be found on the other side of the mountain or the desert or the forest or the sea, he found the answer in legends and tall tales. In the distances danger lurked—dragons, giants, or fiendish half-animal people. He was safe only at home. It would be a mistake, however, to assume that generations of people have remained sessile because they feared to venture into outer regions. Although fear may have had its influence, the more basic reasons for sessility, which are still very much present, are of a more positive kind.

We may grant that there are many people with "itchy feet" who move frequently because they like to; but even with most of them, moving is a choice between staying, and trying for something better.[2] The considerations in favor of remaining where one is or moving are influenced by factors which change with time and place and conditions. It is easier for a man to move if he has already moved once or oftener, or if members of his family have moved. It is easier for a man to move if the road is near and the means of transport available, and easier as he learns more about what to expect at the end of the journey. Moving, if the prospects offered become more enticing and if taboos against it diminish, becomes easier for more people.

But the fact remains that a good share of the population does not move—

[1] Walt Whitman, "Song of the Open Road," *Leaves of Grass*, New York, Modern Library, 1921; Random House, 1944, pp. 165-178.

[2] This writer has written two books on migration in the United States, both published by the University of Chicago Press: *The Hobo*, 1923, and *Men on the Move*, 1938.

and that applies even in the United States—and the idea of finally settling down is hardly ever abandoned even by mobile people.

Whether it is easy or difficult for people to make decisions about moving depends on changes involved in the type of move. If it is only a change of residence from one section of a city to another, less is involved than in a move from one city to another. For urbanites, moving from one city to another may involve less of a problem than for ruralites to move from a village to a city. The decision becomes more difficult if the move involves migration to another country, particularly a country where the migrant must learn another language. The decision becomes even more difficult if one must not only learn a new language but cross an ocean to reach his destination.

## Roads as Urban Lifelines

To avoid ambiguity, we will use the term "road" in its widest meaning to include any type of transportation, from trails traversed by human carriers or pack animals to roads for wheeled vehicles to railroads to rivers and sea lanes to air lanes. All have the same general function of affording access. We must examine briefly the relationship of these means of access to community life, in particular as they affect life and work in the city.

Perhaps in no country have roads played a more dramatic role in the general economic life than in the United States. This is emphasized by the vast expenditures invested by Americans in gaining access to places. Roads had to be improvised that people might reach the new lands. Farmers had to have roads to markets. Cities were growing faster than roads to sustain them could be built. From the beginning of government in the United States, the people were united in their belief that the economic and social interests of the people belonged in the private domain, independent of government. The first exception to this basic principle was made with respect to roads. Shortly after 1800 the federal government gave financial support for the construction of the Cumberland Pike and still later it gave substantial aid for the building of railroads. That issue no longer exists, as is illustrated by the general acceptance of the idea that the federal government must go forward with a highway-building plan for the new state of Alaska. In England, where similar ideas about the independence of private enterprise from public control prevailed, road building and maintenance for a long time were done by toll companies, while local roads remained the responsibility of local authority.[3]

[3] G. M. Trevelyan (*English Social History*, London, The Reprint Society, 1948, p. 386) notes the road-reform needs, felt most in the cities, and the agitation for better communication.

In most European countries, cities grew along routes or at intersections of routes, but roads hardly existed. It was a venturesome undertaking for the king to travel by carriage from one town to another. It was easier to travel by horse. The reverse tended to be true in the United States: The road came through and towns built along it. When railways came, many small towns moved to be near them.

We need not consider here the democratizing and civilizing influence of roads, as the spreaders of ideas, new words, and even diseases, as well as carriers of goods and highways for people. Kipling, for example, was impressed with the cosmopolitanism of the Grand Trunk Road of India, and in his *Kim* he wrote, "All castes and kinds of people move here." The thought that roads are stimulating influences for changing culture needs no elaborating.

Where there is no road, and hence access is difficult, great suffering may result, as in times of famine. Thompson wrote of a famine in Northwest China at a time when rice was plentiful and cheap in other parts of that country. There was no road to bring it to the stricken area. He noted that a coolie with a wheelbarrow of rice "would have to eat his load and a little more making the round trip."[4] Certain western regions in the United States experienced a great drought from 1933 to 1935. Millions of cattle and sheep died for lack of food and water, but other millions were moved away by the railroads. No people starved, but thousands, because they had automobiles and there were good roads, migrated to other states. Without such access by road and by rail, many people might have died in the Dust Bowl.

## Industrialism and the Frontier

Before taking up further the subject of roads and mobility, we will turn our attention to the Industrial Revolution as a change-provoking phenomenon centering in the cities of Europe, and to the frontier, the great stimulator of both industrialism and urbanism. The land frontier itself, the opening of new markets in vast areas of the world, was one type of pioneering. But the development of industry and all the disturbing social and economic changes brought about by it was also a type of pioneering. However crude and humble the technological beginnings of the emerging industries, they introduced new elements into urban living. This was made possible to no small degree by the enlarging land frontier. Trade and transportation assumed a more global

[4] Warren S. Thompson, *Plenty of People,* Lancaster, Cattell Press, 1944, p. 61.

aspect. Besides the revolution in the industries, there was a revolution in transportation, establishing sea lanes and making roads in the new lands.

People also had to be transported to the new lands, and such transportation had already been under way for a century before the Industrial Revolution had really begun. Adam Smith, writing in 1775, called attention to the great progress being made in the American colonies, where a man with small fortune could get good land and quickly become economically independent. He observed:

> But there are no colonies of which the progress has been more rapid than that of the English in North America. Plenty of good land, and liberty to manage their own affairs their own way, seem to be the two great causes of the prosperity of all new colonies.[5]

These growing new colonies, almost wholly agricultural, were a mighty incentive for industrial development in the mother country. Moreover, this beckoning new market related to colonists of European background. Culturally speaking, the American colonists were ready-made customers for English goods. However, they had also inherited with their English culture the idea of independence, a trait which was encouraged by the frontier. They were not unwilling to be in the hinterland of English urban enterprise, but they could not accept the traditional controls to which hinterlands had always been subjected. Colonialism tried to discourage the initiative that the colonists would naturally, and needed to, express. It was, in the apt term of Ayres, a "penetrating phenomenon."[6]

While the plantation frontier attracted persons of higher social and economic status, those who went to the land frontier had to be able and willing to work. They were the poor and the humble whose departure from European countries was not lamented. The frontier demanded much of them and in many ways they were most suited to the situation where class had to be forgotten and where a man had to be able to act on his own initiative, and where he needed to be original and creative.

De Tocqueville, who visited the United States in its early years, noted that a new concept was developing regarding the perfectability of man. He saw New World communities free of Old World traditions.

> When the citizens of a community are classed according to rank, profession or birth and when all men are forced to follow the career which chance had

[5] Adam Smith, *The Wealth of Nations*, New York, Everyman's Library, 1954, Vol. II, pp. 69-71.

[6] C. E. Ayres, *The Theory of Economic Progress*, Chapel Hill, University of North Carolina Press, 1944, p. 113.

> opened before them, everyone thinks that the utmost limits of human power are to be discerned in proximity to himself, and no one seeks any longer to resist the inevitable law of his destiny. Not, indeed, that the aristocratic people absolutely deny man's faculty for self-improvement, but they do not hold it to be indefinite; they can conceive amelioration, but not change; they imagine that the future condition of society may be better, but not necessarily different; and, while they admit that humanity has made progress and still may have some to make, they assign to it beforehand certain impassable limits.[7]

It was not that the common man going to the frontier wished to leave his traditions behind but, since many of these had no meaning in the new situation, they simply tended to pass out of mind. The situations he faced, as Webb put it, were ego-building, and it was not surprising that people on the frontier developed new words to define new realities. Thus new ways of thinking evolved, new attitudes toward risk taking and new conceptions of man-to-man relations.[8]

Moreover, the frontier prompted new conceptions of geographical distance, and the idea of moving from place to place became an ordinary event which, under some circumstances was a virtue. With such changes in thinking the idea came to be accepted that one must also be a road builder. The man occupying a piece of land on the frontier, unlike the Old World peasant who could see only his own field and who had few interests outside his village, saw his location in relation to a region. Others would come this way and people might come from different directions. It was natural to think in terms of roads and access. Perhaps he entertained the dream that his location might become a town, and it would bear his name.

These thoughts apply especially to the American frontier, but there were other frontiers on other continents, each being stimulating in its own way. In this respect the American frontier was unique, called "the greatest of experiments" by Commager.[9] Here developed a type of mentality that later adapted itself neatly to the ways of industrial urbanism. However, long before frontier America was itself to become industrial, it stimulated industrialism in the mother country.

Adam Smith, writing before the American Revolution had gotten under way, mentioned the advantages to Europe of all colonies, and he emphasized

[7] Alexis de Tocqueville, *Democracy in America,* New York, Alfred A. Knopf, 1946, p. 33.

[8] Walter P. Webb, *The Great Frontier,* Boston, Houghton Mifflin, 1952, p. 49.

[9] Harry S. Commager, *The American Mind,* New Haven, Yale University Press, 1950, p. 462.

in particular the advantage to England of her American colonies, as well as the English trade monopoly in the colonies.

> In consequence of this exclusive trade, all that part of the surplus produce of the English colonies, for example, which consists in what are called enumerated commodities, can be sent to no other country but England. Other countries must afterward buy it of her. It must be cheaper therefore in England than it can be in any other country, and must contribute more to increase the enjoyments of England than those of any other country. It must likewise contribute more to encourage her industry.[10]

Smith also noted how industry in England benefited from the produce derived from America. And, of course, he took notice of the importance of the colonial markets for goods produced in the factories of England.

## The Two American Frontiers

In considering the frontier in the United States, it is necessary to distinguish between two western movements. One was in the southern states where slave labor prevailed and where hard work was the lot of the black man only. The other frontier was in the northern states, but the observations regarding it apply also to Canada, except for the areas occupied by the French Canadians. The southern states had a type of plantation economy, while in the northern states it was a family farm economy.

The settlers in the southern states, mostly English, began very early to import slave labor from Africa. The institution of slavery was tried in the northern states but it proved unsatisfactory. The presence of slavery in the South served to discourage later immigrants from Europe who might otherwise have settled there. Thus America as the land of promise for free immigrants was that limited belt north of the Mason-Dixon Line.[11] There was a westward movement in the South, but it was an extension of the slave-labor type of agriculture. Parts of Missouri were in the path of this movement. In Kansas there was contest whether as a state it should be free or slave.

The plantation-slavery economy came to a natural barrier when it reached the drier regions of western Texas, the beginning of the ranching country. Moreover, a suitable labor supply of Spanish Americans was already on hand in the ranching areas. Moreover, slave-labor agriculture had become associated both with certain crops, mainly cotton and tobacco, and to an

[10] Smith, *op. cit.*, Vol. II, pp. 90-91.

[11] For observations on the inability of the South to receive immigrants, see Harvey Wish, *Society and Thought in Early America*, New York, Longmans, Green, 1950, p. 86.

agricultural household way of life that had become identified with a particular geographic and climatic region.

East-west routes developed more slowly along the southern than along the northern frontier. The first railroads westward were through the northern and only later through the southwestern states. City building moved much faster in the North, and the urban development was much more associated with industry than in the South. It was the North that attracted immigrants of many kinds and, in the long run, it was the North that developed faster, as children of these immigrants from different lands and cultures began making their diverse contributions to a single composite American industrial urban way of life.

## Transportation and American Urbanism

The famous Boston Post Road was initiated by the British in the early colonial period because, for military reasons, a land link between Boston and New York was needed. The point of interest here is that the villages along the route were called on to contribute materials and labor, which they did, but few of them at that time had much interest in the road as such. The people in Harlem, for instance, did not then have a road to New York at the other end of Manhattan Island. They usually made that journey with boats, and so were unfriendly to the idea of assisting in building the Boston Post Road. But the road was built and within decades became an avenue of importance.

The first settlers, their communities scattered in a thin line along the seacoast and waterways, had other problems. They were not looking inland, but seaward. Later as their children began moving to new places inland, they had to have roads, and that interest grew with each decade and with each wave of westward settlement. There was interest in canals in order to connect rivers, or in canals following rivers, a type of transportation already far advanced in European countries. These organized canal-building efforts were also seaward looking, except that some were designed to connect cities along the coast.[12]

In 1825 the Erie Canal, connecting the Hudson River with Lake Erie, was completed. This was less a seaward-looking and more an inland-looking venture. Improvised roads were being blazed westward and the Cumberland Pike, built with federal aid, was very much in operation. Settlements were building fast in Ohio, Indiana, Illinois, and points west. Trade was beginning

[12] A chapter on American canals is included in George R. Taylor, *The Transportation Revolution, 1815-1860,* New York, Rinehart, 1951. Canal mileage grew from 1,277 in 1830 to 3,326 in 1840, and then began declining. Many canals were narrow and inefficient.

to move up and down the Mississippi River. New Orleans was becoming a rival of New York for this growing market area. The Erie Canal was New York's answer. Previously the cost of moving a ton of goods from New York to Buffalo, 20 days by wagon, had been $100, and another $100 to take the same goods on to Cleveland. The canal cut this rate by 80 per cent, and it also was a boon to westward passenger transportation.[13]

It was the railroad, however, that opened the way for mass settlement and town building. But railroad building was costly, and private capital was cautious about venturing into such an unproved type of enterprise. The federal government helped indirectly by giving large tracts of public land to the states to help meet building costs. When it was felt necessary in the interest of national defense, federal aid was given directly to the different railroad companies. Thus it was possible by 1869 to complete the first railway line to the Pacific.

It should be mentioned that some of the first railroad projects were started as auxiliary services to canals. In Pennsylvania, canal building was thought to be more important, to be supplemented by connecting railroads. The canal companies apparently used their influence to discourage any interest on the part of states to use public funds for railroad building.[14] With the increase of inland settlement, interest in canals gave way to a rising interest in roads and railways. Railroad companies altogether received more than 159 million acres, becoming the biggest private landowners in history and powerful enough to dominate legislatures and municipalities, to influence the selection of judges, and so forth.

Whether by road or railway, the frontier was made accessible. Whereas between 1800 and 1869 5,900,000 immigrants came to the United States, the number rose to no less than 20,800,000 between 1870 and 1910. This latter period was one of town building and of industrial development over wide areas which had been unoccupied by white men before 1825.

## Mobility within Countries

In European countries canals and railroads also were being rapidly built during the second half of the nineteenth century and, with industry already

[13] Everett E. Edwards, "American Agriculture, the First 300 Years," *Farmers in a Changing World*, Washington, Department of Agriculture Yearbook, 1940, p. 216.

[14] Taylor, *op. cit.*, p. 75. Taylor mentions that the State of New York, because of its investment in the Erie Canal, required railroads paralleling the canal to pay the canal toll rate for any freight handled, but this rate was not collected when the canals were frozen. A similar policy was followed in Ohio for railroads competing with state-owned canals.

getting under way, cities were growing. In 1801 there were less than four million population in French cities of 5,000 or more inhabitants. The number was up to 14,100,000 in 1901. Between 1872 and 1881 the cities of France received 1,699,000 in-movers in excess of out-movers, while for the rural communities the excess of out-movers was 1,238,000. This excess of in-movers for the cities and the excess of out-movers for the country continued. Between 1872 and 1911 the excess of rural out-movers over in-movers was greater than the excess of births over deaths in the rural areas. This meant that the urban labor markets were expanding, but it also meant that people were moving out of as well as into cities.[15]

It should also be mentioned that this new mobility in France was not only between city and country. It seems also to have been between towns or between cities. More recently Chombart de Lauwe found high mobility within the wide area of Paris, between the city and the suburbs or between suburbs. He studied Suresses, one of these suburbs, where the population changed from 26,685 in 1931 to 31,054 in 1936. During this period of five years 41 per cent of the 1931 population had moved away and another 16 per cent changed addresses within Suresses. Only 11,253 of the 1931 population of this community had not moved by 1936. During this five-year period 15,574 persons had moved into this suburb, about half of the 1936 population. This is the record for one Paris suburb for that period.[16]

Unlike France, Germany in 1850 was a federation of independent states, but railroad building was under way. Industrialization, however, was only beginning. There were only 16 cities of 80,000 population or more, Vienna being one of them, and all the 16 had a total of but 1,850,000 inhabitants. France was somewhat more urbanized. In 1939 Germany contained 120 cities of 50,000 or more population. The growth of cities in Germany did not really begin until after 1870, when the many smaller jurisdictions were joined as a single nation. Since then the urbanward trend has been continuous and still goes on. Nellner reported that in 1951 cities of 100,000 or more population received 414,560 in-movers while there were 254,104 out-movers. For German villages in 1951 there were 1,541,355 in-movers and 1,753,959 out-movers, a loss of 222,304.[17] It should be mentioned that many of the moves to

[15] Adolphe Landry, *Traité de Démographie*, Paris, Parot, 1949, p. 463.

[16] Paul Chombart de Lauwe, *Paris et l'Agglomeration Parisienne*, Paris, Presses Universitaires de France, 1952, Vol. I, p. 30.

[17] Werner Nellner, *Mitteilungen auf dem Institut für Raumforschung*, Bonn, Vol. 18, 1953, Table 4. It needs to be mentioned that all German communities keep complete records on all in-moves and out-moves.

villages were made by urban working people finding residence outside the city.

Japan is a country that became urban and industrial within recent decades. Between 1920 and 1955 the population changed from 18.1 percent urban to 51.6 percent urban, while the rural population dropped from 81.9 percent to 48.4 percent of the total. Of the cityward migrants between 1935 and 1940, more than 50 percent were in the six largest cities, which contained 30 percent of the total population. In 1920 all the cities in Japan contained 10,020,000 inhabitants compared with 42,940,000 in 1955.[18]

Isomara, the source of the above figures on Japan, also found a great amount of mobility within urban districts. He made a study of a ward in Tokyo in which there were about 60,000 households. He found that between 1925 and 1955 nearly 97 percent of the householders had changed their places of residence. About 30 percent of these householders had once lived at or nearer the urban center. Movement in general was from the inner urban areas outward. About 70 percent of the householders had their work places outside the ward, mainly in the direction of the urban center. A large share of the outward moves were toward the suburbs of Tokyo.[19]

These few illustrations of mobility in the industrial urban countries could be multiplied, perhaps with even more striking figures. They support the observation that mobility is a basic characteristic of our kind of a society.

## Characteristics of Migration

Except in countries where public records are kept of all mobility involving change of residence as well as change of community or change of country, no very complete picture of the amount of mobility may be had. Perhaps it is sufficient to know that the amount is considerable and that it apparently does not diminish. It is probably sufficient also to recognize that residential mobility is to be expected if there are changes in the labor market, changes in technology which may outmode some occupations and create others, and changes of a social nature. We can say, briefly, that people usually move for what seem to them practical reasons. They find themselves in a situation which holds out less promise than seems to be offered by another situation. Whatever may be involved in such reasoning, the elements are different with people, with

[18] Eichi Isomara, paper on population in Japan presented to the Joint UN/UNESCO Seminar on Urbanization in the Asia and Far East Region, Bangkok, 8-18 August 1956, UNESCO Social Science Department Conference on Urbanization, Document No. 11, July 19, 1956.

[19] *Ibid*, pp. 19-20.

situations, and with time. Some observers have used the term "pull" to describe the attractiveness of a migrant's destination, and "push" for the disadvantages, real or imagined, of the place he would leave.

DeVoto would remind us that migration motivations are not to be arrived at merely in practical terms, by weighing the pull influences against the push compulsions. There may also be involved dream elements, and a challenge.[20] For most migrants the place of destination encourages dreams about starting anew. Certainly this applies to frontiers in new countries, but it applies no less to the city. How much the place of destination needs to offer in order to attract migrants may depend on how little the place of origin affords. This may be seen with reference to the Irish emigrations in the 1840's and after; America was the land of promise, but many were happy to find jobs in the industries of nearby England and Scotland, perhaps going to America later.[21]

That brings us to an old question about which much has been written—how migrants of rural origin finally get established in the city. Ravenstein made a pioneer study in this field and he arrived at certain laws of migration which were tested in the United States recently by Smith. With minor variations, Smith restated the Ravenstein formulations:[22]

1. Most migrants move only a short distance;
2. The process of absorption is like this: inhabitants of the immediately surrounding area flock to the city, creating gaps in the rural population which are filled by persons from more remote districts, which in turn creates other gaps, until the attractive force of the city makes itself felt step by step in the most remote corners of the nation;
3. Each main current of migration sets up a compensating counter current, and the process of dispersal is the reverse of the process of absorption; and
4. Long-distance migrants go immediately to the great centers of trade and industry.

Smith found that these "laws" tend to hold true today in the United States. Atteslander, who made a study of migrations into Zürich, also found that some immigrants came directly to the city, others arrived by stages, as

[20] Bernard DeVoto, *The Year of Decision, 1846*, Boston, Little, Brown, 1943, pp. 46-48.

[21] Thomas Ferguson (*The Dawn of Scottish Social Welfare*, London, Thomas Nelson, 1948, p. 72) gives information on the Irish immigrants in Glasgow, Edinburgh, and other Scottish cities in 1861.

[22] E. G. Ravenstein, "The Laws of Migration," *Journal of the Royal Statistical Society*, Vol. 68, 1885, pp. 167-235. Quoted from, T. Lynn Smith and C. A. McMahan, *The Sociology of Urban Life*, New York, Dryden Press, 1951, p. 308.

indicated by Ravenstein, while others made short stays in the city before settling, going to and from (*Pendelwanderung*).[23] It is necessary to recognize in studying urbanward migrations that rural people in all Western countries are much more urbanized than when Ravenstein made his studies. It is less of an adventure to migrate cityward, for people already know about cities and city life.

Smith, however, was not so much interested in how migrants get to the city from the country as he was in who goes to the city. Some of his conclusions were (for the United States):

1. Younger age groups predominate in rural to urban migrations.
2. In the cities of arrival the median age of migrants is lower than the median age of the resident population.
3. Younger migrants arriving in the cities, in general, come from more distant places of origin than do older migrants.
4. In normal times females migrate to the city in a greater proportion than males.
5. Normally females migrate at an earlier age than males.
6. Of older age groups more males than females migrate to the city.
7. Persons who migrate from farms to the city usually have a higher level of education than farmers who do not migrate.[24]

We must recognize that the types of migrants going to one city or another will depend on the character of the labor market in each place. A steel manufacturing city would attract migrants who would not be attracted to a city of finance and insurance. Clerical workers might be attracted more to a city of finance and insurance than to a city of manufacture, unless it is a large manufacturing city with a varied labor market. It is to be expected that persons in upper occupational groups are likely to be better informed about the possible places of destination than are people of lower occupational groups.

Pons studied the migration pattern for natives who reached Stanleyville in the Belgian Congo. The migrants were unlettered persons from the tribal villages, wholly unacquainted with cities or industry. The following list shows the number of moves by both males and females from the time of leaving their villages until reaching Stanleyville:[25]

[23] Peter Atteslander, "Dynamische Aspekts des Zuzuges in die Stadt," *Kölnische Zeitschrift für Soziologie und Sozialpsychologie*, Vol. 7, No. 2, 1955, pp. 253-279.

[24] Smith and McMahan, *op. cit.*, p. 336 (see footnote 22).

[25] V. G. Pons, N. Xydias, and P. Clements, "Social Effects of Urbanization in Stanleyville," *Social Implications of Industrialization and Urbanization in Africa South of the Sahara*, Paris, UNESCO, 1956, p. 257.

| | (in percent) | | |
|---|---|---|---|
| | *Males* | *Females* | *Total* |
| One move only | 23.7 | 41.9 | 32.1 |
| Two moves | 33.8 | 32.4 | 33.2 |
| Three moves | 21.2 | 13.9 | 17.8 |
| Four moves | 10.0 | 6.4 | 8.3 |
| Five moves or more | 11.3 | 5.4 | 8.6 |
| | 100.0 | 100.0 | 100.0 |
| Number | 1,902 | 1,632 | 3,534 |

Unlike the pattern in an urbanized country, these African males migrate first, and most of them approach the city by a step-after-step movement. Less than a fourth of the number (23.7 percent) moved directly from the village to the city, but many of these who made the move in one step already had relatives in Stanleyville to receive them. After moving and settling, even at one of the intermediate stops, the man may send for his wife or female relatives. This explains the higher percentage of females (41.9 percent) arriving in Stanleyville with only one move.

## Migration under Force or Stress

With the exception of the Negro migrants into the United States, who were initially introduced as slaves, force or stress have not figured to any great extent in American history. There are, however, some examples that need to be mentioned, although they are more related to the frontier than to our subject, the modern urban community. We have heard much, for example, about the practice of Great Britain of using Australia as a place to which minor criminals and other unwanted persons were deported. A similar policy had been followed with respect to some of the American colonies. Davie estimates that between 1717 and 1776 about 50,000 convicts had been sent to America, and of these about 20,000 were sent to Maryland. Many were sent to Virginia. We should add, however, that most of these convicts were poor people arrested for debt.[26]

In the settlement of new lands, not only were the Indians dispossessed, but in some cases entire tribes were moved forcibly to other, sometimes less desirable, locations. As far as the white Americans are concerned, there is apparently only one outstanding example of migration by force, and that concerns the Mormons who now are settled in the intermountain region, mostly in Utah and parts of surrounding states. The Mormons, mainly because of

[26] Maurice R. Davie, *World Immigration*, New York, Macmillan, 1949, pp. 31-33.

religious differences, were first driven from Ohio, later from Missouri, and still later from Nauvoo, their dream city in Illinois.[27] Other examples which might be cited have been minor. We may say with confidence that the United States has been quite free from enforced migrations or migrations due to stress.

But the United States has been the recipient of vast numbers of migrants who have had to flee from oppression—from the religious persecutions that brought the Puritans to New England, the Quakers to Pennsylvania, and the Catholics to Maryland, to the most recent reception of political refugees from Hungary. In large measure, these refugee and escapee peoples have entered American cities and have contributed in developing the American type of dynamic industrial urbanism and all that it implies. On this subject we can do no better than to call attention to the observations of Kulischer with reference to his study of migrations in Europe which, as he concludes, stimulated migrations to America. But the stimulus during the nineteenth century was what he called the "vanguard" of internal trail-blazing migration:

> . . . which opened still further territories and created new resources and new markets for the industrial capitals of Europe. The central column was crossing the ocean. The mighty rearguard was flocking to Europe's restlessly growing industrial centers. These dual currents of expansion and concentration were even more accelerated when industry itself started to migrate, creating new centers nearer the areas of colonization. By its young industry the New World exerted an even stronger attraction upon the migratory stream. The decisive factor was that the process of colonization was in progress.[28]

The lot of most of the people concentrating in European cities during the period cited by Kulischer was one of stress and poverty. While it does not exist in such intensity today, its present counterpart may be seen in such underindustrialized countries as India. Great numbers of people, because of poverty and because, being of lower caste, they are discriminated against, migrate from their villages. For most of them, as reported by Deshmukh, migration is "a leap into the darkness" and the younger people venture first. They are unwelcome in the towns and cities, already filled with the jobless sleeping on the streets or living in shanty towns wherever they find space.[29]

In the towns these Indian migrants endure misery, but they do not return

[27] See Thomas F. O'Dea, *The Mormons*, Chicago, University of Chicago Press, 1957. This writer has also reported on the Mormon migrations in his *Desert Saints*, the same press, 1942.

[28] Eugene M. Kulischer, *Europe on the Move*, New York, Columbia University Press, 1948, p. 27.

[29] M. B. Deshmukh, "Study of Floating Migration, Delhi," in *The Social Implications of Industrialization and Urbanization* (Five Studies in Asia), Calcutta, Unesco Research Center, 1956, p. 172.

to the villages. Quoting Deshmukh, "Many migrant families which took to begging as a subsidiary or main source of income, not only permitted lepers to live with them, but also shared food, clothing, etc., with them."[30] Some of these people, unwanted in the communities of origin and rejected in the communities of destination, do manage to find places in the urban communities. The others keep moving, but the initial stimulation to moving is the illusion of a refuge in the cities.

## Habitual or Chronic Mobility

It is not only the very poor who make up the migrant Indian population. Indians of the upper castes also migrate cityward, but under more favorable conditions. They have the advantage of more education and security at home if they fail in the city, but they are less likely to fail because they can enter the job market at a higher level. Lewis studied a village in North India from which came a relatively high number of migrants from the landowning caste, the Thakurs. They did not lose their property rights when they took employment in the city, even if they remained there fairly permanently.[31] Like the lower-caste Indians, some of these also move from place to place after leaving the village.

Some evidence of chronic mobility is to be found in the industrial countries of Europe, that is, in addition to the urbanization type of moving about. Mention has already been made of the study by Chombart de Lauwe of the high amount of moving about in the Paris region. In some European countries, the amount of moving of people already urbanized is probably much less than it might be if certain limiting conditions did not exist. Italy has tried by law to prevent persons changing places of residence unless they are assured of employment. But the chief barrier to moving about in some European countries is the lack of housing. House building is a slow-moving industry in France, for example. It has become a faster-moving industry in Germany since the war but houses are still too few. Persons moving from one urban area to another a few miles distant or from one city to another may have to wait a year or longer before a suitable dwelling can be found. This makes a lively business

[30] *Ibid*, p. 213.

[31] Oscar Lewis, "Aspects of Land Tenure and Economics in a North Indian Village," *Economic Development and Cultural Change*, Vol. IV, No. 3, April 1956, p. 301. Also see Edwin Eames, "Some Aspects of Urban Migration from a Village in North Central India," *The Eastern Anthropologist*, Vol. 8, No. 1, September-November, 1954, p. 25.

for house-finding agents. Or persons wishing to move sometimes advertise for the exchange of dwellings. Moving in Germany is also made difficult because tenants in new or rebuilt houses must make large cash deposits in advance. It is called a building cost (*Baukost*) and may be the equal of rent for three or more years. This equity, which runs from ten to twenty years, and which is repaid in alleged "reduced rentals," may be sold to another tenant, often a discouraging experience. For this reason urban workers who get new jobs more distant from their homes are forced to commute for months before they can move their families.

Because of such difficulties the mobility of the urbanized European is probably much less than for urban Americans; without these obstacles it would doubtless be higher than at present. According to Census Bureau estimates there were 144,101,000 persons one year of age or older in the United States as of April 1, 1949. Of these, 19.3 percent had changed their place of residence during the previous year, and of those who had moved, about 70 percent had moved from one house to another in the same county. The actual amount of moving was probably higher because such estimates cannot take full account of the rooming-house and hotel-living bachelor population in American cities.

A study of the potential mobility of urban families, as of June 1954, was made by Katona and Mueller. They found that 70 percent of renters and 34 percent of owners, or 48 percent of both groups, were "potentially mobile." This meant that during the ensuing twelve months they expected to move or buy a home, they might move or might buy a home, or they "would like to move (no buying or moving expectations)." Of the renters, 22 percent as against 56 percent of the homeowners said that they "would like to stay" in the present dwelling. Ten percent of the owners and 8 percent of the renters expressed no opinion. On the basis of these expressed "moving expectations," the renters, compared with owners, are more than twice as "potentially mobile."[32]

Between small cities and large ones Katona and Mueller found little difference regarding expectations to move, although of homeowners expecting to move, the percentages were smaller in the very small cities. With respect to the incomes of families interviewed, the potential movers were relatively more numerous among middle-income than among families of very low or very high

[32] George Katona and Eva Mueller, *Consumer Expectations, 1953-1956*, Ann Arbor, University of Michigan Survey Research Center, 1957, p. 85.

incomes. With respect to the ages of potential movers, the percentages were higher for families with younger heads than for families with older heads.[33]

Families living ten years or longer in the same house were found to be potentially less mobile than families living in the same house for less than ten years. Of homeowners living in the same house less than three years, 39 percent expected to move, but of rent-paying families living in the same house less than three years, 73 percent expected to move. Among both homeowners and tenants, the least potentially mobile families were those of which the head had been in the same employment ten or more years, while the most potentially mobile were families whose heads were employed on three or more different jobs since World War II.[34]

Rossi made a study of four residential areas in Philadelphia. These included an area of single homes, the occupants with upper-level occupations and incomes; an area of old houses divided into apartments, but of middle economic level; a district of working-class families, mostly skilled workers; and a type of rooming-house slum area. Low mobility was found in the working-class family district ("blue-collar" artisans), while the highest mobility was found in the rooming-house slum.[35]

As far as Philadelphia is concerned, Rossi's study indicates that mobility is present in all areas, but is highest in the slums. Mobility is a chronic characteristic of the modern urban scene. If it varies with different types of occupation areas, that really means it varies with different groups or categories of the population. All are mobile, but to different degrees, and apparently the degree of mobility of an individual or family is related to social, economic, and other conditions that stimulate mobility or permit sessility. These conditions are subject to change, especially in growing communities.

This moving about which is so native to urbanism, although it is not confined to the city, as Wilson sees it, has its psychological counterpart. Moving about stimulates moving about. The expectations of traditional rural life have little place in the cross currents of urban life. Here one gets status often as he is able with skill to change and shift to avoid disadvantages or to gain advantages. According to Wilson:

> Mobility and rootlessness in urban life acquire additional significance as they imply a transiency in group affiliations and allegiances. Physical mobility is

[33] *Ibid*, p. 87.
[34] *Ibid*, pp. 88, 89.
[35] Peter M. Rossi, *Why Families Move*, Glencoe (Illinois), Free Press, 1955, pp. 19-22.

only the external symbol of shifting, multiple group memberships and the associated ways of thinking and of feeling.[36]

## Migration and Social Mobility

When Morris Birkbeck, an English traveler, visited Ohio in 1817 he stopped at the home of an "old Hibernian" who within a few years after migrating with his family from Ireland had become a leading citizen in the frontier state. They talked during the evening:

> The wife was at a neighbor's on a "wool-picking frolic," which is a merry-making of gossips at each other's houses, to pick the year's wool and prepare it for carding. The son and daughter were married and well-settled, each having eight children. He came to this place fourteen years ago, before an axe had been lifted, except to make a blaze road, a track across the wilderness, marked by the trees, which passed over the spot where the town now stands. A free and independent American, and a warm politician, he now discusses the interests of the State as one concerned in its prosperity: and so he is, for he owns 118 acres of excellent land, and has twenty descendents. He has also the right to scrutinize the acts of government, for he has a share in its appointment, and pays eight dollars a year in taxes—five to the General Treasury and three to his own country; in all about four pence per acre. He still inhabits a cabin, but it is not an Irish cabin.[37]

In his own country this immigrant from Ireland had a very limited status and little or no possibility of rising above it. This is a fairly extreme example of an aspect of all voluntary migration: People hope by moving to improve their economic or social position, or both. This is generally true of local as well as international migrations; at the place of destination the migrant hopes to achieve something that is not possible or hardly possible at his place of origin. The country boy is a diligent student so that he can later find a job and perhaps his fortune in the city.

Lipset and Bendix made a study of the biographies of persons who had achieved leadership in American industry. They found that about three out of four businessmen and industrialists born between 1771 and 1800 began as entrepreneurs, persons who had been workers and later started enterprises; but of business and industrial leaders born between 1891 and 1920 only about one out of five started as entrepreneurs. Nearly half of those born between 1891 and 1920 started their careers as workers in existing establishments and

[36] Everett K. Wilson, "Some Notes on the Pains and Prospects of American City," *Confluence*, Vol. VII, No. 1, Spring 1958, p. 14.

[37] Morris Birkbeck, *Notes on a Journey to America*, London, 1818, p. 31.

had "worked their way up." The record is clear that, in the past at least, those who migrated to the American frontier found more opportunity for advancement than in their own countries.

Today in the United States, as far as business and industry are concerned, Lipset and Bendix remind us that to a much greater extent than formerly position and status are inherited. They point out, however, that there is still a considerable amount of upward movement both socially and economically, especially if generations in the same family are compared. They note that there is still a possibility for the grandchildren of workers "to enter the ranks of the business elite."[38] They might also add other elites: intellectual, political, trade union, etc.

Social mobility, which involves the shifting of people from one social or occupational level to another, is sometimes called vertical mobility. It is a companion term of horizontal mobility which concerns mobility that does not involve change of status. One may move from one area to another or he may move from one job to another at the same occupational level.

Vertical mobility may involve changing from a lower to a higher, or from a higher to a lower form of employment. This is sometimes called professional mobility, but it may also involve change of social status. Such movement may or may not involve a change of residence. Upward social mobility is likely to be most evident in places where industry and population are increasing, giving opportunity for more people to enter new types of work.

A community in which there is little or no new industrial development may offer few opportunities for professional and social advancement. More people may be migrating away from than into such a community. Yet the community may be fairly prosperous. It is only that social and economic change take place slowly. Van Heek made a study of social mobility in a relatively stable city in Holland, a city of about 100,000. His study included 67 managers in textile mills, 1,394 semiskilled and 250 unskilled textile workers in the same mills. On the backgrounds of the 67 managers, van Heek found that:

63 had fathers who had been managers in textile mills;
48 had grandfathers who had been managers in textile mills;

[38] Seymour M. Lipset and Reinhard Bendix, "Ideological Egalitarianism and Social Mobility in the United States," *Transactions of Second World Congress of Sociology*, London, International Sociological Association, Vol. II, 1954, pp. 34-54. The study is also mentioned in Reinhard Bendix, *Work and Authority in Industry*, New York, John Wiley and Sons, 1956, p. 229.

4 of the 67 whose fathers had not been managers were the sons of fathers in public office or other high positions;
10 of the 17 working sons of managers were either textile manufacturers or assistant managers.

Since 1860 not a single son of working-class parentage in the community studied had risen to a management position. Of 2,300 textile workers interviewed, none reported either a father or grandfather in an elite occupation. Some had fathers who were, or had been, foremen. Among the fathers of the textile workers were almost no clerks, teachers, or shopkeepers. Among the textile workers interviewed there had been some between-generation change of professional status, but this had been mainly shifts between the skilled, semiskilled, and unskilled levels; 15 percent of the workers held higher positions than their fathers and 10 percent held lower positions.[39]

In such a stable community as the one studied by van Heek there is little residential mobility and almost no professional mobility; the occupation of a man indicates the social class to which he belongs. This would be only approximately true in a highly mobile and rapidly changing American community where change of status as between generations, called generation mobility, is expected.

Svalastoga made a study of more than two thousand males of 21 years of age and older in Denmark. He was interested in measuring professional and social change as between fathers and sons. He found in general terms that

> . . . out of every ten families, four remain immobile from one generation to the next, two move one step up, two move one step down, while (of the remaining two) one moves further up and one further down. About one-fifth of the variation in status of a filial generation is accounted for by the status of the parent generation. The remaining four-fifths of the variance must be accounted for by other factors than social status of family of origin, but even here the family of origin continues to exercise an influence. . . .[40]

This means that in Denmark, as in other Scandinavian countries, where industry has been expanding and cities growing, people are able to rise socially and professionally through their own efforts. This may explain in part why people in these countries are much less anxious than formerly to migrate to America. They can migrate to their own cities.

[39] F. van Heek, "The Method of Extreme Types for the Study of the Causes of Vertical Mobility," *Transactions of the Second World Congress of Sociology*, London, International Association, 1954, Vol. II, pp. 391-395.

[40] Kaare Svalastoga, "The Family in the Mobility Process," see Nels Anderson, Ed., *Recherches sur la Famille*, Göttingen, Vanderhoeck and Ruprecht, 1958, p. 295.

Urbanization and industrialization have moved rapidly in Japan, especially since 1920 when only 18 percent of the people were urban compared with 52 percent in 1955. The Japan Sociological Society made a study of social mobility in six cities of the million class, the largest of them Tokyo with 6,700,000. The first step was to obtain by interview the prestige status of occupations. The ratings ran from professional or technical person to manager and down to the common laborer. Information was obtained from some thousands of persons about their occupations as well as the occupations of their fathers and grandfathers. Only 6.3 percent of the fathers and 4.7 percent of the grandfathers had the same occupations as the interviewees. In the clerical occupations 15.6 percent of the interviewees were in such work compared with 7.6 percent of the fathers and 1.7 percent of the grandfathers. Industry and handwork claimed 26.9 percent of the sons, 16.5 percent of the fathers but only 6.1 percent of the grandfathers. As a measure of the rate of social and occupational change in Japan, it is necessary merely to mention that 56.1 percent of the grandfathers had been engaged in agriculture compared with only 2.1 percent of the grandsons and 27.6 percent of the fathers.[41]

## The Idea of Assimilation

It is generally true that when people migrate for motives of settlement they entertain ideas of becoming identified with the social and economic life in the places of destination. This is probably true even if they are not welcomed at these places. However, if people migrate from a country with a more advanced civilization to one having a less advanced civilization, they may try to displace the people already in the place of destination or to hold them in a subordinate position. The position of migrants entering towns and cities is likely to be a subordinate one, in the beginning at least. Often their presence is tolerated because they are willing to accept subordinate and menial roles.

But migrants who accept subordinate roles in places of destination more than likely do so with the idea that this will be temporary. They do not migrate with the thought of entering and remaining in subordinate status. If they are not themselves able to rise to a higher social and economic level, the rising may be done by their children or their grandchildren. But rising means overcoming barriers, some of which are socially defined. The dominant groups may be unwilling to accept people of certain religions or of certain national

[41] Japan Sociological Society, "Social Stratification and Mobility in the Six Large Cities of Japan," *Transactions of the Second World Congress on Sociology*, London, International Sociological Association, 1954, II, 414-431.

origins. In time, however, these barriers tend to be lowered and they are likely to be lowered sooner in great cities than in small towns. Assimilation does take place.

In some countries skin color may be a seemingly permanent bar to assimilation; especially is this true of countries where the dominant people are of white skin. In such countries where the assimilation of racial groups is not possible, another type of adaptation takes place: The racial groups tend to be accommodated in that these unlike groups settle "naturally" into fixed relationships. But accommodation is a relationship that may be disturbed in countries where people move about and where the lines between groups or classes are ever-changing. Where assimilation is not permitted and where accommodation does not take place, the situation may be one of conflict.

Le Liepvre and Bousquet made a study of assimilation of immigrants in France. The sample included 4,000 Austrians, Germans, Italians, Poles, Russians, Spaniards, and others. Although many of the men were married when they arrived, 610 had married while in France and of these, 382 married French women; 57 foreign-born women in the sample had married French men. The highest number of marriages was on the part of Polish immigrants. The problem of assimilation was minor.[42]

Bogart in 1952 made a study of the assimilation of Algerians in Paris and the suburbs of Paris. In this case assimilation involved persons of a dark-skinned race. He found that, although the French have varying views about mixed marriages, there was little prejudice against it. Bogart summarizes in Table 6 the answers of the men interviewed when asked about their marriage

Table 6. Marriage status or marriage preference of Algerians in Paris

| Answers to question: "Do you think you will marry here or in Algiers?" | *Immigrés* $N = 51$ | *Migrants* $N = 89$ |
|---|---|---|
| | (in percent) | |
| Married, an Algerian woman in Algiers | 20 | 28 |
| Married, an Algerian woman in France | 0 | 2 |
| Married, a French woman in France | 20 | 1 |
| Living with a French woman | 28 | 9 |
| Thinks he will marry a French woman | 20 | 3 |
| Thinks he will marry an Algerian woman | 4 | 36 |
| Not sure or no preference | 8 | 10 |
| No answer | 0 | 11 |

[42] Thérèse Le Liepvre and Marie H. Bousquet, "Etude de 4,000 Dossiers du Service Social d'Aide des Emigrants," *Français et Immigrés,* Paris, Presses Universitaires de France, 1954, p. 248.

G

status or marriage plans. Bogart found it necessary to divide Algerians into two groups on the basis of their attitudes toward France and the French. The *immigrés* are Algerians who are making a favorable adjustment and are friendly. The *migrants* are Algerians who are unable or unwilling to adapt to the situation in France, keeping mainly in the company of Algerians. If racialism is present, they are most aware of it, and they may, by their own hostility, stimulate it. The difference between the two groups is reflected in Table 6.[43]

France and the United States are wide apart with respect to assimilation across the color line. Negroes were among the first settlers in the American colonies, and Negro families have been for several generations on American soil, but they have never been permitted more than a marginal type of social assimilation. Much of the change in this respect has taken place during the past two decades; many feel that assimilation will move at a faster rate now that enforced segregation is an issue which more and more is a subject of court decisions.

As Negroes were among the first settlers on the Atlantic Coast, so Chinese were among the first settlers on the Pacific Coast. They came, or were admitted, for the same reason—to provide a labor supply. Their numbers increased in the coast states from 13,100 in 1854 to 280,000 in 1882. In 1858 California passed a law barring Chinese immigration, and in 1862 a California law imposed a head tax of two dollars per month on Chinese. Later this tax was raised to four dollars per month, and applied to all immigrants, but it was imposed on Chinese only. During a period of unemployment in 1873, many Chinese were robbed and beaten. Davie reports that the most fierce opposition to them was by the Irish immigrants.[44] In time the Chinese settled in towns and cities and came to be identified with the restaurant and laundry businesses. In the cities they established their own communities, called Chinatowns.[45] With their immigration restricted, and with their total number since 1900 never reaching 100,000 in the entire country, the old bias against the Chinese vanished. It was a type of accommodation. Chinatowns were found in the

[43] Leo Bogart, "Les Algerians en France, Adaptation, Réussie et Non-réussie," *Français et Immigrés,* Paris, Presses Universitaires de France, 1954, p. 73.

[44] Davie, *op. cit.,* pp. 310-311.

[45] Rose Hum Lee, "The Decline of Chinatowns in the United States," *American Journal of Sociology,* Vol. 54, No. 6, March 1949, pp. 422-432. "A Chinatown is a 'ghetto' of Chinese persons, a community within a non-Chinese community, having no independent economic structure, but attached symbiotically to the larger economic, political and social base. Thus its prosperity and decline are interrelated to that of the larger community." P. 423.

larger cities only and with time they have tended to disintegrate. For residence and work, Chinese are able to disperse without their presence being resented.

When the Japanese began arriving in the United States about 1880 they were better received than the Chinese had been. In West Coast states they turned to gardening, and they lived up to the American ideas of industry and resourcefulness. Before 1900, however, prejudice against them began to rise, but it did not reach the point of brutality as earlier with the Chinese.[46] They were not segregated as the Negroes have been. Their children born in the United States may become citizens. A considerable number of American white soldiers stationed in Japan have married Japanese women and have been able to bring their wives home without fear of discrimination.

Here are two opposite attitudes regarding the marriage of Americans across race lines. If a Negro soldier stationed in Germany marries a German woman, he faces a difficult problem if he brings his wife back to the United States, although he is not molested in most cities of the North. If a white American soldier marries a Japanese he is not confronted with such bias. In both types of interracial marriage, attitudes are mellowing, but the mellowing with respect to Negro-white unions is slower. This bias is less in the more mobile urban community. Assimilation which does not involve such marriages also moves faster in this type of community.

## Summary

This chapter considered some of the social implications of mobility, the moving about of people. We recognized that although many people, moving by choice, are easily stimulated to move, most people tend to be more disposed to sessility. Nonetheless, with the emergence of the industrial urban society, mobility has been increasing. As the means of transportation increase and improve, mobility becomes easier.

Two factors which have had great "pull" influence in this increasing mobility have been the rise of industrialism in the cities, and the discovery and opening of wide new regions for settlement. Industry and the land frontier, in fact, have been twin influences, each depending on, and supplementing, the other. Industrialism resulted in unprecedented mass migrations from rural to urban communities, while the land frontier stimulated a smaller, but equally important stream of migrants to the regions of new land and other natural resources.

[46] Davie, *op. cit.*, p. 321.

There were different types of frontiers and each had its importance for urban industrialism, especially in Europe. One type of frontier involved a minimum of migration by the occupying countries, but made use of native labor or of slave labor. Another type involved a larger migration of persons from the occupying countries, who became settlers and tillers. Both types were found in the United States—the plantation type with slavery in the South and the farming type of individual family farms in the North. Both frontiers moved westward, but the northern frontier was the more cosmopolitan, as it included people of different cultural background. Industry and town building moved faster in the North, and here was more mobility, more inventiveness, more progress.

The movement of people to far places, the massing of people in cities, and industrial production stimulated demand for access. The result was more roads, more and improved types of transportation, and improved methods of communication, all of which stimulated greater movement of goods and greater mobility of people. Networks of transportation and communication, centering in the cities, served both to increase the economic power of cities and to enhance their social influence over outlying areas. In the cities the migrants from different places mix and the social changes taking place there radiate outward, both to stimulate change over wide areas and further to stimulate a kind of continuous mobility which might be called chronic. Although in the history of human migration force or stress have been compelling factors, most of the migrations associated with the rise of the industrial society have been, in great measure, migration by individual choice.

A person who migrates may need to change from one type of work to another (professional mobility), and while learning new work he may need to move from one work place to another, even changing his place of residence. If moving to new work means economic advance, that may mean moving from one level of social status to another (social mobility). As one moves upward or downward socially that may also call for residential mobility. In many ways moving about may force changes in the community class structure. Physical, professional, and social mobility are interrelated, and this is especially evident in growing urban places.

Following the move into a place, if the migrant adapts to and is accepted in the community of destination, he has become assimilated. But there are degrees of assimilation. If his presence is tolerated and he finds a role in which he is accepted, a sort of special status, that is sometimes called accommodation. An accommodation relationship is more likely to be found in a more static society, less likely in a changing city. Barriers of skin color may prevent assimilation in some countries and in some communities more than others. However, in any country the attitudes which oppose interracial assimilation tend to change. The rate is slower in some places and at certain times than others, but it tends to be faster in urban than nonurban places.

CHAPTER 8

# Urbanism and Its Incongruities

THROUGH THE GENERATIONS different thinkers have visualized one type of ideal community or another. However these utopias have differed, they have had at least one common characteristic: These dream cities had al. reached a state of perfection. Law and order prevailed and lawful behavior had become habitual. The struggle for space was absent. Unsightly areas were not to be found. The tearing-down and rebuilding process was no longer a problem. Even the traffic problem had been permanently solved.

The cities that we know, however, the actual cities of every man's experience, are characterized by continuous change. They must change because they are dynamic and sharply competitive. This change may be sporadic, as in boom times, or slow, as in times of depression, but it can hardly be described as ordered or orderly. Whether fast or slow, it usually involves both growth and decline. In the dynamic urban life the least planned-for events and the most surprising ones are likely to occur at any time, which for many people makes the modern urban community more intriguing as a place to live. For such people, and their number is legion, Utopia would be dull.

Thus, the real city is a place of contradictions. Its avenues may be clean and attractive, but there is dirt and disorder in many other places. Opposed to its quiet sections are areas of noise, confusion, and tension. These negative aspects of urbanism are the subject of this chapter.

## The Wake of Urbanism

The post office in a city of the Middle West was erected in the 1890's. When it was completed there was a big open-air meeting with speeches, and every street was decorated. It was then very modern, a stout fortlike structure with thick stone walls. The outside stone facing, adorned with hand-chiseled decorations, was something the townspeople liked to point out to strangers. But that was two generations ago. In the meanwhile many new buildings have been built in the central area of the town, each a little more advanced than the others. Gradually the "progressive element" began to be a little less proud of their one pubiic building, calling it the "old post office." The idea began to circulate that a new post-office building was necessary, something in keeping with the spirit of the times. The government should do something. Senators and congressmen were appealed to. "It's cheaper to put up a new building than to modernize the old one," was the feeling of the townspeople. So after about 55 years of service the old building was replaced, although when it was erected people had said it would stand there for generations.

In that same city the argument, "cheaper to build new than to modernize," had been the reason for replacing most of the central city buildings that were serving satisfactorily when the post office was built. They had all disappeared before the post office yielded to a more modern public edifice. Such tearing-down and rebuilding is characteristic of all modern cities throughout the world, but the process moves faster in American cities. Seen from one point of view, it is a form of waste, although from the business point of view the replacement of old buildings with new ones is economically sound. Insofar as such replacement involves waste, it is a type of waste that assumes different forms in different areas of the city. And waste is a many-sided term.

One may speak of agricultural land as being wastefully used if it is not fertilized and guarded against erosion. But urban land may also be used in a way that involves waste. Vacant pieces of land may lie idle and be filled with rubbish, or be occupied by wrecks of buildings. Not only does the community derive no good from such areas, but they may be social liabilities.

Urban congestion, whether of people in work places, in their homes, or in the streets, may involve waste of time and of energy in reaching places or getting things done. This is no better illustrated than in the traffic jams found in some urban areas. We can say that such congestion is normal for the busy city, and we can also say that no matter what is done to relieve congestion in one form, it tends to appear in another form; which means only that the problem must continually be worked at.

Familiar complaints against the city, some cities in particular, are that there is too much noise, or that the air is polluted with smoke and gases. Especially are these charges made against the more industrial cities. It is true that there is too much noise and too much smoke; and it is also true that it is possible to diminish much of the menace in both. That calls for much more control and regulation, which means more cost than people are generally willing to pay.

Much of the complaint against the modern city has to do with appearances. Slums, for example, may be unsightly as well as uninhabitable. Vacant lots filled with wrecks of automobiles and other refuse not only offend the eye, but may convey a negative impression about community housekeeping.

## Appearances and the Reality

Many critics of the city are in some respects naïve and uninformed persons. This may be said of a learned European traveler who asked, "Why does one have to pass miles of railroad yards, warehouses and factories before he arrives in the American city?" The same question might be asked on approaching either Liverpool or Manchester in England, Liége in Belgium, or one of the cities in the Ruhr district of Germany. The answer that never satisfies the questioner is that these cities are work places, and while some work places are bright and clean, others are not. Yet it is not difficult to agree with Professor Brogan, whose observations seem applicable to other countries as well as England.

> Most English towns are drab, and some parts of all English towns are drab. And it is not enough to say that in most parts of most English towns the majority of the inhabitants are poor, which, of course, is true, or that the climate makes it hard to be picturesque and easy to be down-at-the-heel at the same time. There is a lot of unnecessary drabness that is a reflection on the taste, the energy, and the public spirit of the people.[1]

He adds that there is much "unnecessary urban dirt, dullness and boredom," and he concludes that too many Englishmen are too complacent about these conditions. The great majority of people living in such areas of drabness are not at all disturbed by it; in fact, a family may live two or more generations in the same drab district, and may react with attitudes of contempt for persons who may be disturbed because of such drabness.

Not all of the people who live in unattractive or antiquated houses in drab streets are necessarily poor. Chombart de Lauwe found such "islands" in Paris during his ecological study of that city. These clusters of upper-class, often aristocratic, families living in a kind of archaic isolation had earlier at-

[1] D. W. Brogan, *The English People,* New York, Alfred A. Knopf, 1948, p. 278.

tracted the attention of Halbwachs who tried to find an explanation for their stubborn resistance to change. Although other families of their class moved away, these would not. While the area declines and looks neglected, these few families cling to the spot that figures in their memories.[2] To them the area is not a run-down slum, but a place of charm.

Realistically speaking, this observation applies to other areas of the city that are complained about; for example, the docks, warehouses, and railway yards upon which the economy of the city depends. They may be a most efficient network. To the transportation engineer the many thousands of miles of tracks in and about Chicago is a thing to be admired, and he is not likely to be moved by complaints of their drabness. He will ask if they can be used effectively. And to the many thousands of men who operate these essential facilities it is important that these networks can be and are used safely, for railroading is a dangerous work.

Certain industries, furthermore, are dirty by nature. Iron ore makes the hands and clothing red, as grease makes the hands and clothing black, or as one working in a cement plant is likely to be covered with white or gray dust. The plants used by these industries to some may seem to be disorderly places, but workers employed in such plants have a different opinion. Moreover, these industrial plants must be managed effectively and safely; if they are not it may mean loss of efficiency in production and financial losses.

Perhaps no form of enterprise is more complained about in the city than the junk business, yet no other industry does more to keep the city clean. Here every form of waste material—paper, rags, old metals, bottles—is accumulated, sorted, and packaged for different markets. Junk is an indispensable salvage activity in the modern community. The junkyard is usually a well-arranged work place, however much of a jumble it may seem to the uninformed.

## Incongruities of Urban Growth

While any urban community is likely to have its unsightly places, its noises, its smells, and some forms of congestion, these negative phenomena are likely to be more in evidence if the community is growing. And by growth we mean that not only is population increasing, but there is also increased construction and the area of the place is widening. In ecological terms, growth

[2] Maurice Halbwachs, *La Memoire Collective,* Paris, Presses Universitaires de France, 1950, p. 139.

means also that changes are being made in the occupancy of space. Some occupation areas are being encroached upon and the space users must shift their activities. Internal changes in the use of space may give rise to transportation problems that did not exist before. Structures here and there must be replaced by new ones, usually for another use. Perhaps streets must be torn up and widened.

Community growth is usually a bit-by-bit process; building here and tearing down there, while necessary activity elsewhere is delayed. Houses for one social class may be going up out of proportion to the needs of other classes. The provision of public utilities may be lagging behind the needs. The community center may be overbuilding, while measures to deal with the traffic problem are being neglected. Here, according to a United Nations report, is how the situation looks in a booming South American city:

> Caracas may be taken as somewhat typical, though certainly an exaggerated case. . . . Caracas is perhaps the only city where the freeway (all traffic using the same street) is being applied within the urban area. . . . There is the old colonial center with its narrow streets, now trying desperately to handle the automobile and pedestrian traffic resulting from ten and twelve story buildings. Mass transportation is still poorly developed.[3]

This report indicates that public administration has been slow in coming to grips with even such apparent problems as building control and the provision of street space. These Spanish-style cities with their central plazas and indifferently laid-out back streets were designed for less urgent purposes than they are called upon to serve today.

San José, Costa Rica, in 1927 contained 50,580 inhabitants, many of whom lived in the suburbs. By 1950 San José contained 86,901 inhabitants with another 75,736 in the suburbs, and this 200-year-old city is still growing.

> Any expansion was to take place by adding new blocks in a checkerboard pattern. This orderly layout established the need for community living of the time. Even today the smaller villages enjoy most of the advantages of such a pattern. In San José, however, the indiscriminate addition of square blocks in a central built-up area has proved to be a straight-jacket for the functions of a modern city. Traffic congestion has reached a critical point in many sections. The narrow streets, originally designed for an entirely different mode of transportation, are now clogged with buses, automobiles, pushcarts, oxcarts, bicycles and pedestrians. . . .
>
> The central commercial area; approximately 70 blocks of the 700 that comprise the central built-up area, is in the oldest part of the town. The blocks,

[3] *Urban Land Policies*, New York, United Nations Secretariat, Document ST/SCA/9, April 1952, p. 173.

originally divided into four big lots, have been subdivided without limitation, and today their irregularity and the poor relation between width and depth make the construction of an efficient building almost impossible. In order to use every possible square foot, the buildings crowd the lots, and ventilation and light are poor.[4]

Before considering cities nearer home, we should consider the Asian cities, most of which have space-crowding problems in the worst form. Bombay, for example, grew from about 1,500,000 in 1941 to about 2,500,000 in 1951. On the conditions in Indian cities, the United Nations report quotes the following from a 1949 report on environmental hygiene by the Ministry of Health of India. It concerns tensions as well as space problems.

> Thousands have no shelter . . . and live literally in the streets in filthy surroundings. They are the dangerous elements of society—underfed, underclothed, undersheltered and unemployed, or insecure about their future. The refugees have also poured into every town. They have suffered so much that it is no wonder if many of them are surly and uncooperative. Every municipality finds it difficult to supply water, remove refuse, and render the usual municipal services efficiently, owing to the increase of the population. . . . The public transport system has also been strained to the breaking point. Buses, trams and trains are all chokeful of hanging humanity. People lose their temper, balance and health in these trying conditions of urban life.[5]

In these cities population not only exceeds by far the number of houses and jobs, but it exceeds also the capacity of all public services—water supply, transportation, garbage removal, etc.—to meet community needs.

The difference between the incongruities of the overcrowded Eastern cities and overcrowding in Western cities puts the latter in a favorable light, but it is a difference in degree. Perhaps no Western city has ever solved its central area congestion of traffic, but that may be a problem that is never finally solved. There is probably no Western city that has finally ridded itself of slums, although the worst of these slums today are much more habitable than the best three generations ago. During this century the Western cities have been living through a strict and self-imposed learning process in municipal law and order which Versluys missed in Eastern cities.

[4] *Ibid*, p. 210.

[5] *Ibid*, p. 213. This writer knew Bombay in 1945, a time the city was much less crowded than later when it received many refugees from Pakistan. Not only were many thousands sleeping in the streets then, but families in improvised shelters occupied large areas outside the city, as well as vacant areas within the cities. For these shanty towns there was an utter lack of any sanitary facilities. The municipality seemed helpless to do anything to relieve the problem.

> The impression of disorderliness of the Asian city is not only created by the great variety of its buildings or the overcrowding of its narrow roads, or the irregularity of its traffic and deficiency of a number of services, but also by the lack of rigidity with which several rules are maintained. Here again we find a problem of discrepancy between the theory based on the experience of Western cities where street and building alignments are sacred, traffic rules are rigidly applied, and construction laws govern with supreme power; and the practice here which has to cope with ignorance of drivers of bullock carts about traffic rules, which requires that still another family should be housed whatever the overcrowding may be.[6]

## Transport and Congestion

The land area in town or city, however it may be divided, remains fixed, a definite amount of available space. Some of it is occupied for work, residence, or other purposes, but some of it must be set aside for access purposes, for the streets, alleys, and boulevards for common use. Most of the buildings along a street, say in 1850, doubtless served the needs of the community at that time, but with population increase and greater demand for space, these old buildings would not be adequate for the demands of 1900, and still less adequate for the demands of 1950. We may be sure that for a growing American city few of the buildings of 1850 were still being used in 1900, and if standing by 1950 they would be regarded as landmarks. The new buildings in such an evolution are usually higher and provide more space, and can be more efficiently used than the old ones. New buildings can go higher in the air, at least within limits, and as they rise higher they afford more space in which people can reside or work. This means that more and more people must have access through the streets. But the width of the street cannot be changed without moving buildings. The more crowded becomes the street, the less efficient it becomes for access purposes.

The demand for access, especially to the urban center, may be expected to increase with the increase of population. The urban center serves many purposes for many people both within the city and outside. If one expects to get there with his own automobile, he must expect not only traffic congestion but great difficulty in finding a place to park his car. This is not a new problem and present-day transport problems are probably no more frustrating than were those of the days of horse-drawn vehicles. According to Nevins:

[6] J. D. N. Versluys, *The Social Implications of Industrialization and Urbanization* (Five Studies in Asia), Calcutta, Unesco Research Center, 1956, p. vi.

> Without exception, the cities still depended upon horsepower for transportation. In some instances this elementary means of transit was highly developed. The New Yorker, for example, had at his disposal in 1866 some 1,400 hackney coaches, seven lines of omnibuses, employing perhaps 500 vehicles on regular routes; 16 separate lines of horse railway, with 500 cars and not far from 8,000 horses, and a few hansom cabs. . . . Yet, slow-moving horse vehicles crowded the main thoroughfares of the largest cities to suffocation, they littered the pavements with dirt, and cruelly overtaxed many dumb beasts, so limited was their range that outlying districts remained totally undeveloped while the central streets were congested.[7]

There was congestion not only in the central areas, but along all main streets. Even after electric cars replaced the horse-drawn street cars, this new mode of transport was often hampered because of the heavily loaded wagons always blocking the way. For this reason factories had to be near the crowded residential areas within walking distance of most of the workers. The urban poor were confined to the densely populated areas. Being able to get out of the city of a Sunday was an event. Only the rich could afford carriages, but even the carriage owners could not conveniently live more than five or six miles away from the downtown areas. The work life of the horse in the city was usually much shorter than in a town or village. Thus many horses, after a few years of urban service, had to be sent out of the city to be sold for less strenuous work. The point of interest is that, while the horse was a more satisfactory work animal than the donkeys, oxen, and camels used in older cities, he had reached the limit of effective utility.

It was the automobile that came to the rescue, but before that we must mention the bicycle, the social importance of which has been somewhat neglected. Man had been using wheels for many centuries, but it was after the middle of the nineteenth century before he learned to place two wheels in alignment and to use them as a vehicle. There had been various predecessors, but the modern bicycle was not designed until after 1870.[8] It became immensely popular, first as a novelty and then as the utility which it still is on the Continent.

The bicycle was the first conveyance in the history of the city that afforded transport for large numbers of low-income people. Light in weight and using very little street space, the bicycle enabled people to live farther from their work. It gave some relief to slum congestion, but most important,

[7] Allan Nevins, *The Emergence of Modern America, 1865-1878*, New York, Macmillan, 1927, p. 81.

[8] See two articles in the Encyclopaedia Britannica, "Bicycle," and "Cycling."

it enabled urbanites by the thousands to leave the city on holidays. The League of American Wheelmen, formed in 1880, became the first pressure group for improved roads, at least for side paths. What the bicycle started in the task of getting people out of the slums was later done more extensively and effectively by rapid transit lines.

Automobiles, which were becoming increasingly effective and numerous beginning in most cities about 1910, permitted more rapid movement through the streets. More people in less time could enter or leave the city, but cities were growing and more people needed to use the streets. An auto truck was able to move more goods a greater distance in less time than was possible with the most efficient horse-drawn vehicles, but the volume of goods to be moved was increasing. The horse was pushed off the streets, but the horse-age congestion was replaced by an equally exasperating new congestion. The streets were rid of horse litter, but the urban air became polluted with the fumes of motor vehicles. Traffic moved infinitely faster, but crossing the streets became more dangerous.

In some large cities schemes have been employed of putting transport lines underground, thus more than doubling the capacity of passenger movement on common carriers. Undersurface transport permits more people to move in and out and about the city with greater speed, but it often aggravates rather than relieves pedestrian congestion on the street surface.

## Characteristics of Slums

Birmingham, England, is a city that reflects the story of the Industrial Revolution with all its blundering as well as inspiring advances, and with all the mistakes that forward-looking men can make in building a metropolis. This story has been told by Gill and Briggs (a volume by each), and here we see how from village to great city in less than two centuries Birmingham was ever confronted with problems of growth and change. Gill reports how haphazard and seemingly planless this growth was, how conflicting interests were involved with the laying out of each street.[9] One-time areas of growth became areas of neglect in later decades. Always there were in-between habitation districts of the very poor, many of which were removed or changed as the growth process went forward. These slum areas were not peculiar to Birmingham.

[9] Conrad Gill, *History of Birmingham*, London, Oxford University Press, 1952, Vol. I, p. 367.

A definition of the slum is offered in the report on urban land policies of the United Nations:

> . . . a building, group of buildings, or area characterized by overcrowding, deterioration, unsanitary conditions or absence of facilities or amenities which, because of these conditions or any of them, endanger the health, safety or morals of its inhabitants or the community.[10]

While the slum is all this, it may be something more, as when it is seen in relation to urban change and growth. It may be an interstitial area of apparently static character, surrounded by areas in process of change. Or, in time perspective, it may be a habitation area in an interim position between a former better use and a coming new use of the space. Most of the writers who placed the word in our vocabulary loaded it with various evil connotations. Actually the slum is not defined in all its implications in terms of evil, or in pointing accusingly at certain persons, usually the landlords and politicians, alleged to be responsible for it. Without doubt, individuals may be blamed for some aspects of the slum, but it is too much to blame them for its existence. This becomes somewhat apparent when we enumerate some of the characteristics of slums.

1. *Appearance.* This may be called a universal mark of the slum; its aspect of neglect and disorder with respect to buildings, yards, and streets. The appearance is generally one of structural over-age and decline.
2. *Economic status.* Generally a slum is inhabited by people of the lowest income, although there may be occasional buildings of equally run-down appearance inhabited by families that are not so poor. In general, however, the slum is a poverty area.
3. *Overcrowding.* We may find that the space is overcrowded with buildings or the buildings may be overcrowded with people, or both. If the slum is "retreating" many buildings may have declined until they are unsafe for habitation, but there may be overcrowding in the buildings still occupied. The uninhabited space may then be occupied by such unwelcome occupants as junkyards.
4. *Population.* In a slum of heterogeneous occupancy, many of the inhabitants there are persons not welcome in other residential areas, or they cannot afford to live elsewhere. Thus, it may be a refuge area for the aged, the chronically sick, the homeless, and the socially maladjusted. But such "odds and ends" may not be welcome even in some slums where a "sense of community" exists. If the slum population is racial or cultural, it may have a degree of social organization. It is a poverty area, slumlike in appearance.

[10] *Urban Land Policies, op. cit.,* p. 200.

5. *Health and sanitation.* For understandable reasons, when compared with other areas of residence, the slum is characterized by low standards of sanitation. The slum is often most neglected by the public services for sanitation. For mixed reasons, it may also be an area of high sickness and death rates.
6. *Morals.* The slum may be an area of delinquency, crime, and vice, but this is more likely to be true of the socially disorganized slum. While such a slum may not be the habitat of "successful" criminals, it may be the habitat of marginal types or the hiding place of fugitive criminals. Vice may be found in the slum, but is by no means confined to the slums.
7. *Way of life.* Slums differ widely with respect to the social organization of their inhabitants. They range from the slum in which the inhabitants are strangers to one another, and wish to be; to the family slum in which there is wide acquaintance between the inhabitants. Slums inhabited by immigrant groups may have a firm social organization.
8. *Social isolation.* While every residential area within the modern city tends to be socially isolated from others, partly by choice and partly by location, the slum is especially so. It is the area of lowest status and this is known to slum dwellers. Their chief link with the rest of the community is their identification with the labor market, but there may be an additional link through politics. Slum dwellers function equally with others as citizens and as they are able to identify themselves with political groups.
9. *Mobility.* The slum is usually an area of high residential mobility, but a family-occupied slum may have a low rate of residential mobility. The high mobility of slum inhabitants is more true of American than European cities. Perhaps the most mobile is the slum occupied mainly by homeless males, especially itinerant workers.
10. *Slum permanency.* In the rapidly growing American city some areas once occupied by slums are later occupied for other purposes. The "removed" slum emerges elsewhere, perhaps to be replaced still later by another occupancy. These shifts may be in response to related changes due to urban growth.

Regarding the final item above, slum permanency, it must be said that the removal of the slum is not merely a matter of tearing down old buildings. Standards of living must also be improved. Spate made this observation about slum removal in New Delhi:

> Nor have matters been improved by "improvements" of the slashing type denounced by Patrick Geddes. The southern edge of the town—that facing New Delhi—has been given a face-lift. In most walled towns, East or West, the strip immediately behind the wall becomes a locus of the worst slums; here the wall and the hovels behind it have been torn down and replaced by a long frontage of bright and brash shops and offices, modernistic rather than really modern. But it is a fake façade, one block deep, and one can stand on the roof of the

building which houses the Delhi Planning Organization and look straight down onto mud huts and cattle stalls plastered with cakes of the dried cow-dung used for fuel. And it is to be feared that the displaced dwellers by the wall have exchanged one squalor for another: they have gone to swell the fantastic sprawl of new suburbs which has flowed round the old Civil Lines and right over the Ridge.[11]

## The Slum and Cosmopolitanism

It may very well be asked if there can be a cosmopolitan city into which all kinds and classes of people gather without there being a cosmopolitan slum. The poorest people must have some place to live, and of these people there are many kinds. The slum may be for some the place for beginning the upward climb, while for others it may be the last stop on the way down. Perhaps this thought may have been in the mind of Zorbaugh when he wrote the following about the cosmopolitan slum, the slum where many types of the poorest are found in Chicago.

> The slum is an area of freedom and individualism. Over large stretches of the slum men neither know nor trust their neighbors. Aside from a few marooned families, a large part of the native population is transient; prostitutes, criminals, outlaws, hobos. Foreigners who came to make a fortune, as we used to go west, and expect to return to the Old Country as soon as they make their "stake," who are really not a part of American life, and who wish to live in the city as cheaply as possible, live in the lodging houses of the slums. Here, too, are the areas of first settlement, the foreign colonies. And here are congested the "undesirable" alien groups such as the Chinese and the Negro. The slum gradually acquires a character distinctly different from that of other areas of the city, through a cumulative process of natural selection that is continually going on as the more ambitious and the energetic keep moving out, and the unadjusted, the dregs and the outlaws accumulate. . . . The city, as it grows, creates about the central business district, a belt of bleak, barren, soot-begrimed, physically deteriorated neighborhoods. And in these neighborhoods the undesirable, and those of low economic status, are segregated by the unremitting competition of the economic process in which land values, rentals and wages are mixed.[12]

[11] O. H. K. Spate, "Aspects of the City in South Asia," *Confluence*, Vol. VII, No. 1, Spring 1958, p. 25. This writer for his doctoral dissertation at New York University studied the history of slums on Manhattan in New York City. As the city grew, many slum areas were replaced, but other slum areas emerged, each time a little farther uptown. In the course of this migration the quality of slum housing improved. This moving was quite independent of antislum reform efforts.

[12] Harvey W. Zorbaugh, *The Gold Coast and the Slum*, Chicago, University of Chicago Press, 1929, p. 128.

Whether such areas are a permanent fixture of the modern city, for the present at least, they do exist as incongruities in the city. The very poor must live as they can and where permitted. Versluvs wrote of the heterogeneity in the cities of Asia:

> Somewhere in a fashionable street one may see a sort of hole in a wall, or in some corner where a tailor or a cobbler is busy doing his work on a surface of a few square feet, or a tiny shop may be installed. Next to tall modern buildings, low and dirty shacks seem to wait for their being demolished, which may take a long time. In the heart of the city one may find semi-rural islands of clusters of small houses in tiny fruit gardens, just outside a busy street with modern stores and offices.[13]

In the United Nations report on urban land policies mention is made of congested shanty towns in and near the center of Singapore. "Their shanties are built of old boxes, corrugated iron, palm thatching or whatever else is available." The city has tried to meet the housing need, but the pressure for dwellings is such that the authorities have accepted the minimum standard of no less than 140 square feet per family and *not more than five persons per room.*[14]

It has probably always been true that most newcomers to the city take their first stand at or near the urban center. Here are the natural "ports of entry," a term used by Robert E. Park. Thus the so-called "Left Bank" in Paris, a wide area across the Seine from the principal central area, is the district of strangers of every sort, from poor art students to political refugees. Whatever the size of their purses, here they can find some place for shelter. Perhaps there would be difference of opinion whether this street or that may be called a slum, but in this general area can be found all the types of marginal people. They are a cosmopolitan agglomeration and as a group they lend color to the life of the city, which is less true of the less cosmopolitan mixture found in the Chicago slum described above by Zorbaugh.

## Slums within the Cities

Much less is written today than several decades ago about the slums in American cities. At the turn of the century they were a distressing problem to many people. Organizations such as the settlement houses were very active in trying to do something about ridding cities of slums and all the evils asso-

[13] Versluys, *op. cit.*, p. vi.
[14] *Urban Land Policies, op. cit.*, p. 126.

ciated with them. Prominent in these efforts were Jane Addams of Hull House in Chicago, Lillian D. Wald of the Henry Street Settlement in New York, Robert A. Woods of South End House in Boston, and many others. This anti-slum movement was inspired by the work of Samuel A. Barnett, founder of the first social settlement, Toynbee Hall in London. Many American private movements for housing reform, care of children, home nursing, etc., emerged out of the social settlement movement. The story of the battle to rid the cities of the congested and unsightly tenements has been told by DeForest and Veiller. The following is a single item from their two-volume report.

> [There are] thousands of poor persons, but comparatively few buildings suitable for their accommodation; most of the houses are those which were formerly occupied by the wealthy who have moved uptown, and now in their dilapidated state many of them are tenanted by poor Irish and German immigrants. Large rooms have been divided into dwellings for two or three families (each, perhaps, taking boarders), where they wash, cook, eat, sleep and die—many of them prematurely, for the circumstances in which they live make fearful havoc of health and life, and in addition, night lodgers consisting of homeless men are allowed temporary shelter.[15]

This report concerned New York mainly, although sections were given to tenement conditions in other American cities. One value of their study is historical. For example, what use is made today of some areas where the slums stood, slums described half a century ago by DeForest and Veiller?[16]

Another type of slum that has passed into history was the fairly permanent shanty settlement, so much associated with the growth of American cities. To mention them is to remind ourselves that a great amount of progress has been made as far as this type of urban incongruity is concerned. In some of our cities so-called squatter settlements endured a generation or more, the owners of shanties paying ground rent. They located on hilly land, along river banks, on marshy spots; land not in demand for other uses. Such a slum was "Dutch Hill," located where Forty-second Street in Manhattan meets the East River. The site, adjacent to the United Nations building, is now occupied by the Tudor City skyscraper housing development. It was formerly a rocky hill,

[15] Robert W. DeForest and Lawrence Veiller, *The Tenement House Problem*, New York, Macmillan, 1903, Vol. I, p. 77.

[16] An example is Gotham Court, a tenement which in the 1850's housed 120 families in cubicles averaging 196 square feet each. Families used group toilets and took water from a common tap. 71 families were interviewed, including 504 persons. There had been 138 deaths in the 71 families in two years, including 59 infants. In spite of complaints against it, this tenement continued in use until 1896 when it was replaced by a nonresidential structure. *Ibid.*, Vol. I, p. 133.

a workingman's shanty settlement occupied by families unwilling to live in tenements. To reform groups it was a disgraceful sight. Here is a description of the place sometime after 1860, by Brace.

> The houses are little board or mud shanties scattered like the wigwams of an Indian village, with most perplexing paths winding among them. Some are of the primitive block form with a hole in the roof for a chimney; others are arched, others have a sharp Gothic gable. Occasionally something entirely new in architectural design will meet you in the shape of a rectangular box with diamond lattice work, which on nearer approach you discover to be a railroad car banked in and made into a house. Some of the better class of cabins have gardens, others are built on the edge of the quarry; others almost undermined by the digging of streets, and must soon be removed. Each house has a retinue of dogs, goats and pigs; the dogs usually being overcome by their labors in the streets dragging ash carts, are a peaceful race, only their progeny evincing a quarrelsome disposition.
>
> The goat seems always an unhappy animal unless he can mount on an inclined plane, an upturned cart, or the roof of a cabin will suffice. The pigs and hens are usually found making excursions into the shanties and carrying off whatever may be lying under the bed or table. All the inhabitants of these houses are squatters. They have a plot of ground and they have built their cabins on it to remain until the rightful owner turns them away. When they move they sell their houses for $5 or $10. They are all Irish and German laborers, many working in the quarries nearby; others, especially the German women, live on the sale of rags and bones which they and their children gather all the day long through the streets of the city.[17]

Such semipermanent squatter settlements, common to many American cities before 1900, and so much a part of American urban growth, may be called slums in the physical sense only. Although they were lacking in the common amenities and were considered as unsightly, they were frequently well integrated with respect to their community life. In fact, because these squatter clusters were often regarded with disapproval and suspicion by the larger community, the squatters were probably more united than might have been otherwise the case. In some cities when the time came to remove such settlements the squatters were so well organized and were able to offer such resistance that strong police action was needed to oust them. Behavior norms within the shanty town were not always the same as those in the rest of the community. It would have been regarded as reprehensible, for instance, for a shanty-town man to call in the police in case of trouble, and the coming of a policeman into the community tended to put everyone on the defensive. Some-

[17] C. Loring Brace, *Dangerous Classes of New York*, New York, 1872, p. 151.

thing of this attitude was found by Haynor in his study of slums in Mexico City.[18]

## Slums outside the City

While some of the squatter-type shanty slums were in the cities, many were not. They were at the edge of the city, and this is true of cities in some countries today. Miner reported some slumlike clusters near the center of Timbuctoo, but mainly the central areas were inhabited by the better-situated families. The poor lived in small communities of makeshift structures outside the city.[19]

This seems also to be true of even the most modern cities in the Latin American countries. The United Nations report on urban land policies mentioned that there were different names from country to country for both the slums within the city and those outside:

> In Mexico, Central America and Venezuela the tenements which occupy old dwellings near the center of cities are called *casas de vecindad;* in Cuba, *solares;* in Chile, Argentine and Uruguay, *conventilles* and in Brazil *corticos* and *villas.* This slum-type dwelling consists of a series of one- or two-room dwellings surrounding a long narrow courtyard where water supply, wash tubs and toilets are located. As many as sixty families may share the same courtyard, which is often reduced to a mere alley. Water supply and sanitary facilities are generally inadequate; there is no clean spacious area where children may play safely; buildings are structurally unsafe; and domestic privacy among family members is lacking. . . .
>
> The other slum type in Latin America is similar to the shanty town found in the outskirts of southern and western cities in the United States, and the squatters' settlements now prevalent in other parts of the world. These are the *ranchos* of Venezuela and Colombia, and the *fravelas* of Brazil; they tend to be built beyond the limits of the developed areas, or closer in on hill tops and in ravines which are otherwise unsuitable for building purposes. Surprisingly enough, public land, rather than private, is often used for this purpose. It is estimated that ten percent of the urban area of Caracas is occupied by these *rancho* towns. The one-room shacks and shanties are built of adobe and waste materials from the large homes and apartment buildings at present under construction in the cities. The districts which the buildings form are generally without paved streets, sanitary sewers or water pipes to the dwelling. Community

[18] Norman S. Haynor, "Criminogenic Zones in Mexico City," *American Sociological Review,* Vol. XI, August 1946, pp. 428-438.

[19] Horace Miner, *The Primitive City of Timbuctoo,* Princeton, Princeton University Press, 1953, Chapt. I.

> water spouts are provided by the authorities, and sometimes electricity, improvements which only add to the permanency of the miserable dwellings.[20]

It is hardly necessary to dwell upon the health hazards for the entire urban community where a large share of the low-income workers live in squatter settlements under such conditions. The hazards become evident if we consider that many of the workers are servants in the best homes or are food handlers. These makeshift settlements were described by Fitzgibbon in a report on rural communities in Uruguay, but he referred to them as *rancherios,* and he estimated that there were no less than 500 such in that country. He noted that the way of life in these settlements of landless workers is characterized by an undue amount of illegitimacy, prostitution, drinking, and gambling.[21] The problem is in part due to overcentralization of the population in Montevideo and an overuse of the land for large ranches.

Another variant of this type of squatter slum is found in and near the rapidly growing cities of North Africa. LeTourneau describes them thus:

> Many new arrivals cannot find housing and have settled in vacant areas, made a small payment to the owner, put together precarious shelters of tent cloth and odd pieces of lumber with walls of oilcans fitted end to end—hence the name "oil-can city," *bidonville* has been given to them. The public authorities have not succeeded in abolishing this blight on the North African cities, for the inflow of population has always far outstripped their measures. Thus in Casablanca, which holds an unenviable record, there is a *bidonville* of more than forty-five thousand persons. Sordid as they are, these miserable habitations are not worse than those which their occupants, once country people and now become proletarians, left behind. Structurally they are actually an improvement over the rural dwellings in North Africa and are, moreover, markedly less unhealthy, in spite of appearance, than the overcrowded districts of the old medinas.[22]

## Social Implications of the Slum

Like any other area within or adjacent to the city, the slum changes and moves. It is perhaps the most sensitive of urban habitation areas because it reflects quickly and in different ways, mostly negative, various social, economic, administrative, and other aspects of community life. Whatever the

[20] *Urban Land Policies, op. cit.,* p. 178.

[21] Russell H. Fitzgibbon, "Uruguayan Regional Problems," *Social Science,* Vol. 29, No. 2, April 1954, pp. 80-81.

[22] Roger LeTourneau, "Social Change in the Muslim Cities of North Africa," *American Journal of Sociology,* Vol. 60, No. 5, March 1955, p. 530.

problems are that are reflected by the slum, they are not necessarily solved by slum removal, important though that may be. Usually the slum, or particular slums, are removed naturally in the course of urban change and development.

One fairly universal characteristic of the slum is the poverty of its inhabitants. Therefore, as the levels of living of increasing numbers of people rise, the number of slum inhabitants diminishes, although there may remain a certain "hard core" minimum, even in cities of very high levels of living. In many rapidly growing underindustrialized cities, however, the number of people trying to enter the urban community far exceeds labor market capacity; hence, as we have noted, in Asia, Africa, and Latin America unemployment contributes to the existence of the shanty slums. Poverty is linked with unemployment or underemployment, but it is only a partial explanation of the existence of slums.

There are many slums, whether of the inner-city tenement type or the outside shanty-town type, in which a good share of the inhabitants are employed. They live fairly permanently in these squatter communities, one reason being that better housing is not available. Private capital does not venture into this type of construction, and public housing programs may be lacking. The squatter slum comes to be complacently accepted as the way of living to which low-income people are accustomed and seem to prefer.

Describing the residential distribution of the population of Kampala in Uganda, Southall noted that the rich live at the center of the city, but only 10 percent of about 30,000 unskilled workers live in the town. The rest live in a ring of shanty towns just outside the city. Most of the clerks and better-paid workers live in a belt outside the worker settlements, and ride bicycles to and from their homes.[23]

The distribution of habitation areas described by Southall for Kampala is not unique. But it presents an incipient suburban movement in which the bicycle is the first means of rapid transit. The workers in the first fringe of habitations, while outside the city, are within walking distance of their work places. The better-paid workers are able to live a greater distance out because they have their own transportation.

The more efficient methods of rapid transit in the Western cities not only enabled more people to live a greater distance from their work places in the city, but enabled people living in crowded areas, including the slums, to escape

[23] A. W. Southall, "Determinants of the Social Structure of African Urban populations, with Special Reference to Kampala, Uganda," *Social Implications of Industrialization and Urbanization in Africa South of the Sahara,* Paris, UNESCO, 1956, p. 573.

to suburban areas. Because of this, especially in some American cities, certain slum areas have become partially abandoned. They are called areas of blight. (See following section.)

## Urban Decentralization

It is well known that most great cities of a million inhabitants or more, and smaller cities to a lesser degree, are in process of decentralization. If they are not losing population, they are not growing at as fast a rate as are towns and cities suburban to them. This does not mean that the great city is declining; its importance as a work place may be increasing, but more and more people who work in the city prefer to live outside, in the "ring." And the rings are enjoying a more rapid population growth than the larger cities. Bogue found that of the 147 principal Standard Metropolitan Areas in the United States in 1950, the central cities claimed 30.7 percent of the national population increase between 1940 and 1950, while ring areas outside these central cities claimed 48.6 percent of the national increase. The 71 Standard Metropolitan Areas of 1910 had a reverse growth relationship; 46.8 percent of the national population increase between 1900 and 1910 was in the central cities and 20.8 percent was in the towns and cities of the rings.[24]

Decentralization imposes new problems on the central cities. For example, Boston is one of the slow-growing American cities (748,060 in 1920 and 801,444 in 1950), but in the 40 suburbs of Boston are more than a million population. Boston is the work place for many of these people, and it must provide services for them. They do some of their shopping in Boston, but they vote and pay their taxes outside of Boston. In their citizenship relations they may oppose Boston on many issues. If they have automobiles, they drive to the central city, which adds to the traffic congestion in downtown Boston. In a study of Boston's decentralization problem, Ballard and Rockwood found that the number of persons entering the central area daily by private automobile rose from 332,770 in 1927 to 488,568 in 1938, a gain of nearly 49 percent.[25]

[24] Donald J. Bogue, "Urbanism in the United States, 1950," *American Journal of Sociology*, Vol. 60, No. 5, March 1955, p. 480. A Standard Metropolitan Area (except in New England) includes the central city, the county in which it lies, and contiguous counties if, according to certain criteria, they are socially and economically integrated with the central city. See Henry S. Shryock, Jr., "The Natural History of Standard Metropolitan Areas," *American Journal of Sociology*, Vol. 63, No. 2, September 1957, pp. 163-170.

[25] W. H. Ballard and M. L. Rockwood, *A Survey in Respect to the Decentralization of the Boston Central Business District*, Chicago, Urban Land Institute, 1940, p. 27.

The movement of large numbers of people to the suburbs, with consequent decentralization of the city, has in the past created a further urban incongruity; the blighted area. It can still be found in some American cities. Normally a blighted area is one formerly used for residence, perhaps occupied by low-income renters, families or single persons. An exodus to the suburbs leaves these least desired dwellings vacant; or, perhaps only an occasional building is occupied. Some of the buildings may be in a state of disintegration; others may have been removed. Although the area, when it was occupied, was called a slum, in its state of blight it is something less than that. The land still has value, but it yields no income. It waits to be occupied for some other use. Until this transition takes place, the area is not only an economic loss, it is also unsightly.[26]

Decentralization also deprives the municipality of certain sources of revenue. If one of the functions of municipal housekeeping is to improve the appearance of the city, this becomes more difficult, since, while the costs of public administration rise, public income does not rise at the same rate.

Briefly, decentralization leaves the central cities unable to cope with the incongruities about which people complain. But there are incongruities outside the city limits which may also be embarrassing.

## Unintegrated Public Authorities

Not all the incongruities of modern urbanism, at least in American cities, have to do with unsightly slums or areas of blight or with the various equally evident types of overcrowding. In order to perform its administrative work effectively, the city must have access to the water supply which comes from places far outside the city. The city must be able to dispose of its sewage, garbage, and rubbish, and this may give rise to protests from other communities. In the interest of health, the city may need to be concerned with various undertakings beyond its boundaries, such as draining a swamp. For the convenience of the urban population the municipal corporation may wish to establish back-to-nature areas in the mountains or along water fronts. These many needs can be met only if and when the outlying areas are willing to cooperate.

These outside communities, although not always willing to cooperate with one another on mutual problems, usually join forces to resist any efforts toward improvement initiated by the central city. Around the city of New

[26] See the stimulating analysis by the Editors of Fortune, *The Exploding Metropolis*, New York, Doubleday & Co., 1958.

York for example, 397 local authorities are to be found in three states. Pittsburgh is surrounded by 225 local authorities, Philadelphia by 174, and Chicago by 162. A good share of the residents in the outlying communities are migrants from the city where they still work. As described by Gordon, they are persons "who ran away."[27]

These near and far suburban communities—villages, towns, and even large cities—are economically and socially identified with a single metropolitan area, but their ultra-independence renders difficult, especially for the central city, the implementation of metropolitan planning and administration.[28] The problem is sometimes solved piecemeal as the city is able to annex one after another of these outlying communities, but occasional towns or villages refuse to be annexed, or oppose annexation for years. Perhaps as they grow, and acquire more people and a mixture of viewpoints, the community attitude toward annexation may change. It depends in each case on the advantages that annexation seems to offer. Nonetheless, some places may resist annexation even though other adjacent suburbs may have been added to the central city. Gist and Halbert conclude that in time most of these "embattled satellites" yield. In their opinion, annexation is merely a matter of "giving political recognition to the social and economic coalescence that has already occurred." They would agree that some places are exceptional, and two of these stubborn exceptions are Hamtramck, almost surrounded by Detroit, and Evanston, surrounded in large part by Chicago.[29]

In the case of Salford, England, an *enclave* of Manchester, resistance to annexation has become a deep tradition. In 1792 these two towns, separated by a small river, had a combined population of about 60,000. Beginning about 1840, Manchester became increasingly industrial, whereas Salford remained

[27] Charles Gordon, "A Modern City's Transportation Needs," Washington, *Annual of the American Planning and Civic Association,* 1944, p. 132.

[28] Ted F. Silvey at the Corning Glass Conference mentioned living in a New York City suburb located in New Jersey. In the region are 60 towns and villages none of which is able to function as efficiently for public service as they might if all were joined into a municipal area, but the idea is resisted by local business and political groups in each place. See Eugene Staley, Ed., *Creating an Industrial Civilization,* New York, Harper & Brothers, 1952, p. 22.

[29] Noel P. Gist and L. A. Halbert, *Urban Society,* New York, Thomas Y. Crowell, 1948, p. 132. The attitudes of suburbs toward the central city are expressed in other ways as well. Suburbs may resist various types of legislation that may be beneficial to the central city, joining forces in state legislatures in opposing laws proposed by large cities. Such attitudes may reflect political differences. Suburbs tend to be politically more conservative than central cities. Often, if the city is Democratic, the suburb is Republican. See Gordon E. Baker, *Rural versus Urban Political Power,* New York, Doubleday, 1955, p. 18.

quietly residential. Manchester was growing and many outlying places were being annexed. Salford, too proud, refused to be annexed, certainly not by such a grimy city as Manchester was coming to be. Although most of that proud leadership is gone, the old attitudes remain alive. In all respects except that of political identity, Salford is part of the larger city, and in fact today is being decentralized as is Manchester and many other cities. While the population of Manchester between 1931 and 1951 dropped from 766,378 to 706,400, that of Salford dropped from 223,458 to 176,036. Through more than a hundred years these two cities have been mutually inconvenienced in their public services and, although in many respects they are a single community, they remain politically separated.

## Social Disorganization and Deviation

Cultures tend to remain stable unless disturbed by changes in the physical environment, by technological change or perhaps through contact with other cultures. The cultural change is in the direction of the complex; it is also disorganized to some degree. As Freedman states:

> Cultures tend to change in the direction of greater complexity, the degree of complexity being directly related to the amount and intensity of culture contact. This sort of change involves the decline of folk characteristics. Related aspects of a single culture frequently change at different rates, the resultant "lag" being an important source of disorganization.[30]

We are primarily interested in culture in the urban society, and what we find, particularly in the most urban communities, is not an integrated culture but often a mixture of cultures oriented or being oriented to an overall secondary way of life. Within the embrace of the impersonal and secondary urban life may be found a great variety of subcultures represented by various clusters of people from as many different places and backgrounds. Disorganization results because all of these subcultures are in process of change, perhaps in the same direction, but at different rates.

Faris describes social disorganization as "a disturbance in the patterns and mechanisms of human relations." The parts of a society work at cross purposes and there may be confusion and disagreement.[31] The term would have various applications in relation to the behavior of individuals or groups,

[30] Ronald Freedman and Associates, *Principles of Sociology*, New York, Henry Holt, 1956, p. 355.

[31] Robert E. L. Faris, *Social Disorganization*, New York, Ronald Press, 1948, p. 49.

but only insofar as this behavior deviated from expected norms. In the complex modern society, however, there are many groups, and the norms may differ from one group to another. A boy may be in full accord with the norms of his gang but the gang may not be in accord with the norms of the wider community. Perhaps because the boy is so integrated into the gang he may be somewhat less in accord with his family. In relation to the family norms and the community his conduct might be called deviant.

What is normative behavior and what is deviant may be differently defined by different groups in the community, and these definitions tend to indicate the points at which different groups may be in the process of social change. This is readily seen when the views about the behavior of teen-agers are judged by the older generation or the behavior of a striking trade union is judged by the different sections of "the public." Deviant behavior may concern those in the very front of the change process, but it may also concern those at the rear. The idea is illustrated by the traffic on a high-speed highway. If the median speed is 60 miles an hour, one motorist may be arrested for racing more than 90 miles an hour, but at the other extreme a driver may be deviant and a hazard to traffic if he drifts along at less than 30 miles an hour.

Precisely because change in the urban community is more rapid than in less urban places, there are many individuals who cannot adjust. They may be the "lone ones" who have few or no group associations. While many of these people are no problem to themselves or others, the exceptional one may be a problem. The worth of a man in the modern community depends in large part on his ability to work. There are individuals who, for various reasons, are unable to meet the competitive requirements of work. They may be sick or crippled or aged and weak. Perhaps they need to be trained for other kinds of work. Other individuals may be willfully deviate and predatory in their behavior. Still others may be mentally deficient, or mentally ill and in need of psychiatric help. These different types of deviate individuals are symbolic of the social incongruities of the urban community.

We need not here consider the issue of whether or not these social incongruities, including crime, are more characteristic of urban than rural communities. The statistical evidence indicates a greater prevalence of these types of deviation in the cities, but for different reasons these statistics are not comparable. On the contrary, we can well be impressed that the degree of social incongruity in the modern city is not greater, and that so many people of so many diverse backgrounds are able to meet the adjustment demands imposed by urbanism. This may be evidence that urban life is less anonymous and chaotic than has sometimes been assumed. In this book we have emphasized the extent to which the urban man identifies himself with the larger community through his membership in secondary groups. This should not divert our attention too much from the prevalence in the urban community of untold numbers of small primary groups. One becomes integrated into various types of groups, and:

> Normative integration is a property of groups rather than of individuals. While it is observed in the conformist or nonconformist behavior of individuals, this quality of behavior can be ascertained only if one determines the group memberships of the individual. The norms of a particular group are the criteria used to establish conformity or nonconformity.[32]

We need to know more about the function of informal groups and their role in facilitating the adjustment of individuals to urban life.

## Summary

The purpose of this chapter was to present some of the negative elements of urban development and change, the things that people complain about: smoke, noise, dirt and disorder, congestion, ugliness, and so on. It has to be recognized that much of what is objected to is hardly to be avoided so long as cities are the workshops of the world; yet much that is objectionable can be remedied. We noted that there is much waste in the changing and growing city, that in the course of change areas of neglect and decline will appear here and there; but these may disappear as development proceeds, although other such areas may appear. The more advanced Western cities are in a better position to cope with their incongruities than are the underindustrialized, overcrowded cities in some countries.

Problems of transport and access seem to be ever-present, regardless of what measures are devised to cope with them. Where more effective transport has been achieved it has opened the way for relieving central city congestion. There has been an exodus from the slum. Actually, there has been an excess of outward migrations, leaving behind vacant areas and new administrative problems for the cities. The transit problem, too, changes with each plan for solving it, but it may also change with technical advance in transport facilities, as when motor vehicles displaced horse-drawn vehicles.

The slum was considered not only as a continuous challenge to the urban community, but also, when seen in long-term perspective, as a barometer of

[32] Freedman and Associates, *op. cit.*, p. 173.

urban evolution. As family dwelling places, most slums in American cities have greatly diminished, although a type of slum inhabited by certain kinds of individuals—homeless persons, the maladjusted, rooming-house dwellers of low income, persons down in their luck, and the like—still exists. While slums as marginal areas of residence tend to disappear in one location or another, they may appear elsewhere, but the minimum conditions of the later slums tend to be higher than those of earlier ones.

In the earlier decades of American cities the shanty-town slum of semi-permanent character was a familiar sight. These were out-of-town squatter communities (not to be compared with the auto camps of today). Such squatter slums are a familiar sight around cities in the less developed countries, especially in cities where industry is growing and people are gathering faster than housing can be provided. The size and number of these outside shanty-town slums indicate the extent to which the city is the refuge for those who can no longer sustain themselves on the land.

Cities as corporations are also faced with growth and supply problems which bring them into competition with places beyond their borders. Because of this opposition a city may have difficulty opening roads, finding recreational space for its people, securing a water supply, obtaining an adequate free space area. Some of these problems are met in part by annexing outlying areas, when that is permitted. While the city and its immediate hinterland are one socially and economically, the many political subdivisions which comprise the metropolitan area prevent effective regional planning and administration.

The incongruities of the modern city are, in the main, peculiar to the urban way of life, and they relate not only to congestion, slums, apparent disorder and ugliness, but to the problems of people as well. Many of these are generally identified as problems of social disorganization, and they are found in proportionately greater measure in urban than in rural communities. While these social problems are continually being worked on, events and change may bring new problems to attention and aggravate existing ones. The equilibrium of the community, however, tends to prevail even though certain problems may at times assume challenging importance. It may be said that the processes of disorganization are neutralized by processes of organization, that the urban community has certain inherent qualities of balance and self-discipline. These qualities of the urban way of life have been designated as part of the generations of accumulated experience through which the evolving city has lived.

CHAPTER 9

# Collective Community Behavior

AT THE CLOSE of the preceding chapter the modern city was described as behaving in a self-disciplined manner, and the urban population as containing within itself the heritage of generations of city-dwelling; a heritage that is modified, added to, and passed on. People speak of communities as if they were entities capable of being proud or indifferent, gay at times or sullen at other times. It is more than a figure of speech when a city is referred to as if it were "a behaving personality" capable of taking a position in favor of this or opposed to that.

## The Community as an Entity

A city or a town is an entity, on the one hand, with respect to its social organization. It is a social structure, embracing a variety of social substructures: families, informal groups, formal groups, neighborhoods. On the other hand, the city or town is a civic entity, a corporation organized under law with a government. The structure of the government includes the various functioning branches: police, fire protection, water supply, health, education, and so on. In this corporation the people as citizens are shareholders.

The social structure is a network of relationships, some of which may be friendly, others unfriendly; and these may involve degrees of dependence or independence. This structure is "real" and can be described. It has continuity through time, "a continuity," according to Radcliffe-Brown, "which is not static like that of a building, but a dynamic continuity, like that of the organic structure of a living body. Throughout the life of the organism its structure is being constantly renewed; and similarly the social life constantly renews its social structure."[1] As a continuing entity it is also a nucleus of social habit, carrying with it the social heritage and extending it into the future.

As a civic entity, the urban community forms another network of relationships, between the impersonal organization called the corporation or government and the individuals who comprise the constituency. The breath of its life is the law and the regulations formulated in the name of law. In different ways that change with time and circumstances the relationships of the social structure are marked by inequalities, but inequality is not natural to the civic organization. And the civic organization has a continuity of its own, which in some of its manifestations is a separate type of heritage extending into the future.

Between these two aspects of community entity, the social and the civil, there is such an infinite interweaving of influence that the two are often indistinguishable. In a relationship of networks, often delicate and sometimes invisible, each is dominated in some measure by the other. What may be called a particular kind of community behavior is generally participated in by both the civic and the social. When, in particular situations, one is acting and the other may appear to be standing aside, this standing aside may be in fact tacit participation. Thus, when the community behaves or acts, the action is an interplay of the social and civic entities. Of this more will be said in Chapter 12.

## Community Consciousness

The "soul of the city" (l'âme de la cité), as the term is used by Bardet, is a colorful way of identifying an essentially local group. It is made up of shared sentiments and experiences. It is knowledge that is shared and social values that are mutually understood and accepted.[2] This concept seems to be hardly more or less than the concept, "social heritage," used above. Halbwachs recognizes that something remains to be carried forward when people have collective experience, and if these people continue together in the same place through generations they acquire a "collective memory."[3] Collective memory, although a useful term, turns out to be another synonym for social heritage.

[1] A. R. Radcliffe-Brown, *Structure and Function in Primitive Society*, London, Cohen and West, 1952, p. 191.

[2] Gaston Bardet, *Mission de l'Urbanisme*, Paris, Ouvrières Economie et Humanisme, 1949, pp. 164-167.

[3] Maurice Halbwachs, *La Mémoire Collective*, Paris, Presses Universitaires de Paris, 1950, pp. 133-136.

Like "soul of the city," it implies recognition that the agglomeration comprising the community is a social entity.

Hoselitz, in dealing with phenomena involving the behavior of the urban agglomerate has used the term "sentiment of community" and again "city consciousness," this latter term being descriptive of the individual acting in a civic role. Through city consciousness a bond is forged "between rich and poor, old and new families, native sons and immigrants. What they had in common was their joint interest in the government of the city, its civic improvement, finances, embellishment and growth."[4] In other connections we hear the term "sense of community," generally with respect to attitudes of community responsibility that people have or should have.

These and similar terms have the common characteristic of being descriptive of community life as organized and the community as a behaving entity. It is because we recognize the entity character of the community that such terms come into use. We think of the community both as a behaving entity and as one that is conscious of itself. Whether we call it city consciousness or civic consciousness, it relates to community self-awareness and finds expression in types of community behavior. It may be evident in a crisis, or in elections, or in the moods of people as they go about their work or other activities.

## Manifoldness of Community Behavior

With respect to what does the community behave? At the moment we are interested in the community behaving as an entity; behavior concerned

[4] Bert F. Hoselitz, "Cities in Advanced and Underdeveloped Countries," *Confluence* (The Relation of Advanced to Underdeveloped Countries), Part II, 1955, p. 328. Wohl and Strauss hold that people must visualize the community in terms or symbols, thus the chaotic whole is reduced to something orderly and descriptive. One may call it a mechanism, or use an array of adjectives, or: "The entire complex of urban life can be thought of as a person rather than as a distinctive place, and the city may be endowed with a personality or—to use a commonplace—a character of its own. Like a person, the city then acquires a biography and a reputation. Personified cities can be described with personal pronouns and, through the use of appropriate verbs, conceived of as having capacities for action and possession. And, following this fashion of speaking, we make the same allowances for and judgments of cities that we are ordinarily inclined to make of people." Thus by the use of symbols for the complexities of the urban life, the individual is able to function in this environment and to express himself about it. R. Richard Wohl and Anselm L. Strauss, "Symbolic Representation and the Urban Milieu," *American Journal of Sociology*, Vol. 63, No. 5, March 1958, pp. 528, 531. For a similar application of the term "personality" to conceive of the city as a whole with "its smells, its noises, its people, its clothes, its vehicles, its animals" and so on, see John E. Burchard, "The Urban Aesthetic," *Annals of the American Academy of Political and Social Science*, Vol. 314, November 1957, p. 113.

with performing certain work and achieving various projected goals will be considered in later chapters. A young and growing city or town may go about whatever it does in a vigorous manner, while a fairly sedate and static community may act in a hesitating manner. Places may be compared in terms of their manner of behaving. Carl Sandburg in his poem about Chicago described it as brawling and boisterous, a railroad builder, hog butcher, and so on; a a place where everything was being done in a big way. More than fifty years earlier an article in the *Atlantic Monthly* described Chicago (boasting about its 300,000 people, 60,000 buildings, and 60 miles of cedar block streets) as the great Babylon of the West, growing but not content.[5]

We may think of a community as behaving with respect to its own internal affairs. If a citizen in a European community reaches the age of one hundred, it would be surprising if the local authorities and leading citizens did not go as a body to honor him or her. Here the key people in more or less a routine manner do an honor in the name of all the people. The same key people on another day may make a call expressing sorrow to a family that has suffered some tragedy. On another day the same authorities and representative key personalities will walk in a parade honoring the birthday of the city. The personalities leading such a parade would be the leaders of that phase of the community which we have designated the civic organization, others would be leaders in the various nongovernmental associations representing the social organization of the community. Perhaps on another day the community will behave as an entity in celebrating the opening of a new bridge; the occasions for such activity are numerous.

Community behavior with respect to internal matters may be very different from its behavior toward other communities. According to the code of the primitive community, more of the virtues are found in the local group than among the outside groups; the local people are braver than those in outside places, and those in other places are not to be trusted. To varying degrees, traces of such attitudes may still be found in the difference between the behavior of a community in local matters and its behavior toward other communities. However, these attitudes, inimical or friendly, are more likely to be found in smaller communities than in larger ones, or they may be evident in the relations between very small communities and large ones.

[5] The article, "Chicago," in the *Atlantic Monthly* (Vol. 19, 1867, 325-245) is mentioned by Allan Nevins, *The Emergence of Modern America, 1865-1878,* New York, Macmillan, 1927, p. 79.

## Attitudes toward the Next Town

Bernot and Blancard made a study of Nouville, a French village of about 800 inhabitants, and here they found deep-seated attitudes about the neighboring places. Near Nouville in a forest region is a smaller village not as old and somewhat less advanced than Nouville. In the forest village, Coutrain, is a glass-making plant which is smaller than the one in Nouville. Generations ago Coutrain was merely a hunting lodge in the forest, apparently closed to Nouville people. Today Nouville people seldom go there and they can think of no reason why they should. There are occasional marriages between Nouville and other villages, but rarely between Nouville and Coutrain. Nouville's self-image is expressed by the way Nouvillers say, "we of Nouville," and in contrast how they say "those of Coutrain." There is a small city nearby concerning which both Nouville and Coutrain have the same negative opinion. Each village favors itself when villagers speak about "those of" the city.[6]

In his study of the youth of Elmtown, a small city in the Midwest, Hollingshead reported that Elmtown people had unfriendly feelings toward Diamond City, a small neighboring city. While these attitudes were probably not of the elemental kind found in Iranian villages, where the traveler is usually warned against the people in the next village, they were, nonetheless, held with some degree of conviction. The two towns did have considerable contact, but it was in sports that the attitudes, at least on the part of Elmtown, found expression, particularly in contests between the high school basketball teams.

> Community pride is at issue in basketball games. Elmtown is approximately two and a half times larger than Diamond City, and it reflects the usual belief in American culture that bigger is equivalent to better. Besides, Elmtown considers itself a "cleaner" town than Diamond City. Elmtowners for generations have associated immorality, vice and crime with Diamond City. Thus when the Indians (Elmtown) and the Jewells meet on the athletic field more is at stake than winning or losing the game, particularly if the game is basketball.[7]

While Elmtown people, especially the elite, are very class conscious, the top people are ready to forget social rules if a poor boy from "the other side of the tracks" becomes an ace in local sports. Hollingshead noted how the

[6] Lucian Bernot and René Blancard, *Nouville, un Village Français,* Paris, Institut d'Ethnologie, 1953, pp. 360-362.

[7] August B. Hollingshead, *Elmtown's Youth,* New York, John Wiley and Sons, 1949, p. 194.

local businessmen are the movers in these bursts of community spirit, and privately they are sometimes not averse to offering rewards to players if they make good scores, especially against Diamond City.

## Bias against the City

The bias of the rural man against the city and everything urban has a long history. It began, as Turner reasons, with the rise of cities.

> The original cities were clusters of peasants under priestly rulers. The social base of urban cultures was, therefore, a peasantry, whose members, as tillers of the soil and keepers of cattle, created and replenished the economic surplus that went into the hands of their priestly rulers. In every respect, except one, these tillers of the soil lived as had their primitive peasant-village ancestors; the one alteration in the primitive manner of life was the imposition of the social controls that drained away from them the surplus their labor produced.[8]

The controls listed by Turner meant that the peasants were obliged to give of their surplus to support the priests and the temple or to pay interest on loans. And so gathered at this control center the craftsmen and others who lived by trade. Much has happened between those beginnings and the present relationship between rural people and the city. Feudalism was an advanced outgrowth of earlier relationships between those who held the land and the authority and those who tilled the land. Even today in some preindustrial countries most of the tilled land is owned by families who live in the cities, or the tillers are held in bondage to money lenders who live in the cities.[9]

Through these many generations there was almost nothing in common between the interests of urban communities and those of rural villages. The position of the rural man was always one of giving what was asked and bowing low, and he developed to a fine point the arts of evasion and deception. The modern farmer in Western countries is far removed from peasantry, and the modern American farmer is still farther away. Possibly in the long history of rural-urban relations no rural people were ever more free from urban exploitation. Nonetheless, the American farmer in his primary-group village is likely to have his reservations about the city. This was observed by the

[8] Ralph Turner, *The Great Cultural Traditions*, New York, McGraw-Hill, 1941, Vol. I, p. 294.

[9] During the Great Depression farmers in states of the Midwest staged demonstrations against mortgage sales. Mortgages were held by city banks and insurance companies. The sales in many cases were frustrated. Civil authorities were in sympathy with the demonstrators.

Englishman Lord Bryce, who was writing about American life about 1900. This was before the arrival on the farm of automobiles and the radio.

> Yet he is not free from the usual defects of agriculturalists. He is obstinate, tenacious of his habits, not readily accessible to arguments. His way of life is plain and simple, and he prides himself in its simplicity, holding the class he belongs to the mainstay of the country, and regards city folks and lawyers with a mixture of suspicion and jealousy, because he deems them inferior to himself in virtue as they are superior in adroitness, and likely to outwit him.[10]

Bailey, a leader in the Country Life Movement, who was writing later than Bryce, and whose sympathies were rural, regarded the rural man as superior in matters of work. The rural man, he argued, works for the joy of creation; the city man has lost interest in production, but he joins a trade union, and the trade unions, Bailey reasoned, place emphasis "on rights rather than on duties."[11] If Bailey could see these farmers four decades later as they organize to obtain price support from the Government, he would find that they behave not unlike trade unionists. Farmer organizations, however, are differently oriented from the trade unions, which are oriented to a very non-rural time-measured and money-valued industrial concept of work. The modern farmer is apt to articulate his urban bias by pointing to trade unions, as earlier he pointed to the "slick" lawyer as the symbol of urbanism. On this attitude Edwards writes:

> Much of the trouble came from the fact that higher pay and shorter hour agitation by labor unions sometimes offended the rural mind, which out of its own experience had acquired a deep respect for long hours and hard work for humble rewards. The enforced dependence of the urban worker has never been sympathetically comprehensible to the farmer with his traditions of independence and individualism. In considering industrial disputes, country people have tended to look upon work as a moral duty, to regard insistence upon conditions and terms of labor as a partial abrogation of that moral duty, and to project their own moral and non-exploitive outlook into the industrial situation.[12]

Atkeson, who had been active in the 1880's trying to further the organization of farmers and industrial workers into a single trade union, came to think differently about this idea later. In fact, fifty years later he wrote auto-

[10] James Bryce, *The American Commonwealth*, New York, Macmillan, 1909, II, p. 294.

[11] L. H. Bailey, *The Country Life Movement in the United States*, New York, Macmillan, 1911, p. 139.

[12] Everett E. Edwards, "American Agriculture, the First 300 Years," *Farmers in a Changing World*, Washington, Department of Agriculture, 1940, p. 146.

biographically, "The farmers want all they can get for what they have to sell and, of course, the laborer-buyers desire it as cheaply as possible." This provides little common ground on which to stand economically. He added, however, that both groups do have a common interest "in the stability of our government and in the perpetuity of our institutions."[13]

## Antirural Bias

It is difficult to associate bias with urban life in general, although bias in many forms can be found among urban people. It must be found where it is, and when it is found it does not reflect the whole urban agglomerate. Because this agglomerate is secondary and impersonal, we can associate tolerance with urbanism in general. Wilson wrote:

> The city seems to generate that freedom for variant behavior which some call tolerance, some merely the sophistication of those who, knowing all the answers, all the variants on the human theme, are either sated or immunized, accepting the most bizarre as in the urban course of things. Some call it callousness. Some call it evil. But all would agree that here the range of choice is greater. In this setting philistine and bohemian work out a symbiotic accommodation.[14]

Urbanism as it is reflected in the most urbanized man may have this appearance of tolerance, aloofness at times or neutral response, in the crowd manners of the urbanite. How does he think and behave with respect to ruralism? We can get part of the answer in the urban writings about rural people. The farmer as depicted in urban-centered radio programs, for instance, is a negative stereotype, a very backward person. He must talk with a nasal twang. His humor is expressed in stale "cow jokes." Bias of this kind, and much of urban reaction to ruralism is of this type, takes the form of a kind of fun-poking humor.

This characteristic urban attitude may be called contempt for small places, or contempt for things that seem backward, or impatience with the ways of rural people. It may lead to name calling, and the ruralite becomes a "hick," "rustic," "clod-hopper," "hill-billy," "an old Silas," or is called by some other disparaging name. Such terms are not necessarily loaded with negative meaning, even though they may be cynically expressed. It does not

[13] T. C. and M. M. Atkeson, *Pioneering in Agriculture, One Hundred Years of American Farming and Farm Leadership*, New York, Judd, 1937, p. 113.

[14] Everett K. Wilson, "Some Notes on the Pains and Prospects of American Cities," *Confluence*, Vol. VII, No. 1, Spring 1958, p. 4.

bother the urbanite that his stereotype of the ruralite does not square with reality. As he makes the farmer a comic stereotype, he makes of the equally nonexistent cowboy a more inviting and heroic stereotype. In either case the urbanite may be naïve but he is not emotionally involved in these expressions, for he is too occupied with other things.

These urban attitudes are not organized, nor are they the result of any concerted meeting of minds. Apparently they are attitudes that have been developing for generations, becoming articulate through one medium or another. Landis would call the school teacher such a medium. He complained, "The rural school teacher, because he is usually trained and prejudiced in favor of the urban, sees little that is fascinating in the rural environment; it is drab, colorless and dead."[15]

Urban biases against what is not urban may develop very quickly, as Pons discovered when he interviewed Belgian Congo natives who had been a few years in Stanleyville. Urban living had weaned most of the 38 men interviewed by Pons from much that had been important when they lived in the villages. Most of them had returned to their villages to visit their kinsmen, but they found these visits to be painful experiences. They were accused in the villages of "following the Europeans." Some found they were no longer trusted.

> An analysis of the statements made by the 38 men in explaining their hopes and intentions revealed that life in Stanleyville is frequently appreciated by Africans, not only as attractive itself but as a refuge from many difficulties and hardships in the village. "Kizungo," or life with the Europeans, offers an escape from the "harsh" authority of tribal chiefs, from sorcery, from the drudgery of work in the fields, from obligations to demanding kinsmen and from the hostilities and jealousies of the village. The town also has its own positive attractions.[16]

These Africans had developed antivillage attitudes even before they had themselves become fully urbanized. These attitudes, however, would be much more peripheral and emotional than attitudes of bias on the part of urban people who have not known rural life but recoil from it.

But the city as a corporation or a particular organized section of the urban population may formally give expression to antirural bias. This the trade unions may do in times of unemployment. In some states the urban people in their capacity as citizens often become outspoken against the

[15] Paul H. Landis, *Rural Life and Process*, New York, McGraw-Hill, 1948, p. 189.

[16] V. C. Pons, "The Changing Significance of Ethnic Affiliation and the Westernization of the African Settlement Patterns in Stanleyville," *Social Implications of Industrialization and Urbanization in Africa South of the Sahara*, Paris, UNESCO, 1956, p. 669.

"upstate apple knockers" or the "downstate clodhoppers" or the "cow counties" when the city seeks legal permission in the state legislature for some program of development and the proposal is defeated by the rural legislators.

Local governments must secure permission of the state government for almost every new activity they wish to undertake. With few exceptions, the state legislatures are controlled by rural and small-town majorities, so that often the vote of the urban man is worth no more than half the vote of the rural man. It has always been so, since American legislatures began as rural bodies. Extensive urban growth is a development of the last several decades, and reapportionment of representation in state legislatures usually lags far behind urban growth. Rural legislators yield power only when they must, but seldom to the point of rural-urban equality, which is understandable.

The following 1954 figures from Baker apply only to the lower houses of the state legislatures, the so-called democratic assemblies. The six most urban counties in Georgia with 32 percent of the population had 9 percent of the state representatives. The 9 most urban counties in Florida with 60 percent of the population had 23 percent of the representatives. The Wilmington urban area with 59 percent of Delaware's population had 23 percent of the representation. The 10 largest cities in Connecticut with 46 percent of the population had 7 percent of the representatives (but 46 per cent of the senate seats). New York City with 53 percent of the New York population had 43 percent of the state representatives, but pays about 55 percent of the state taxes. Only two states, Baker notes, are exempt from the inequality charge: Massachusetts and Wisconsin. He adds that the same discrimination against urban areas is found in the United States House of Representatives; rural interests predominate.[17]

Similar information is presented by Anderson and Weidner, and they also mention the fear among rural voters of what evils would befall them if the cities gained equal political power. It must be recognized that in countries where cities have political dominance the rural areas fail to develop. Urban leaders, say Anderson and Weidner, feel that state governments in many cases are weak and they frequently complain about the

> . . . allegedly rural viewpoint of many state legislatures and the resultant failure or refusal to give adequate consideration to urban needs, coupled with frequent direct legislative interference in city affairs. It is a part of the traditional

[17] Gordon E. Baker, *Rural versus Urban Political Power*, New York, Random House, 1955, pp. 16-17, 42ff.

American working philosophy of life to believe that rural people and rural ways and conditions are inherently better than urban.[18]

This kind of discrimination gives rise to much of the urban bias against the hinterland, while the rural justification is in part the argument that the city is not to be trusted with political power, that city politics are corrupt. (This argument ignores the fact that corruption is often charged to county governments.) Many rural people still hold that cities are immoral places. These rural attitudes toward the city may be found in all urbanized countries.

## The Pride of Communities

Motivations of pride are evident in much of the collective or corporate behavior of communities. This is the theme of a London song which contains the line, "London pride was handed down to us," a sentiment that is found in many songs about cities. Or pride may be expressed in legends which are proudly treasured. Because London was a proud and self-sufficient city when William the Conqueror arrived, it was the only English city that received from him a special charter, which added no powers but took none away. "London Pride," Brogan wrote, is more than the name of a flower in that city:

> It is the perfect description of an attitude which at a turning point in world history determined how it should turn. It is easiest misunderstood by the visiting foreigner. The Londoner may be mildly gratified at being complimented on the excellence of his subway system. But he would not be more than mildly irritated if he had to admit that it was no better than New York's. He tolerates the chaos of the South London waterfront and the snail-like progress of the new high-speed roads on the outskirts. He likes, but takes for granted the comparative abundance of parks and squares, but in those parts of London which are least well off in this respect local pride is as deep and complacent as it is in Richmond or Battersea.[19]

Stewart had something to say about community pride in his book about the evolution of cities. He mentioned Herford, England, a city that bought its freedom from Richard I in 1189:

> The little town of Herford had a burgher tradition of its own. It was an intensely local pride, constantly immersed in the preservation and extension of those rights and privileges which it had bought from the king and coerced from the bishop. The first duty of the burgher was no longer to his king or country—he

[18] William Anderson and Edward W. Weidner, *American City Government,* New York, Henry Holt, 1950, pp. 114-117.

[19] D. W. Brogan, *The English People,* New York, Alfred A. Knopf, 1948, p. 289.

> had freed himself from these duties by money payment—but to his own community. If the country were at war, the defence of the city walls was more important than support of the Royal Standard. If there were peace, the everyday work of the town, however menial, was infinitely preferable to any traditional feudal tasks. It was better to serve the town by repairing the walls or digging a drain for the common weal than to undertake the servile duties on the Lord Bishop's demense.[20]

In that time the citizens of Herford had their independence, a collective property of which few other towns could boast. That experience is part of the history of the place, and one of many reasons for community pride.

One city may express its pride with reserve, while another may be boastful about it. In the course of his social study of Coventry, England, Kuper mentioned that the city had passed through many crises since it had been saved by Lady Godiva, who on a challenge to save the town from a tax levy rode naked on a white horse through the streets. The city weathered other crises and each encouraged the common idea that Coventry can meet any misfortune. It rose in pride from the World War II bombings. At a postwar public occasion, as reported by Kuper, the mayor of the city declared in an address, "We are the finest, most industrial, and most historical city in Europe."[21] That may be called forgivable pride.

Following World War II, when Louis Adamic visited Yugoslavia and came again to his native village, there was a ceremony in his honor at the village school. The welcoming speech was read by a teen-age pupil. Besides telling the honored returner of the villagers' admiration for his writings, the pupil recited the names of other writers who had come out of the same small district. Adamic remarked, "I was grateful for the girl's delivery; it was nearly as awkward as mine when I replied to her address."[22] This is an example of perhaps the most genuine expression of prideful behavior, the home town honoring the local boy or girl who "made good," although he made good in the city.

## The Display of Community Pride

In his *Autocrat of the Breakfast Table* Oliver Wendell Holmes wrote, "The axis of the earth sticks out visibly through the center of each and every

[20] Cecil Stewart, *A Prospect of Cities*, London, Longmans, Green, 1952, p. 69.

[21] Leo Kuper, Ed., "Blueprint of Living Together," in *Living in Towns*, London, The Crescent Press, 1953, pp. 30-31.

[22] Louis Adamic, *The Eagle and the Roots*, New York, Doubleday, 1952, p. 146.

town and city." The truth of this statement is illustrated in various types of display behavior evidenced by cities. We might use for examples cities that recently acquired Major League baseball teams: Baltimore, Milwaukee, Kansas City, San Francisco, and Los Angeles. Or, as another example, we could cite the excitement in Milwaukee when its team won the 1957 World Series. In large measure, these displays, which are not peculiar to American cities, evidence systematic efforts of local groups to "put our city on the map." While the new American city may feel that need, London does not—or, rather, the expression of it is different. London has its high, Lord Nelson monument that stands as if it were the axis of the earth; Paris has its Eiffel Tower.

This display urge is also evidenced in the folders which are prepared for visitors; not infrequently such guide material will contain a map showing the strategic position of the city. Whether it be Kansas City, Chicago, Brussels, or Hamburg, the city's position is placed in the center of the map. In such behavior, which seems fairly universal, feelings of pride and various local practical interests become intertwined.

In his book about American cities, Perry described Kansas City as being a leading "bread and meat town" at the geographical heart of the nation, and most representative of America in taste and culture; a town where things are more important than ideas, and where the emphasis is always placed on the superlative and "biggest" is a favorite word. Here are the biggest stock yards for handling range cattle, the biggest railroad station in the United States, the biggest market for winter wheat, and the "biggest single grain elevator." Not only the elite in Kansas City, but the people on the street, are ready to tell anyone that their city is the heart of an "empire."[23]

Perry also reported on the rivalry between Dallas and Fort Worth, thirty miles apart. They have already had their "skyscraper race." They could not agree on a single airport midway between the cities and so, to the everlasting inconvenience of the traveling public, there are two airports. Dallas, a cosmopolitan city, reflects the industrial cities of the eastern states. Fort Worth faces the West. Perry describes it:

> Fort Worth, "Where the West Begins," is a cow town, a big-hearted, loud-mouthed, "Howdy, stranger" town. It is the capital of that magnificent realm of western range lands that stretch west to New Mexico and northward almost to Kansas and Colorado. In much of this land rain is a curiosity and everything either "sticks, stings or stinks." And the sturdy western folk with the rawhide hands and the high-heeled boots rub off in a town where they bank, buy and

[23] George S. Perry, *Cities of America,* New York, McGraw-Hill, 1947, p. 244.

> frolic. Fort Worth probably comes nearer expressing what the word "Texas" signifies to most outsiders than any city in the state. In its effort to make a hit with the outside world, its problem is simple: it has simply to be itself, exuberant and whooping and friendly, since nearly everybody loves a western story.[24]

A city may take pride in not being industrial, in not being a boom town, and it may avoid all the striving to become "biggest" in this or that. Santa Barbara, California (1950 population, 44,973), is such a place. It is a coastal city of residence and recreation, a retreat from the world for rich people and intellectuals. Its avenues and beaches are free of billboards, and there is no "night life." Woe to the homeowner who does not have a perfect lawn and a clean backyard. In competition with hundreds of cities in California, Santa Barbara has won the Better Homes Award many times. It calls itself "The most civic-minded community in the West." The city takes pride in its voluntary isolation, a luxury that many of the residents can afford; but the money was gained in cities that are in the thick of industrial and commerical competition, cities that are too occupied with work to be civic-minded in that way.[25]

## Symbols of Community Pride

Tunnard observed that "pride of place" is characteristic of most people when they speak of the home town. Thus a Frenchman, especially if he lives in Paris, is proud of the monuments in his city, because he knows that all the world admires them, and he takes some of that admiration to himself.[26] For this reason, apparently, a city places its monuments in the central areas where the largest number of strangers must pass.

In the city of Arles, France, is a statue of Frederic Mistral, dialect poet of the Provence and Nobel prize winner. He was born in Bouches du Rhone in 1830, but Arles is proud to claim him. Ludwig von Beethoven, the composer, was born in Bonn, Germany, in 1770. At the age of twenty-two he went to Vienna, where he became famous, and lived there until his death. Yet in Bonn there is a monument to him, a street and a small park are named for him, and the house where he was born is maintained as a museum. The facts in neither case prevent either Arles or Bonn from having the pride that finds expression in monuments.

24 *Ibid.*, p. 88.

25 *Santa Barbara, A Guide to the Channel City and its Environs*, American Guide Series, New York, Hastings House, 1941.

26 Christopher Tunnard, *The City of Man*, New York, Charles Scribner's Sons, 1953, p. 361.

Whether it be monuments, fine new buildings or historic old ones, a church of distinction, a museum, or whatever, every town and city has its landmarks that strangers must not fail to see. Many are symbolic of events in the history of the community or honor the memory of local personalities.

The Statue of Vulcan in Birmingham, Alabama, is a symbol of local pride that acquired an importance different from its original purpose. Originally it had been designed to "sell" Birmingham, steel city of the South, at the St. Louis Exposition, and it was then laid aside. Here is more of the story:

> The Statue of Vulcan, on top of Red Mountain overlooking the city, is a 60-ton statue of iron painted a brilliant aluminum, and portraying the blacksmith of Roman mythology. Fifty feet high, it stands on a tower 120 feet high and is visible from all parts of the business district. The statue was erected by Guiseppi Moretti for Birmingham's display at the St. Louis Exposition in 1904. When the Exhibition closed, it was shipped back to Birmingham and remained neglected for years at the State Fairgrounds.
>
> In 1936, through the efforts of civic and industrial leaders, the Iron Man was moved to Red Mountain and placed in the present position. On the base of the tower, built by WPA funds, is an inscription which reads in part, ". . . o'er and o'er Nature has flung her treasures with a generous hand and Birmingham sits enthroned."[27]

Not infrequently the symbol of local pride may not be merely local in its meaning. This thought is illustrated by a statue to the popular comic strip character, Popeye the Sailor Man, who, when he eats spinach, is able to perform heroic deeds. The statue for this champion of all the solid virtues is found in Crystal City, Texas, which calls itself "The spinach capital of the world," and thus is honored its chief salesman. It conveys a touch of lusty humor and serves a practical purpose as well, but it would be unfair to call it merely an advertising device, for this monument may be as close to the heart of local pride as many others seen elsewhere.[28]

Popeye as a symbol is at home in Crystal City as William Shakespeare, a national symbol in England, is at home in Stratford-on-Avon, or as the monument of George Washington is at home in the national capital. While many of the monuments in Washington seem not at home there,[29] this is not true of

[27] *Alabama, Guide to the Deep South,* American Guide Series, New York, Hastings House, 1941, p. 177.

[28] *Texas, Guide to the Lone Star State,* American Guide Series, New York, Hastings House, 1940, p. 506.

[29] Some of the monuments in Washington are gifts of states or societies in the nation. They are to generals or political leaders. While they have a right to stand at important places in the national capital, they are not associated with the life of the city. They may have their place in American history, but they are not symbols of local pride.

the 555-foot obelisk to the first President. Although the Washington National Monument Society began in 1832, the centennial of Washington's birth, the collection of funds did not begin until sixteen years later. The inner wall up the stairway is lined with inscribed stones, the gifts of hundreds of contributors, among them states, cities, religious and other groups. There were many disputes, some of them political, most of them unworthy. But, as Floyd Dell tells us, "The conflicts are forgotten, the Monument remains, a serene symbol of national unity, and it would be difficult to imagine Washington without it."[30]

## Community Behavior in Crises

However we may describe that quality of unity in collective action, it may not be evident in the usual day-to-day life; in fact, the life of a community under normal conditions may seem anything but collectively organized. But if something happens—a flood, fire, earthquake, military attack—then the collective action very soon appears, although it is often preceded by panic. Some of these events of the disaster type become heroic episodes in local history; the great Chicago fire (1871), the Boston fire (1872), the Johnstown, Pennsylvania, flood (1889), the Galveston flood (1900), the Baltimore fire (1904), and the San Francisco earthquake (1906). Not infrequently the great disaster may mark a turning point in local history, as did the London fire of 1666.

For cities of the Middle Ages and ancient times, the great events of local history were frequently of a military nature, as when a walled city withstood a seige. Such an event was the seige of Vienna by the Turks in 1683. Vienna in 1679 had already lived through a devastating plague and the city had seemed to be "fast vanishing beneath the hecatombs of the dead." Now came an army of 300,000 that had destroyed everything en route. The Turkish ruler sent the following surrender order:

> We, Mela Mohammed, Glorious and Absolute Sovereign Emperor of Babylon and Judea, of the Orient and the Occident; King of all earthly and heavenly kings; Sovereign King of Holy Arabia and Mauretania; Heir and fame-crowned King of Jerusalem; Lord and Master of the Grave of the Crucified God of the Unbelievers, pledge to thee, Emperor of Rome, and to thee, King of Poland, as well as to all thy followers, our Holy Word by the Mercy of God, Who rules in Heaven, that we are about to plunge your little countries into war.
>
> We have with us thirteen kings with 1,300,000 warriors, infantry and cavalry. With this army, the like of which thou and thy followers have never seen, we

[30] From an unpublished manuscript prepared by Floyd Dell with the cooperation of this writer. It concerned an uncompleted study of public works and community behavior.

> will crush thy little country without mercy or compassion beneath the hooves of our horses, delivering it to fire and sword.
>
> Firstly, we command thee to await us in thy capital, Vienna, so that we may behead thee, and thou also, little King of Poland, do likewise. We will exterminate thee as well as all thy followers, causing that very lowest creature of God, which is a giaur, to disappear from the earth. We shall first expose great and small to the most terrible tortures and then hand them over to the most ignominious death. I shall take thy little Empire from thee and sweep away the entire population, allowing thee and the King of Poland to live only so long that you may satisfy yourselves that we have carried out everything we promised.[31]

Plague-weakened Vienna did not surrender, but the defense would have failed had it not been for the relieving armies of Poles, Bavarians, Saxons, and Swabians who joined in putting the Turks to flight. The Cross had won over the Crescent, and Vienna entered a new era. It was more than a military achievement; a cultural threat had been removed. Vienna became a boom city, attracting people from all lands. It was the beginning of the Baroque period of art to which Viennese still point with pride.

Better known is the failure of Rothenburg to hold the walls against the forces of Charles V, led by the Belgian general Johann Tilly in 1631. Because of this resistance, Tilly swore to liquidate all the leaders of the city. While the executioner was being hunted to do his duty, wine was brought for the victorious officers. Somewhat mellowed, Tilly offered to spare the lives of the condemned if someone in the city could drink in a single draught a two-liter tankard of wine. The burgomaster accepted and drained the tankard. The occasion is honored by a festival play, "Der Meistertrunk," by Adam Hoeber. Mainly because of this story Rothenburg was not shelled by the American armies during World War II. On ceremonial occasions, the old tankard, symbol of a great event, is brought to honor distinguished guests, who join the burgomaster in drinking from it.

The life of a community may sometimes be endangered by economic changes which may call for collective action of another kind. The situation may be discouraging and benumbing, not heroic and challenging as in disaster or war. A depression is such a situation, and often the behavior of a community in this condition is marked by inertness. But not always: The story of Ellwood City, called "Steeltown" in a book by Walker, illustrates how even in a depression threat a community can organize itself defensively.

In this city of 14,000, not far from Pittsburgh, the chief source of employment for many years was a tube mill. From 1930 on it became increas-

[31] Ernst Marboe, *The Book of Austria*, Vienna, Staatsdruckerei, 1948, p. 68.

ingly evident that this mill was outmoded and for various reasons was not a good investment. A more modern plant would be built at Gary, Indiana. It was cheaper and wiser to close the mill than to remodel it.

Ellwood City had become to a considerable degree a home town. Even immigrants from overseas were quite integrated into the life of the community. The workers were well organized in their unions, and union leaders were also leaders in many community organizations. The threat to close the mill, which came in 1946, was really a threat of depression for Ellwood City. Even the promise of the company to move the workers to Gary was not welcome because so many workers were homeowners. Tradespeople faced a gloomy future; the entire community stood in economic danger. The situation was not depression itself, but the threat of it.

It was the strong trade union that took the lead in organizing the "save Ellwood" campaign. Appeals were made to the mill's parent company, the United States Steel Corporation. Various situations were developed which gave the plight of Ellwood City national publicity and which placed the steel corporation in an awkward position. After some two years of this organized community action the firm decision to close the plant lost some of its force. Implementation of the decision was "postponed."

This does not mean that the mill will not move one day. The point illustrated here is how a community when faced with a crisis can organize itself and take collective action and how afterward the community feels a sense of achievement. It does not mean, as in the case of Ellwood City, that the problem is finally solved; it does mean that a community faced with a threat, unemployment in this case, may be stimulated to behave collectively.[32]

## Sentimentality and the Community

Sentiment and fantasy may be found in American community behavior, but it is not in keeping with practical American traditions. The American way of life is still dominated by the earnestness and matter-of-fact thinking of the frontier. Communities in general do not reach back to the times of legend making, although certain Minnesota towns celebrate Paul Bunyan Day, for which men prepare for months in advance by growing beards. Mormon communities celebrate July 24 as Pioneer Day to commemorate the day when Brigham Young and his band entered Salt Lake Valley. New Orleans is

[32] Charles R. Walker, *Steeltown*, New York, Harper & Brothers, 1950, pp. 133-141.

an American city that celebrates carnival at the beginning of Lent. Decoration Day is a day of sentiment but it is not a community occasion.

Most Old World cities, on the other hand, have had their events of deep emotion which are remembered periodically. Leyden, Holland, in 1574 was relieved of the Spanish seige after months of near starvation. As a reminder of its liberation in October of each year the city has a celebration during which people march in the old costumes and eat bread and herring as a reminder of the hunger endured by Leyden's defenders.

How a community may behave sentimentally is illustrated by the Heiner Fest in Darmstadt, Germany. The heart of this city was 80 percent destroyed and much of the rest was from 50 to 60 percent demolished by World War II air raids. About 50,000 people, nearly half the population, had been dispersed into the hinterland. In 1950 a good share of these people were still crowded into nearby towns and villages. Some had decided to settle there and become suburbanites. About 20,000 had returned, if not to the same streets, then to nearby streets, waiting until their old neighborhoods could be restored.

Darmstadt in the old days had always had a summer festival but, because it was associated with the Grand Duchy, it had been discontinued. While not wishing to revive the old festival, local leaders in 1950 thought that some new one should be started, something close to the hearts of the people. Since every Darmstadter speaks of himself as a "Heiner," as a New Yorker is a Gothamite, the planned fest was given that name, and the theme would be "Old Darmstadt," the prewar city. Three days at the beginning of July would be given to a grand homecoming.

Everyone in and near the city was asked to bring out any relic recalling the good old days. With these borrowed articles, and with relics and reminders owned by the city, the stores put displays in their windows. Artists were put to work. None of the planners dreamed that in addition to people living near the city another 30,000 would come from more distant places. Fifty thousand "Heiner" badges had been prepared for sale, all of which were sold the first half day and thousands more had to be hurriedly made. For many it was three days of sentimental looking-back. Groups paraded through the old bombed streets. Around the old castle at mid-town a number of dancing platforms had been erected. Everywhere was the city's new slogan, "Darmstadt Lives."

The old bell tower at the city center had been rebuilt, and many thought that this carillon of 21 bells, a landmark since the fifteenth century, could not be restored; but a bell-maker near Limberg performed the feat. The new

*Glockenspiel* was ready for the Heiner *Fest.* A woman wrote to the newspaper, "When I heard the bells again I had to cry. I depend so much on old Darmstadt." Poems were written to the bells. One of these "from the heart of a Heiner" began:

> Be happy! We have again our Glockenspiel.
> The beautiful tower rose proudly from the ruins.
> Joyfully ring out the tones and the songs,
> So long silenced by the fiery storm of war.

Although it was started primarily with the idea of stimulating business during a quiet season, the Heiner *Fest* turned into a gala celebration reviving the city's past. The ego of the community burst into almost incoherent expression, a demonstration of sentiment for the home town which was shared by all classes.

European communities often have their own stories of legendary figures —stories which, as they are handed down, become interwoven with myth. An example is the "Trumpeter of Säckingen." Säckingen is a German town on the Rhine opposite Switzerland, and no visitor there can fail to hear the story or be shown the small, bronze statue of the Trumpeter. There had been a larger statue also, but that was destroyed by the Nazis because of their dislike for such fantasy. After the war the story was revived with new vigor. It concerns a young musician who worked for the Freiherr and ruling family of the village. A secret love developed between the Trumpeter and the daughter of the Freiheer. When the father learned of this he was irate, and the youth, a commoner, had to flee, but before going his way he played a parting serenade to his sweetheart from the Swiss side of the Rhine and, so the legend goes, this was a pledge to meet again. The Trumpeter joined a band of musicians and sometime about 1650 arrived in Vienna where he found a position in one of the churches. The Freiherr, to solace his daughter, and see to it that she would meet the proper people, obtained a place for her in the Emperor's court at Vienna. One day in church, where the maid from Säckingen went with the Emperor's party, she saw among the musicians her Trumpeter. As would be expected in that day of a good lady, she fainted. When the Emperor wanted to know why, she told her story. The Emperor approved the marriage, first knighting the Trumpeter. The Freiherr had been overruled, and the pair returned to Säckingen and lived happily. Near the church is the double grave and nearby is the home where the Trumpeter lived with his widowed mother. These are shown to visitors, and the legend of Säckingen lives on. Years later the story inspired someone to write the famous trumpet piece, "The Trumpeter of Säckingen."

## Rivalry Behavior of Communities

Mention has already been made of the rivalry between the Texas cities, Dallas and Fort Worth. For many years there was a similar rivalry between Chicago and St. Louis and for a time St. Louis was in the ascendancy. It is well known that Los Angeles and San Francisco are rivals for the dominant position in California. Because of this rivalry it was found expedient during the Great Depression for the federal government to establish two relief administrations in California, one centering in each of these cities. Similar rivalry relations exist between certain cities of Europe: Bremen and Hamburg in Germany, Zürich and Basel in Switzerland, Liverpool and London, and Liverpool and Manchester in England.

There have always been rivalries between cities, and sometimes the ancient city lived or died as it was able or not to defend itself with armed force. If the city, as in the case of Carthage, existed mainly because it was militarily strong, its defeat in battle usually meant extinction. But if for economic and geographical reasons the ancient city was favorably situated, as Delhi, India, it would rise again, even though completely destroyed in battle. Rivalries between modern cities are mainly economic and technological, and these are the bases in terms of which cities compete. Thus many American towns that existed in the days of wagon roads lost their importance if they were not touched by the railroads.

The case of Corinne, Utah, illustrates how the fate of a town depends on lines of transport. When the Union Pacific Railroad was built, Corinne was merely a construction camp located where the railroad crossed the Bear River. Other such camps along the Union Pacific faded as soon as construction was finished, but some of the leaders in Corinne believed the location was ideal for a town, for the wagon road northward to Idaho ended here and they believed that later a railroad to the north would join the Union Pacific here. Furthermore, because of its location where the Bear River enters Salt Lake, people believed the town would become a lakeport for steamers.

But there is another side to this story. Utah in the 1870's was almost completely dominated economically and politically by the Mormon Church. The small non-Mormon (and violently anti-Mormon) minority gravitated to the only non-Mormon town in the state, Corinne. It was a mixture of characters and adventurers, for Corinne was a gambling and saloon town. The one anti-Mormon newspaper in Utah was printed here. For all the power Brigham Young and the church leaders held, they took no action, at least no violent action, against the "Berg on the Bear," the "Sodom by the salten sea." But when the railroad-to-Idaho plan got under way, the Mormons, having already built a section of railroad at their own expense, were in a bargaining position. The new line terminated in Ogden, not Corinne. The fate of the place was so quietly sealed that few people in Utah today know it ever existed.[33]

## Intrametropolitan Contest

A community achieves a sense of self-realization as it engages with some measure of success in competitive activity. Collective effort is needed to meet needs of which appreciable sections of the population are aware. Thus intercommunity rivalry that finds expression in various forms of competition, from developing new lines of transportation to winning baseball games, may have wholesome consequences, especially for a growing community. But if the modern urban community grows enough to enter the million class, its growing may take the form of decentralization, which introduces a type of rivalry that may be damaging to community unity, at least for a time.

*Newsweek* in 1957 printed a number of articles about the challenging effects of suburban growth. Special attention was given to Los Angeles with its dozens of suburbs, each of which is a growing, ego-strong city. A good share of the higher-income families who earn their money in Los Angeles spend much of it in the suburbs, although they buy luxury items in the central city.[34] The Chamber of Commerce of the United States in 1952 published figures showing that for cities under 250,000 inhabitants the central-city stores did 70 percent of the business in the metropolitan area, but for places of 500,000 to 1,000,000 inhabitants the central cities received but 40 percent of the trade in their respective metropolitan areas.[35]

Although separate politically from the central city, suburbs are linked socially and economically with the central city. However, insofar as they are

[33] Dale L. Morgan, *The Great Salt Lake,* Indianapolis, Bobbs, Merrill, 1947, p. 279.
[34] "Suburbia, Exurbia, Urbia," *Newsweek,* April 1, 1957.
[35] Quoted by Richard U. Ratcliff, "Efficiency and the Location of Urban Activities." See Robert M. Fisher, Ed., *The Metropolis in Modern Life,* New York, Doubleday, 1955, p. 148.

separate economic units with respect to trade, they compete fiercely with the central city. In some residential suburbs can be found a strong sense of community, particularly in the socially exclusive suburbs; usually such examples of community unity are largely motivated by a strong anticity bias.

Duncan and Reiss describe the suburb as a "political artifact," a part of the metropolitan area that belongs to the central city and that, although not self-sufficient, enjoys a temporary position of civic independence. Ultimately it may be absorbed by the city. The authors cite recent examples of cities that have taken in large areas of the metropolitan district; Dallas, Texas, for instance, was able to increase its population from 294,734 in 1940 to 434,462 in 1950.[36]

Ratcliff, who takes the position that, while congestion and other inconveniences of the central city have encouraged the growth of suburbs, and while the city is disadvantaged by this development, the inconvenience is likely to be temporary. The frictions and inconveniences are very much what is to be expected in the redevelopment process.[37] But meanwhile, according to Anderson, the metropolitan family of communities is in a state of strife.

> Within its roughly-defined area, each agglomeration may have a certain unity of social and economic structure. But from the political and legal point of view, it is a congeries of many overlapping, conflicting, often selfish and disputatious units of government, including counties, cities, villages, boroughs, towns and special districts. The entire metropolitan community, therefore, is politically amorphous, without structure or unity. It is not a corporate person. In the eye of public law it does not exist; even in the field of politics, it speaks with many contradictory voices. It is a place of divided loyalties and conflicting allegiances. In short, it is a community that is not really a community.[38]

Each suburb, viewed separately, acquires out of these rival relationships a high degree of internal civic, economic, and social unity. While it cooperates very little with neighboring suburbs, it is likely to join with them in resisting real or feared encroachments of the central city. Perhaps the central city in the long run will win. In the meanwhile the metropolitan area suffers the many inconveniences that might be removed by establishing some regional civic authority. This is a problem of chaotic intercommunity behavior that is largely peculiar to American cities.

36 Otis Dudley Duncan and Albert J. Reiss, Jr., *Social Characteristics of Urban and Rural Communities, 1950*, New York, John Wiley and Sons, 1956, p. 7.

37 Ratcliff, *op. cit.*, pp. 145-147.

38 William Anderson, "Political Influences of the Metropolis." See Robert M. Fisher, Ed., *The Metropolis in Modern Life*, New York, Columbia University, 1955, pp. 57-58.

## Summary

The community, as presented in this chapter, is an entity capable in various respects of behaving collectively and of exhibiting types of emotion in its collective behavior. It is identifiable in a social sense, in that people are joined in a network of formal and informal groups of different kinds (families, social clubs, religious organizations, business and professional associations, political parties, cultural societies, neighborhoods, and others), and each of these is oriented in its behavior to the total community. The history of these groups is interwoven with the history of the community. They are the carriers and the transmitters of the social heritage.

The modern community is also a civic organization with a government, and as such it is a political entity. Members of the groups which comprise the social organization are also members of the community as a civic organization. The social membership is generally personal and private, while the civic membership is quite impersonal. The civic community is a legal entity and it behaves in accordance with law. Between the two entities, however, there is an overlapping and intertwining of functions and characteristics, until the line between the two is sometimes quite indistinct. The civic organization is an instrumentality of the people in their collective activities.

We speak of the community as "behaving"—resisting or supporting, making decisions or showing emotion, as if it were a personality—and it is sometimes so described. Personality characteristics are often attributed to communities, as when they are described as hard working and earnest, active and gay, sluggish or depressed or sleepy; since a community is a place of both work and living, these terms may not be inconsistent. Often community behavior depends on whether there is much work or little, high average income or a low wage level; whether it is venturing into new enterprises or is retreating from old ones. The morale of a community is affected by its status (gaining or losing) in its competition with other places.

It was noted that between urban communities and their rural hinterlands there has always been an interdependence which through the centuries has frequently operated to the disadvantage of the rural people. This relationship has varied in nature from the extreme of the rural place's being parasitically dependent on the city to its being, as in some parts of the United States, in a position to exercise restrictive influences over cities. Regardless of whether rural people are exploited by the city, they have generally entertained a bias against cities on grounds of morality. They regard rural life as being more wholesome. Urban bias against the country rises to anger rarely, but it continually finds expression in fun poking and other forms of depreciation. The rural people are pictured as backward and dull, and the farmer is often portrayed as a comic character. Rural or urban community behavior of the bias-motivated type tends to diminish as countries become more industrialized and rural areas more urbanized.

One fairly universal characteristic of community consciousness and cohesiveness is seen in collective behavior motivated by pride. This may be more true of the larger community, as it behaves angrily if offended or boastfully after completing some great endeavor. It may build monuments to honor its heroes and to commemorate great events in its history. Its statues and monumental buildings are conspicuously placed for visitors to see, and the citizens take pleasure in pointing them out to strangers. This pride in its history becomes part of the local heritage of story and glory given by one generation to the next. It is part of the "picture" the individual has of his home town.

As was noted in the chapter with respect to the Heiner *Fest,* a community may also behave sentimentally. Small communities may become sentimental during home-coming festivals. Large communities may become sentimental during founding day celebrations or memorial events.

Whatever form collective behavior may take, in the urban community especially it is likely to be very closely related to the serious business of earning a livelihood. If a city is to grow, it must buy more, make more, and sell more. Even if it is not growing, it must remain active and resourceful, else it may become less important and slip backwards. These efforts to live, to live well, or to live better involve competition with other places equally determined to grow, and equally fearful of slipping backwards. Competing ancient cities often settled the rivalry issues by war, and some cities, once destroyed, did not rise again. Others, favorably situated, did revive. The modern city must meet its rivals in the markets. It must gain advantage by superior technological means or by being quicker in reaching new supplies of raw materials. Sometimes a community in these efforts may have heroic experiences. But whether these experiences are heroic or not, they call for various forms of collective community action, of cooperative effort on the part of the community as a civic organization working with the network of nongovernmental groups.

Growing modern cities, particularly those of the million class, are now confronted with a new kind of problem: decentralization. Large cities are becoming metropolitan districts, but metropolitan districts outside central cities are frequently a dispersal of noncooperating independent suburban communities. They are socially and economically a single unit with the city, but politically they are divided. What seems to be needed is a new political unity which includes the city and the suburbs. Doubtless this will come in time. Until it does, however, large cities so situated are threatened with various disintegrating influences over which they have no control, and these may affect negatively both the civic and the socioeconomic organization in these cities.

CHAPTER 10

# Groups and Class under Urbanism

WHETHER A COMMUNITY IS RURAL or urban, it is divided into status groups and, whether this division is formalized or not, it will be recognized generally by the people, and there will be what may be called an approximate conformity to it. Such a social stratification which places people in a status relationship with one another is an essential part of the social order of the society in which it is found. The class structure may be simple or complex; in either case it will reflect the ways of life and work in that particular society and locality, as it will also reflect the different developmental phases of the people.

Our interest in social classes is primarily concerned with the class systems in urban communities and with the social groups that are peculiar to urban life. Class structures under urbanism may assume different forms from those found in nonurban societies, and these various urban class systems have much more in common with one another than any of them has with the nonurban type of class system. Social class under urbanism is more complex, and it is also less stable, since its milieu is one of change. Urban social class adapts to the demands for change with a facility that is rarely found in nonurban class systems, which tend to disintegrate under the expanding influence of urbanism. While an urban class system is unstable with respect to a particular form, it persists even though it changes from one form to another. Urban class systems change but do not disappear.

## Identity of Classes and Groups

One may belong to a group and be quite aware of it, especially if it is a closed group held together by personal acquaintance. One may also belong to a loosely integrated group of scattered members held together by recognized common interests: occupational, religious, national, and so on. Especially in the urban community, one may belong to a group and not be aware of it, or he may be made aware of it under certain conditions. Large numbers of teen-agers, responding separately to the same stimuli, may find themselves behaving in concert. Members of the older generation may find themselves thinking in harmony with other people in their age group elsewhere in the community and without collusion they may vote the same way in an election. Normally, the group is a collectivity of a few or many persons; it may be a family, a gang, a club, a trade union, an association, etc.

The terms "group" and "class" are sometimes used interchangeably. For our purposes, however, the group is specific and can be identified. It may have a name, a formal organization, and a locale. The class is more diffused. One may identify himself with a class, but he is a member of a trade union, a church, a sport club, and so on. One may belong to different kinds of groups, but he is identified with but one class; in fact, his group memberships may help to identify his class. Say Vidich and Bensman:

> By the word "class" the authors mean typical configurations of social and economic behavior which make it possible to distinguish groups of individuals from each other. That is, classes are identified in terms of productive activity, patterns of consumption and other forms of social and economic behavior. The term does not necessarily imply a recognition of "belonging together" by the members of the class, though in some cases such a recognition may exist.[1]

The meaning of social class depends largely on when and where and by whom the term is used, as well as the circumstances attending its use. It is a term that came to the fore with the emergence of industrial urbanism when population was being reshuffled and masses were submerged in poverty around the factories. This stimulated the concept of two classes, the exploited and the exploiting. For our purposes it is not necessary to examine the history of the term "social class" or to examine the varied meanings of the term today. We are primarily concerned with one basic question: How is a person identified as belonging to one or another social class? Bendix asked himself the same question and found this answer:

> It is awkward but unavoidable to speak of "members" of a social class. Strictly speaking, individuals merely share certain social and economic characteristics; but such individuals may also join to act in common and only then are they

[1] Arthur J. Vidich and Joseph Bensman, *Small Town and Mass Society*, Princeton, Princeton University Press, 1958, pp. 51-52*n*.

members without quotation marks. Since most references to social class, however, relate shared characteristics of individuals to their capacity and actual readiness for collective action, it is not possible to keep the two meanings verbally separate.[2]

If we identify a class in terms of its characteristics, then it is necessary to identify persons having them. The characteristics are such factors as amount of wealth, occupation, way of living other than work ways, and ways of thinking, insofar as thinking relates to social values. We find that most people accept the assumption of a lower, middle and upper class, yet the individual may not be able to identify the class to which he belongs. Useful though they may be, statistical efforts to arrange people into class categories are not always helpful. Cole reminds us that even if a person identifies himself with a class, that may not be the class to which he really belongs.[3]

The idea of social class as an approximation, however, has its uses. Whereas the people of a community may be divided into hundreds of groups, it is a convenience to arrange these groups into a classification of three or four levels, by whatever name these levels are called. In saying this, we cannot ignore the importance of history. The ideas people have had about social rank, and by which they have lived for generations, may have much to do with their social stratification today, and with their ideas about social stratification as it affects them.

## *Edelstand* and Caste Systems

In the earlier decades of the United States there was frequently a cynical unconcern about Old World class concepts, perhaps naturally so since most of those who migrated to the new country came from the lowest and most oppressed classes. The European gentleman who could supervise, mix in politics, fight duels for his honor, and who knew the social niceties, although he did no work, was only occasionally, as Commager notes, transplanted to American soil.[4] On the frontier, "gentleman" was often a sneer word, but in time the ways of the gentleman were consciously copied and in many respects Old World class concepts, however modified, did take root in American soil.

[2] Reinhard Bendix, *Work and Authority in Industry*, New York, John Wiley and Sons, 1956, p. 14. Bendix reviews the literature on class in countries of the West and East, especially Russia, and relates the idea of social class to industrial development.

[3] G. D. H. Cole, *Studies in Class Structure*, London, Routledge and Kegan Paul, 1955, pp. 3-4.

[4] Henry S. Commager, *The American Mind*, New Haven, Yale University Press, 1950, p. 15.

On European soil upper class culture was not uprooted; rather it was diffused and tended to seep down. It was the nobles who first began using the knife and fork, and it was they who for years had a monopoly on opera going. In the German-language countries these were the *Edelstand*—the noble class, the aristocracy, the "high and well born"—who, according to Reigrotzki, comprised two traditional although interrelated types of authority: the ecclesiastical and the civil-military, or the sacred and the profane. *Stand* was a hierarchy of several levels, usually with the king at the apex. It was, moreover, a closed order into which one gained admission by birth, although occasionally by adoption. One did not leave his class; there was no place to go. He might be expelled, but that was unlikely. This anointed segment of society was so accepted by all the rest who stood without class.[5]

The citizen of the noble class would consider himself dishonored if his name were associated with work. He had property by inheritance and he might get more by marriage, or he might win it by the sword or by gambling. He might collect tribute from merchants, but he did not engage in business. In time this class, especially in England, began admitting to membership certain merchants and industrialists, a dilution process which became general.

The caste system represents another type of social crystallization which took place through long periods of time, and it suggests the extent to which a group may be assigned to a specific hereditary place on the basis of the type of labor it performs. India is considered to be the principal caste-system country, but some type of caste classification of people may be found in other regions. Such classification is a natural development in a relatively static social order. Said Warner, "Caste belongs to an order of human relations which places people in superordinate and subordinate relations," and he added that the elements of the caste-making processes are found even in America.[6]

However, as we have said, the caste system tends to be a crystallization of a division of labor, a division of labor based on hand work, and such a crystallization could take place only in small places where work ways are traditional. Lewis found the following castes in a North Indian village of 150 families. The figures indicate the number of families in each caste: Jat, 78; Brahman, 15; Camar (leather workers), 20; Bhangi (sweeper), 10; Kumhar (potter), 7; Jhinvar (water carrier), 5; Khati (carpenter), 4; Dhobi (washerman), 4;

[5] Eric Reigrotzki, "Sozialstruktur," *Handwörterbuch der Sozialwissenschaften*, Stuttgart, Fischer, 1955, 10th Lieferung, p. 579.

[6] W. Lloyd Warner, "The Significance of Caste and Class in a Democracy," *The Social Welfare Forum, 1953*, New York, Columbia University Press, 1953, p. 292.

Nai (barber), 3; Chipi (tailor, calico printer), 2; Lohar (blacksmith), 1; Baniya (merchant), 1. The Brahmans are the ruling families generally, and by occupation they are officials or they may follow the upper professions such as medicine or teaching. In this village, according to Lewis, the Jats own most of the land and are tillers and they, more than the others, are on the money economy.[7]

In such a social system where one is born to his caste and his work, the introduction of urban influence may be expected to have disturbing effects. While the caste concept does follow rural migrants to the city, its influence there declines, as was much the case with the preindustrial class system in the countries of Europe.

## Social Class and Citizenship

Although these systems of fixed social classes did exist in the ancient cities, and even in cities of the Middle Ages, apparently they do not fit as comfortably as in the rural places. The fixed class systems became less secure as cities became more commercial and as more of the leading families depended for their income on trade and industry. The old systems were more directly oriented to land owning and to an exploitation of agriculture. This does not mean that the idea of a social class system was set aside, only that a type of social class system (rural-oriented) became transposed into another type (urban-oriented).

With growing industrial development, with people moving from one income level to another, and with changes in work that diminished the importance of old occupational groups, even the most exclusive hereditary social class was not able in the city to hold its lines. The process of diluting the old social class systems rendered the lines between classes more and more indistinct. However, as Marshall reminds us, the idea of social class, and its acceptance, is still present:

> There are many principles by which a society may be divided into groups. There are several on which it may be divided into layers. But social class is a single principle which can only produce one result. It permeates the whole community, so that its application yields a single scheme of location which, in theory, assigns a place to every component part of the whole. It may appear as though clearly defined class groups exist only at the top and the bottom of

[7] Oscar Lewis, "Aspects of Land Tenure and Economics in a North Indian Village," *Economic Development and Cultural Change*, Vol. IV, No. 3, April 1956, pp. 280-282.

> society, and not in the middle. But we know that class enters into the life of every member of the community, because our society is possessed of a class system. It recognizes this form of differentiation as a force affecting social behavior and social opportunity.[8]

In addition to the influence of social and economic change on the firmness of a social class system, another influence which did not exist in earlier cities must be mentioned. This we may designate as citizenship. Social class operates on an assumption of inequality, and by it one attains social status. Citizenship, on the other hand, operates on the assumption of equality, and by it one attains civic or political status. What citizenship stands for was once the monopoly of the nobility, and it was later shared by the "barons" and "princes" of industry and commerce who came to be adopted into the dominant social class. But they were unable to hold this monopoly against the demands of the masses for political equality.

Social class status is somewhat, although not wholly, private in its exercise. Civic status is somewhat more public in character, and its exercise disregards class lines. Within the individual each status is related to and influenced by the other. Civic status cuts across all class groupings: A lower class man's vote counts no less than the vote of an upper class man, although the latter, because of his prestige, may have more influence in civic affairs. While we must admit that the ideal of civic equality is not always realized, so long as the ideal is recognized in the law as a right, the attempt to achieve it can go on. The idea of citizenship—as we are using the term here—as a right is new; more than that, it is only now in process of being recognized and firmly grounded. As this process moves ahead the old class systems and their authoritative position in community life are weakened and replaced by other concepts of social class.

The evolution toward this citizenship, largely urban-centered, even in the more advanced countries, was a step-by-step transition. The process may move faster in some communities and regions than others. For example, where race is a basis for class distinction the dominant racial groups are sometimes able to deny full civic status to the subordinate racial groups. As rural Americans once argued that the foreign-born in the cities were not ready for responsible citizenship, many urban as well as rural people argue that Negroes are not ready for this citizenship, or that they might use its rights to disadvantage white people. Similar attitudes are reported from many sources regarding

[8] T. H. Marshall, *Citizenship and Social Class,* London, Cambridge University Press, 1950, p. 94.

the dominant white colonial peoples now establishing industrial urbanism in Africa.

Southall, for example, notes that "The population in nearly all of the larger African towns are stratified into layers which are very largely determined by race. This gives the urban community many of the characteristics of a caste structure." He notes further that there are firm class divisions of traditional nature within the African population.[9] Balandier concludes, however, that the dominance of the white minorities tends to unite the African groups in opposition to the whites.[10]

## Acceptance of Class Systems

An engineer was once visiting in the southern part of the United States to locate a site where a factory could be established. Most of the little towns in the county under consideration were friendly to the idea. The factory would make business. He was asked by an old Negro if the factory would bring more white people into the county. When the engineer answered that it would, the Negro made this observation. "Ma 'pinion, suh, we's got 'bout all th' white people in dis county we c'n care fo'."

This unwitting acceptance of a class system is not unique. This Negro had lived his life in relation to a particular conception of a class system, and he was adjusted to it. If the work and life of a community change, there may be disagreement about what the class system is, or should be. However, even in the course of social change, some conception of social rank will be accepted by most people. This is what Hollingshead found in Elmtown after talking with people of different social and economic groups:

1. They asserted the existence of classes and gave them names.
2. They assigned different prestige to the several strata.
3. They identified particular persons as members of specific classes.
4. They mentioned some criteria for placing persons in each class.
5. They thought of themselves as members of classes.
6. Finally, and this is important from our viewpoint, they associated behavior with classes.[11]

[9] A. W. Southall, "Determinants of the Social Structure of African Urban Populations, With Special Reference to Kampala, Uganda," in *Social Implications of Industrialization and Urbanization in Africa South of the Sahara,* Paris, UNESCO, 1956, p. 557.

[10] Georges Balandier, "Urbanism in West and Central Africa, The Scope and Aims of Research," *Social Implications of Industrialization and Urbanization in Africa South of the Sahara,* Paris, UNESCO, 1956, p. 501.

[11] August B. Hollingshead, *Elmtown's Youth,* New York, John Wiley and Sons, 1949, p. 74.

It seems that in Elmtown a favorite preoccupation of people in the course of conversation is identifying other people with classes. It is probably a preoccupation of people in all societies, and such conversations can hardly go on without people also identifying themselves with one class or another on the basis of whatever the criteria used. However, the criteria are apt to differ from one person to another, even though the individuals may not consciously think in terms of a systematic set of criteria. These criteria may be different for old people and young people, different again for rich people and poor people, again different as between communities, and different from time to time.

Mayntz made a study of social stratification in a small industrial city in Germany (about 18,000 population). A thousand adults were interviewed, and supplementary information was secured about each. The study was also concerned with social mobility. The persons interviewed were asked about the existence of social class, and there was no disagreement as to its existence, but there were different views about the number of classes. Only about two-thirds of the respondents were willing to name the number of classes. Of the 692 who had opinions, 4 percent divided society into two classes, 46 percent into three classes, 16 percent into four classes, 15 percent into five classes, 9 percent into six classes, while the rest named more than six.[12]

When these respondents were asked to give names to the classes, the answers seemed to depend on the social and economic status of the speaker, and sometimes on his political views. For example, there was no question about there being an upper class, but there was great difference of opinion regarding who belonged to it. The difference of opinion did not concern the factory owner, but the high public official, the professor, the physician, and the businessman. Many were disposed to place doctors, professors, public officials, and important businessmen (storekeepers) in a level under the upper class. There was much more disagreement about who belonged to the middle class (*Mittelstand*), and still more disagreement about the social levels under the middle class.[13]

Only 637 of the thousand persons interviewed were willing to give opinions regarding the class to which they identified themselves. Here we find many contradictions. For example, 34 percent of the 637 included themselves in the middle class, including five of the ten larger employers who would be regarded by others as upper class. Moreover, 42 percent of the white-collar workers

[12] Renata Mayntz, *Soziale Schichtung und sozialer Wandel in einer Industriegeneinde*, Stuttgart, Ferdinand Enke, 1958, p. 86.

[13] *Ibid.*, p. 93.

and small officials identified themselves with the middle class, although in most of the opinions expressed by others they were rated as under or lower middle class or in a level below the middle class. Even 18 percent of the unskilled workers put themselves in the middle class. Of the higher public officials, 20 in all, only three put themselves in the middle class, whereas they were rated middle class by most of the respondents. In brief, there was little agreement between, on the one hand, the ideas respondents expressed regarding names of classes and the kinds of people in classes and, on the other, their ideas about their own places in the class system.[14]

Respondents were in greater agreement when given the titles of 18 occupations and asked to rate these according to their conceptions of class. The occupations, familiar to all respondents in the city studied, were rated in rank order, from highest to lowest, as follows:

1. Factory owner
2. Bank director
3. School official
4. Physician
5. Pastor (priest)
6. City inspector
7. Teacher
8. Master baker
9. Storekeeper
10. Bookkeeper
11. Locomotive engineer
12. Shop foreman
13. Postal clerk
14. Salesman
15. Textile worker
16. Factory mechanic
17. Railroad laborer
18. Laborer helper

There was not full agreement; for example, about 55 percent of the respondents rated the physician upper class, about 42 percent rated him middle class, and about 3 percent put him under the middle class. While 75 percent rated the railroad laborer lower class, 25 percent put him in the middle class. The bank director was rated upper class by 80 percent and middle class by 20 percent.[15]

These four steps taken by Mayntz—naming the social classes, identifying the kinds of people in the classes, self-identification with a class, and rating occupations according to class—support the conclusion that people think of society in terms of a hierarchy of classes. While ideas differ, there is a strong tendency to associate people in classes in terms of the work they do, but this association also assumes progressively higher education and higher incomes from the lowest to the highest of the groups included in the classes.

Warner and Lunt found in "Yankee City" that people not only think

[14] *Ibid.*, p. 103.
[15] *Ibid.*, p. 119.

I

about social class, but much of their social behavior is oriented to gaining or holding status in one class or another. They experimented with criteria by which people might be placed in classes. While recognizing that the idea of a class system is amorphous, and that people are continually moving into and out of classes, they were able to settle on a six-level class pyramid, arrived at by dividing each conventional class into two groups. Thus emerged: Upper-upper, Lower-upper, Upper-middle, Lower-middle, Upper-lower, and Lower-lower.[16]

Cole summarized a number of studies of the class structure in Great Britain. While not rejecting other criteria that might be used for identifying people with classes (groups they belong to, where they live, what they own), he concluded that for large-scale practical uses occupation is the most dependable criterion, but that it, too, is a faulty one. "Nevertheless, if a single criterion is to be used, occupation is likely to be better than any other, where the purpose is to arrive at a rough estimate of the numbers of persons to be included in the various social classes."[17] He might have added that it is perhaps the measure most often used by people in their thinking about social class.

## The Classes and the Groups

Social classes are concepts in the minds of people. Amory, in his *The Proper Bostonians,* described a particular social class in New England, most of whose members are listed among the 8,000 names in the social register of Boston.[18] Although Boston is a strongly Catholic city, almost no Catholics are in the social register. If we examined this book in terms of the occupations of the registrants we would find them distributed among different classes, but they would be homogeneous if compared on the basis of "best-old-family" criteria, and in that sense might be called a social class. They would be admitted to membership in certain organized groups which would not admit persons not in the social register. But social registrants will also be members of organized groups that may also be exclusive although admitting persons not in the social register, and such mixed groups may be identified by other Bostonians as upper class in character.

One does not belong to a class in the sense that he may be a member of

[16] W. Lloyd Warner and Paul S. Lunt, *Social Life in a Modern Community,* New Haven, Yale University Press, 1941, pp. 83-91.

[17] Cole, *op. cit.*, p. 8. Also see p. 149.

[18] Cleveland Amory, *The Proper Bostonians,* New York, E. P. Dutton, 1947, p. 12.

a social club, a political party, a trade union, a fraternity, or some cultural association. One may by his thinking and behavior *identify* himself with a social class, and he may identify other people as belonging to one social class or other, but the *belonging* is largely subjective. One belongs to a social class and is identified with that class as he associates with and is accepted by other persons who are identified with that class. He must be a member of the kind of groups, especially social groups, in which such people have membership. Warner and Lunt reported 357 associations (They do not use the term, "group.") in Yankee City. These they were able to arrange into a hierarchy of 19 types ranging from those whose members were of what they called upper-upper class to those that included only members of the lower-lower class.[19]

In the modern urban community, the individual can hardly become integrated with the work and life of the agglomerate unless he joins with others having like interests, like objectives, like capabilities. Urban community life functions largely in terms of group behavior, particularly the organized secondary groups; otherwise urban life would be chaos. While civil government with its final authority functions to insure order and provide services, order is in fact realized through the functioning of the many organized groups, as they compete or cooperate, each promoting its own interest. In the caste society where the rural tradition still prevails, one is born to his social status and his work. It is not necessary for him to join special groups, for the kinship group serves any need he may have. In the urban community, except to some degree for upper class families of wealth, that kinship structure cannot function. It tends to be replaced by various special interest secondary groups.

Some organized groups more than others may be associated with what may be called social class interests. Groups organized in relation to the arts, dancing, yachting, or social club activities may be much more preoccupied with social class striving than are businessmen's associations, trade unions, or hobby clubs. This does not mean that an association organized for practical purposes is not influenced in its selection of members by class considerations. The contrary is usually true. This is brought out by Warner and associates in their study of "Jonesville," a small city in the Middle West. They found that social class considerations enter somehow in determining both membership and leadership of practically all organized groups in that place. Some draw members from one class, some from another, and some cross class lines. Thus they become ranked by opinion in the community according to the social level of the dominant members.

[19] Warner and Lunt, *op. cit.*, pp. 307-308.

> Different types of organizations are peculiar to different social levels; certain classes tend to belong to one type of organization, others to different types. It is possible, therefore, to distinguish the pattern of associational behavior of the various classes and to show how this pattern maintains and strengthens the class structure of the community.[20]

## Social Classes and the Individual

Organized groups in which the individual may have membership are a means to an end, conveniences for achieving certain purposes. The individual may join a trade union for ideological reasons, but his ideology may be class-oriented. If his interest in trade-union membership is merely practical, he may abandon his membership should it prove to be of no advantage to him, as many workers do in periods of deep unemployment. A worker may become a foreman and still find the trade union useful to him, but if he becomes a superintendent he may find himself in an income and social class above the level of the trade union. If the individual moves from a lower income occupation to a higher one, it is not unlikely that he will try to join organizations which previously were above his social and economic level.

For an individual to rise from one social class to another may also depend on which is permitted and possible and whether what he does conforms to social expectations. Expectations, quoting Burgess and Locke, are patterns of behavior which the individual "expects from others and which he may expect from himself. Once a person incorporates these group expectations in himself he uses them in weighing and criticising his potential future behavior as well as the behavior of others."[21] In the rural community these expectations are more clearly focused than in the city, but they are no less present in the urban community; there, however, the individual is confronted with other influences. He can follow the course his father took, the logical thing to do if the individual belongs to an upper class family; or, if he comes from a lower class family, he may try charting a new course. That means trying to move into a higher social class. He learns that the social expectations of the higher class may be different from those in his own class. Perhaps he tries to meet these higher expectations by getting more education, which may be helpful.

[20] W. Lloyd Warner and Associates, *Democracy in Jonesville,* New York, Harper & Brothers, 1949, pp. 130-131.

[21] Ernest W. Burgess and Harvey J. Locke, *The Family, from Institution to Companionship,* New York, American Book, 1953, p. 244. The authors add, "The potent effect of family expectations may be seen by posing the question; "To what extent are a person's life activities predetermined before his birth by the status of his parents and by their expectations?' " *Ibid.,* p. 245.

LeTourneau reported on changes taking place in the cities of North Africa, particularly Casablanca. Until a few decades ago, the city had a patriarchal society, but this order has yielded to Western expansion. As the city has grown its class structure has changed. There is the working class and the emerging middle class of white collar workers, specialists, and intellectuals. Many with university training have entered Casablanca, but not all have met with success:

> More often they vegetate for life in mediocre jobs, lacking will or intelligence, or enmeshed in family affairs, for it is one thing to have fine ideas but another to execute them in surroundings so poorly prepared. So a great many young people, torn between their dream career and family obligations, have been forced to give up their studies in order to earn money and help provide a living for a large family, or else have married young and are exhausting themselves in bringing up their children in their own way in the face of family opposition.[22]

The individual, regardless of whether he is in a city of Africa, or elsewhere, may meet with difficulty in rising from one social and economic level to another. He must find a higher-paying position and he must sacrifice to train himself for a better position, but the young man in Casablanca is still bound by his family obligations and may be able to rise only to the upper edge of his own class. He settles with the hope that his children may fare better.

Under the most favorable conditions, such as those found in American cities, an individual within his own working years may move upward several levels through the social class structure. He learns how to take advantage of all opportunities and how to make the best use of all group affiliations. Others do well to make a one-level advance from a lower to a higher social and economic level. Still others are hedged in by restrictions which cannot be overcome in one lifetime. Hill calls attention to the condition of Negroes in a small southern town, "Eddyville," a place where the old serenity is being disturbed by the expansion of industry. The Negro has always been held in a subordinate position, but certain undercurrents of new influence are now appearing:

> From many sources Negroes have obtained a conception of themselves as citizens with inalienable rights of equality before the law, or respect due them as persons. The low position of the Negro on the totem pole of power and respect in Eddyville is becoming increasingly humiliating. Under the system still enforced in Eddyville the lowest white woman may expect the most venerable and aged Negro to yield his rights in her favor. Adulthood is achieved by

[22] Roger LeTourneau, "Social Change in the Muslim Cities of North Africa," *American Journal of Sociology*, Vol. 60, No. 5, March 1955, p. 533.

> marriage by whites, even when the person is still emotionally and economically immature, but Negroes are called "boy" and "girl" until old age. . . . No organized avenues of protest have developed through which orderly action may be taken for redress against abuses of Negroes by whites not covered by law. Even justice under the law must be obtained through power-white intermediaries rather than directly.[23]

Formerly these Negroes were not resentful of the treatment they received from the whites, but now individual Negroes feel a sense of humiliation. There is a growing disposition to move into the status of citizenship, never before permitted. This is an advance quite distinct from moving from one economic level to another, and to achieve it Negroes are increasingly joining organizations.

## Rural Life and Social Class

Urbanites are ever confronted with the need for making adjustment to the man-made environment by which they are surrounded, and no small part of that environment is the social structure. Rural people are not so situated except as more and more they become identified with the urban way of life. Those who live in the most rural of communities, who may be called the most rural of people and the least affected by man-made elements in their environment, will be preoccupied hardly at all with adjustment problems. And the farther we look back into the history of the rural way of life the less it seems to have been concerned with problems of social class. Under feudalism we see a social class system that was both firm and all-embracing. If there were problems later about the social order, they were less the result of internal than of outside changes and they necessitated rural adaptation.

Initially the social class systems in urban places were extensions of rural systems, adapted as circumstances required to the urban situation. The urban situation, always different in complementary ways to rural situations, and being the more dynamic, served as a stimulus to the rural way of life. If the urban class system changed, it was because urban society, in order to live, had to keep the initiative and had to be creative; had to continually find ways of stimulating the rural part of society to produce enough for itself and the urban part of society also. Of this, Turner says:

> The city, of course, formed as a result of the accumulation of surplus which permitted its inhabitants to follow pursuits other than producing the means of

[23] Reuben Hill and Associates, *Eddyville's Families*, Chapel Hill, University of North Carolina, Institute for Research in Social Science, 1953, pp. 414-415. Mimeographed.

their own subsistence. From the first, therefore, production in the urban cultures had two levels: (1) the primary, which, besides feeding the tillers of the soil, furnished the means of subsistence for some who did not work at agriculture, and (2) the secondary, which was carried on by those who could work at industries other than agriculture. The output of the secondary level was all surplus.[24]

If more and more people live in cities, they must ever find new ways of impelling agriculture to produce more. While it is not a conscious or systematic urban effort, still the elements of inducement and compulsion are present in all relations of the city with the country. Rural society may evidence types of creativity in adjusting to urban pressure, but it is not confronted internally with the need of changing its values or its social organization. It does change reluctantly under pressures from without.

At the beginning of this century, for example, Finland was a predominantly rural country with a closed class system which included the clergy, the land owners, and the peasants. Workers in the towns were outside the class system. Since 1906 there have been drastic political reforms with the Socialists gaining more and more power. The country has become increasingly urban and industrial. The influence for change in the direction of a welfare state has been urban, while the main resistance is the agrarian group which apparently is slowly losing ground. At issue are opposing views championed by opposing political groups regarding the social class system and the role of the individual in society. In connection with his study of social class in rural and urban communities, Sariola wrote:

> Since the value structures have become so institutionalized in the party system, it follows that to many Finns the worth of individual assets and capacities has become secondary to collective ideals. This change from individualism to collectivism is gradual. Whereas the Bourgeois holds the belief that the upward and downward change of status depends upon the performance of the individual, and the Social Democrats believe that he should be protected against the hazards of too wide discrepancies between status extremes in society by social planning consistent with individual freedom, the Communists deny the correspondence between achievement and reward and believe in a totalitarian rearrangement of rewards.[25]

What is taking place in Finland, as in other industrial countries to dif-

[24] Ralph Turner, *The Great Cultural Traditions, The Foundations of Civilization,* New York, McGraw-Hill, 1941, Vol. I, p. 270.

[25] Sakari Sariola, "Defining Social Class in Two Finnish Communities," *Transactions of the Westermarck Society,* Abo, Abo Tidnings och Tryckeri Aktiebolag, 1953, Vol. 11, p. 156.

ferent degrees, is an urban political approach to changing the social class system. Actually it aims to give collective security to urban workers corresponding to the security rural people have in their land.

## Urban Life and Social Class

Hutton characterized life in American towns of the past mid-century as "chaotic, fluid, highly mobile, intensely individualistic and competitive, immoderate, intemperate, and above all, materialistic."[26] During such a period of town building one would find only a rudimentary interest in social class, and there was little in such towns of group organization. Individualism was rampant. But after a decade or two organized life (apart from the civic organization) did begin to take form. "Top" society began to find itself, businessmen had their associations, literary clubs became organized; America was beginning to become urban even though much of it was still frontier, and while still a frontier, the embryonic beginnings of a social class system were evident.

The top society that emerged was composed mainly of families that came out of the working population, but now, through hard work, shrewdness, and good fortune, had become rich. This was a new kind of upper class. What it lacked in aristocratic tradition it made up in wealth. If it did not have the cultural background of the European upper classes, it could at least pay for the trappings of aristocracy. Veblen, who observed the performance of this new upper class in the 1890's, was especially aware of its efforts to win esteem through spending and display:

> In order to gain and to hold the esteem of men it is not sufficient merely to possess wealth and power. The wealth or power must be put in evidence, for esteem is awarded only in evidence. And not only does this evidence of wealth serve to impress one's importance on others and to keep their sense of his importance alive and alert, but it is of scarcely less use in building up and preserving one's self-complacency.[27]

Members of this class, at least the founders of it, as Veblen recognized, took pride in having worked their way up, and they talked about it more than their children did later. The idea went abroad—and was widely believed and still persists among Americans—that the road to the top is wide open, although it is a hard one to travel. It is that idea which encourages the clerk to believe

[26] Graham Hutton, *Midwest at Noon*, Chicago, University of Chicago Press, 1946, p. 224.

[27] Thorstein Veblen, *The Theory of the Leisure Class*, New York, Modern Library, 1931 (original 1899), pp. 36-37.

that he may one day be superintendent or manager, and encourages the superintendent to believe that one day he may join the board of directors. It also inspires the farm boy to migrate to the city where perhaps one day he may become rich and influential. For all this striving, the theater of action is the city, where the upper class makes its home and where all the other classes compete for advantage.

It may be argued that the urban arena where the struggle for position takes place is a scene of strain, noise, and confusion. Quite unlike the rural society, as noted above, urban society must be oriented to a man-made dynamic environment which, although it may not be a symphony of motion, does have a rhythm. This rhythm is not timed by the cycles of nature, but is paced by the ticking of the clock. Here the movement of men and machines must be coordinated. Timetables and work schedules are to the urban agglomerate what sheet music is to an orchestra. Working, walking, riding, eating, sleeping, keeping an appointment—the urban man is ever clock-conscious.

No matter what his social class, the urban man cannot go to his work an hour earlier in summer or an hour later in winter except by resort to the fiction of setting the clock forward or backward by one hour. But it is necessary for all clocks to be set one hour forward or an hour backward; the clock-regulated mechanized environment must not be thrown into confusion. If the city alternates between daylight saving "summer time" and "winter time," then the suburbs must also do so, and the rural man must fall in line. This change can be achieved only by enacting a law, and such laws are usually opposed by rural people.[28]

This type of urban environment, a rapidly expanding type, tends to develop its own kind of social class system, one that is oriented both to work and change. Old class systems of rural origin were oriented to work only for the lesser classes and these systems were not prepared for social change. Of the earlier upper classes in England, a type that would be out of place in the American scene, Tawney wrote:

> Because rewards are divorced from services, so that what is prized most is not riches obtained in return for labour, but riches, the economic origin of which, being regarded as sordid, is concealed, two results follow. The first is the crea-

[28] Jean Daric, comparing time in country and city, observed that urban life is oriented to technical time, which he designated "pure," while rural life is oriented to natural time. Work in relation to natural time brings muscular fatigue only, while work in relation to technical time also brings nervous fatigue. He suggested that differences between these two types of time may be reflected in rural and urban sickness rates. See Georges Friedmann, Ed., *Villes et Campagnes*, Paris, Librairie Armand Colin, 1953, p. 416.

> tion of a class of pensioners upon industry, who levy toll upon its product, but contribute nothing to its increase, and who are not merely tolerated, but applauded and admired and protected with assiduous care, as though the secret of prosperity resided in them. . . .
>
> The second consequence is the degradation of those who labour, but who do not by their labour command large rewards; that is of the great majority of mankind. And this degradation follows inevitably from the refusal of men to give the pursuit of industry the first place in their thoughts about it.[29]

## Functions of the Elites

We recognize that social class systems are present in all societies, that modern social class systems under the influence of urbanism are in continuous process of change, and that the urban man is identified with social class most effectively as he is affiliated with groups. If we ask, however, how the individual makes use of groups and how groups serve him, we are confronted with queries about how groups in the urban environment operate and how they are led. In this connection, the word "elite" is coming into use.

Writing about the elites in British society, Cole points out how very competent thinkers have used the word with different applications. These contradictions notwithstanding, Cole finds that the word meets a need, and he defined it thus:

> I propose to consider elites, not as constituting social classes, but rather as groups which emerge to positions of leadership and influence *at every social level*—that is to say, as leaders of classes or of other important elements in the social structure. I am here using the word "leader" in a wide sense, to designate not only *Fuehrers* or officially appointed chiefs but also all persons who, for whatever reason, exercise a substantial influence on the social attitudes and social conduct of any considerable section of the groups to which they are attached.[30]

The present popularity of the word "elite" reflects the need of a term that is not loaded, as is the term "leadership," with moralistic connotations. Sumner used the term more than half a century ago in answering his own question, Who does the thinking for the group?

> The notion that "the group thinks" deserves to be put by the side of the great freaks of philosophy which have been put forth from age to age. Only the elite of any society, in any age, think, and the world's thinking is carried on by

[29] R. H. Tawney, *The Acquisitive Society,* New York, Harcourt, Brace, 1920 (renewed 1948 by R. H. Tawney), p. 37.

[30] Cole, *op. cit.,* pp 105-106.

them by the transplanting of ideas from mind to mind, under the stress and strain of clashing argument and tugging debate. If the group thinks, then thought costs nothing, but in truth thought costs beyond everything else, for thousands search and talk while only one finds; when he finds something, a step is won and all begins over again.[31]

The elite are the leaders in groups; small informal groups of a few members or organized secondary groups that may include thousands of members. These groups, as already noted, are found at all levels of the social class system, and some may cut across class lines. For each group, the elite is the inner circle of key persons, the active members of the group who think for the group and often speak for the group. As groups differ in prestige, depending on what they do, what they stand for, how much wealth they collectively represent, and how many votes they can control, so their elites differ in prestige.

The elite of a group tends to have prestige within the group as it is able to gain prestige outside the group, a prestige that is vicariously shared by all group members. Understandably, within the group there may be sharp competition between persons trying to gain prestige and to have a place in the elite circle. Thus in any group there may be some who rise to higher position while others fall back, but these contests are regarded as in-the-family matters. There are many ways by which leaders attain elite status and many ways by which they hold it, or lose it. Hence there is continuous change in the elite groups.

We need to distinguish between an elite and those individuals who are identified as leaders. The head of an organization may be the member of an elite, but that elite will include persons who are not leaders in the sense that they are heads of organized groups. Speaking for the moment of leaders, Stauffer found in a nation-wide study of tolerance that leaders tend to be more tolerant on political, moral, and other issues than the cross-section of the population.[32] This finding may be regarded as encouraging for social stability. It is also encouraging for the city, where the many groups of rival interests tend to form clusters and, hence, elites. Among these groups are included rival political groups, religious, economic and other organizations, business, industry, the labor movement, cultural and welfare groups, and the elite cluster standing for high society.

[31] William G. Sumner, *Folkways*, Boston, Ginn, 1940, p. 206.

[32] Samuel A. Stauffer, *Communism, Conformity and Civil Liberties*, New York, Doubleday, 1955. See especially Chapt. 2.

In the American urban community are certain neutral spots where the different elites come together. One may be organized by key citizens who personally may be members of different elite groups, persons of cosmopolitan interests and recognized integrity. It may be called "The City Club" or "The Civic Club," and it may have a regular meeting place with kitchen, dining rooms, meeting rooms, library, and even a professional staff. It may be formally organized and may have various standing committees for different local problems. But in the great city, as Mills observed, there are other neutral places of meeting—the high-class cafés and night clubs, for example.

> In café society, the major inhabitants of the world of celebrity—the institutional elite, the metropolitan socialite, the professional entertainer—mingle, publicly cashing in on one another's claim for prestige. It is upon café society that all the spotlights of publicity coincide, spreading the glamor found there to wider publics. . . . Café society is above all founded on publicity.[33]

Obviously, not all elite clusters want to be seen in café society, and some elites do not want to be seen with other types of elite in public. But whether the different elites meet on neutral ground in the night spots, in "The City Club," or elsewhere, they are the core of community life. In them all levels of social and economic status are represented, and through them all issues are brought out into the open. In their relationships, sometimes competitive, sometimes cooperative, the complex organization of community life is reflected.

## Primary and Secondary Groups

No such hierarchy of elites is found in the rural community, nor is it needed. Individuals have no such need for organizing themselves into the many types of formal groups, since rural life is basically informal and primary. In this respect, however, rural communities differ widely, depending on the extent to which the rural people, although still agricultural, have developed urban-oriented secondary organizations—for example, cooperative associations such as political parties and civic bodies.

Stirling studied the social structures of Turkish villages and the relation of these local primary groups to the civic structure of the central government. The entire country is divided into administrative districts called *vilayets,* each of which is divided into lesser districts called *nahiye,* each of which, in turn, includes 15 to 20 villages. The public representative in each village is the

[33] C. Wright Mills, *The Power Elite,* New York, Oxford University Press, 1956, pp. 72-73.

*muhtar,* who as an official of government works with the Council of Elders. He is the civil officer, while the Elders are the traditional, kinship-linked folk government. The Council of Elders is the apex of a primary group class system, while the *muhtar* is a minor local representative of a wide-reaching secondary organization. He is a trained man, and a symbol of the law. As such, he names peoples for public jobs, receives visitors, and imposes decrees in the name of "they," the government. He is supposed to be elected for this post, but he is actually named by the Council of Elders.[34] In some respects, village life could go on without the *muhtar,* but he is turned to when new problems arise.

However, although he stands high in authority, as authority is related to secondary group organization, his prestige is low in the local primary hierarchy of lineage groups. The Council of Elders includes the recognized key personalities from these kinship groups. Each lineage group has its *oda* or meeting place, usually in the home of an important kinsman. Here the men meet, smoke, and talk. The older members have their seats nearer the hearth; others have their seats in an accepted prestige order. There are different scales by which one gains prestige and authority: piety and religious learning, moral reliability and helpfulness, economic position, occupation, age, and family position. One gains his status largely in terms of individual character and effort. Stirling noted little evidence of hereditary stratification. Most honored is the man who made the journey to Mecca.[35]

Such primary group village government in Turkey may be expected with time and technological change to lose importance as the village comes increasingly into contact with the expanding industrial urban way of life and the importance of the *muhtar* concomitantly increases.[36] These problems have already been resolved by small communities in the West. Thus American

[34] Paul Stirling, "Social Ranking in a Turkish Village," *The British Journal of Sociology,* Vol. IV, 1953, pp.41-42.

[35] *Ibid.,* pp. 35-37.

[36] See the reference to the *cuaird,* the council of old men in the Irish village, Chapter 4, Footnote 13. This compares somewhat to the *muhtar* in the less advanced Turkish village. Such a traditional primary group need not be present in order to insure a sort of primary group control and guidance. The control structure may persist even in the face of change, as it seems to in Guatemala, a society of two social classes, the upper level Ladino and the lower level Indians. Each group has its own control system but the two are linked in an interrelated control network of dominance and subordination. Now, according to Tumin, the system does not work so well because the Indians, especially the younger ones, are coming under town influence. This tends to undermine the dominance of the Ladino. Melvin M. Tumin, "The Dynamics of Cultural Discontinuity in a Peasant Society," *Social Forces,* Vol. 29, December 1950, p. 140.

farmers may join an organization of farmers: the Grange, the Farmers Union, or some other group of the secondary type. The farmer's wife may join the Red Cross or the Women's Christian Temperance Union or the Daughters of the American Revolution, and be as much at home in these organizations as the wife of a suburbanite. The local storekeepers and the lawyer may form a local branch of a secondary organization such as Rotary International or the Lions Club. Young men in the village may have a branch of the American Legion. The young girls may have a Girl Scout group and the boys a Boy Scout group. Such secondary organizations have branches in thousands of places.

While the primary group type of government may be fairly nonexistent in American communities, the primary group is very much present; and, although it may be most at home in the village, the primary group belongs to the city as well. Its function is social control by personal contact:

> Seen positively, the primary group is a structure of mutual affection that generates spontaneous informal acts. We value the good opinion that others have of us. Our behavior seeks to gain and maintain this response. Thus, to its members, the primary group becomes a value in itself, and not merely a means to some other goal. If a friendship is made for a purpose, say for political advantage, or to close a business deal, then it probably would not have those characteristics which define a primary group relationship in its purest form. A primary group relationship is not contractual or formal. It is personal, spontaneous, sentimental and inclusive. Human society relies heavily on such relations for its maintenance.[37]

Although membership in the family, the gang, or the clique may entail obligations which may be extremely exacting, it must be emphasized that the primary group is not contractual. In the rural community these obligations are more enforceable by group opinion than in the urban environment. Precisely because the primary group obligations do lose force in the urban environment, contractual obligations tend to replace them. We need to recognize that primary groups differ widely. The traditional village may be called a primary group: everyone knows everyone else (and something about him); but within this village group is a pyramid of groups of higher or lower importance, although the important group for each individual is his family. Within the extended family the men may be joined into one primary group and the women in another. Such primary groupings are not contractual, to be

[37] George A. Lundberg, Clarence C. Schrag, and Otto N. Larson, *Sociology*, New York, Harper & Brothers, 1954, p. 426.

sure, but relationships involve obligations. Even membership in the group, however satisfying, is not voluntary.

The primary groups, as described by Lundberg and associates above, are in large measure of the voluntary type, the family being the main exception. In the urban-oriented society circumstances may place the individual in one primary group or another, but circumstances also allow him a measure of freedom to leave a primary group as he becomes identified with others. In the course of the urban man's lifetime he may be identified with a great variety of different primary groups. To some he may remain permanently attached, but with others the attachment may be determined by a variety of circumstances. Through change in education, work, or with new experience one may outgrow his groups.[38]

### *Characteristics of Secondary Groups*

Even the most urbane individual makes his way more satisfactorily in the modern community as he is able to be in person-to-person contact with agreeable associates in primary groups. As Katz and Lazarsfeld have observed, the primary group has been "rediscovered," by which they mean that, although we have known about the primary group, we now see its relevancy for research and experimentation. It is the "anchorage" point for attitudes and values and a center for interpersonal communication. It is something that can be used as a medium of education.[39] On the common-sense level, politicians and leaders of organizations have developed their own skills in the use of primary groups, but social science research has put this knowledge into a more systematic form.

The secondary group in many respects is a cluster of primary groups, the latter forming and re-forming while the former continues performing as another kind of instrument. A person behaves in one way as a member of a secondary group, but he behaves in another way in a primary group whose

[38] Prabhu interviewed 523 persons who had migrated from Indian villages to urban life in Bombay. While all grew up in the caste system, he found that 266 now opposed caste system and 281 favored intercaste marriage. Sixty-nine percent would eat in the house of a lower caste person, if invited. Seventy-one percent would invite lower caste persons to their houses. Prabhu was surprised that so few, in answer to his questions, gave "I don't know" answers. These people had not changed their social class, but their attitudes toward caste had been changing, and they behaved accordingly. Pandhari Nath Prabhu, "Social Aspects of Urbanization on Industrial Workers migrating from Rural Villages to the City of Bombay," *The Social Implications of Industrialization and Urbanization* (Five Studies in Asia), Calcutta, Unesco Research Center, 1956, p. 54.

[39] Elihu Katz and Paul F. Lazarsfeld, *Personal Influence,* Glencoe (Illinois), Free Press, 1955, pp. 41-42, 44.

members may also be fellow members of the same secondary group. His relations to the secondary group are more utilitarian.

The doctor joins the medical association or a professor joins a sociological society for some of the same practical reasons that impel the factory worker to join a trade union or the factory owner to join the manufacturers' association. The young lawyer who joins the bar association may also contrive to be invited to membership in some exclusive social club where he can "meet the right people."

Secondary organizations are formal groups created for getting things done. They enjoy prestige as they are able through the wealth and prestige of their members or the sheer number of members to wield influence. Whatever their individual purposes, one of the characteristics they have in common is that their objectives are usually stated in vague and idealistic terms. They have still other characteristics in common, and some of these are listed below. We may say of the typical secondary organization, for example, that:

(1) It has a written constitution and written rules, also a purpose which is stated in the constitution.

(2) It has elected officers who serve for prescribed periods and have duties which are defined. It may also have standing committees or special committees for performing assigned functions.

(3) It may charge the members fees or dues which are periodically paid to meet organization expenses. It keeps financial records of income and expenditures. It may even impose fines on members for breaking accepted rules.

(4) It keeps a membership list. It also keeps a record of all meetings and of all actions taken at meetings, and into the record may go any information pertinent to the history of the organization.

(5) While such an organization may have a local character in a community, it may be affiliated with like groups in other urban places. It may, in fact, be a unit or branch of a nation-wide organization.

(6) Although a particular secondary group may be short-lived, it is more usual for such organizations to continue through time, adding new members as other members resign or die.

(7) Within its special field of interest, it may cooperate with other organizations of somewhat similar interests, or it may use its strength to compete with them. It may cooperate with any organizations in the community in coping with common problems, provided such cooperation does not jeopardize its own special interests.

The secondary group, organized under law, tends to become somewhat impersonal in its operations, acquiring a character and purpose—we may say "an existence"—apart from the individual member. Sometimes the secondary group may seem to become the special property of the central elite which wields the power (according to the constitution, of course) and speak for the organization. Certain very large organizations, national in their scope, sometimes seem to assume a sort of private-government character, and not infrequently civil action must be taken against them in order to protect the interests of the membership. This observation, however, does not apply to most of the hundreds of secondary organizations found in the modern large city.

### *Joiners of Secondary Groups*

Not all people in the urban community who might do so become members of secondary organizations, nor is it necessary that they should. An individual's joining depends on his status and situation. Persons who do not join trade unions may enjoy some of the advantages gained by trade union action. A storekeeper may not be disadvantaged if he does not join a businessman's organization; in the small town it would be difficult for him not to join, but it may not matter so much in the large town. A doctor, however, can hardly afford to remain outside the local medical association. Many veterans who have served in the Army, Navy, or Air Force may have no interest in joining a veterans' organization. Two or three housewives out of ten may be interested in joining organizations, while the others are too preoccupied with their home duties.

Reigrotzki, in his study of the social participation of adults in the German Federal Republic, found that 37 percent of the females and 73 percent of the adult males belonged to organized groups. But the percentages for both sexes were somewhat higher for the middle-age groups than for persons under 25 or over 60 years. For males with more than high school education, 78 percent held memberships; the percentages were lower for those with less education. Of public officials, 90 percent had group memberships, compared with 80 percent for farmers, 76 percent for the self-employed, 75 percent for hand workers, and 70 percent for white-collar workers. (These figures are for males only.)

The size of the community in which the individuals lived made little difference in the percentages who were members of organizations. For both males and females the percentage in organizations was slightly higher for Catholics than for non-Catholics. Following are the several organizations and the percentage of males and females holding membership:

| | *Percentage males* | *Percentage females* |
|---|---|---|
| Trade unions | 23 | 3 |
| Sport and athletic groups | 20 | 4 |
| Professional organizations | 14 | 4 |
| Singing and music groups | 13 | 3 |
| Fire brigades, hunters | 9 | 0 |
| Church and religious groups | 8 | 6 |
| Social associations | 7 | 2 |
| Women's groups | 0 | 8 |
| Refugee organizations | 5 | 4 |
| All other | 12 | 6 |

It should be mentioned that church membership is not included, nor is membership in political parties. However, memberships in political parties were strikingly low, 9 percent for males and 4 percent for females, although most of the Germans interviewed identified themselves with one or another political party ideology.[40]

Dotson made a study of participation in voluntary associations in Guadalajara, a city of 378,000 in Mexico. In interviewing 230 males and 185 females, he found that only 42 percent of the males and 32 percent of the females had memberships in organized groups. Of those who had memberships, 16 percent of the males and 50 percent of the females were in religious organizations. Twenty-eight percent of the males and 15 percent of the females were members of sport or athletic clubs, while 24 percent of the males and 13 percent of the females were in trade unions. A direct relationship was found between memberships and income: There were 24 memberships per 100 persons with less than 250 pesos per month, and the number increased to 163 memberships per 100 persons with monthly incomes of 3,000 pesos or more.[41]

Komarovsky interviewed a sample of New York adults (2,223) and found that only 32 percent of the males and 9 percent of the females were members of voluntary organizations. She found that for both sexes the percentage of persons in organizations was greater for skilled than for white-collar workers and the percentage increased progressively by level of income. In the professional groups the percentage of memberships for women was equal to

40 Eric Reigrotzki, *Soziale Verflechtungen in der Bundesrepublik*, Tübingen, J. C. B. Mohr (Paul Siebeck), 1956, pp. 59, 164-172. The sample included 4,000 cases with 3,476 completed interviews.

41 Floyd Dotson, "A Note on Participation in Voluntary Associations in a Mexican City," *American Sociological Review*, Vol. 18, No. 4, August 1953, pp. 383-385. On Dotson's American study, see "Patterns of Voluntary Association among Urban Working-class Families," *American Sociological Review*, Vol. 16, No. 5, October 1951, pp. 687-693.

or higher than for men.[42] In "Jonesville" Warner found that 50 percent of the upper-lower classes had such memberships compared with 55 percent for lower-middle class persons and 60 percent for the upper-middle class persons, while almost all members of the two upper classes were members of voluntary associations, most having multiple memberships.[43]

Cauter and Downham in their study of Derby, England (under 200,000 population), found that voluntary association, or "club," memberships were confined to 47 percent of the adult population, 62 percent of the males and 33 percent of the females. Trade unions and professional organizations were not included. Club memberships were highest (51 percent) for the 16-24 age group and progressively lower for the older groups. The percentage with memberships was lowest (42 percent) among persons of elementary school level, and it was highest among those above the secondary school level (66 percent). With respect to social stratification, excluding the upper class because of the smallness of numbers, they found that 58 percent of the respondents were middle-class holders of club memberships, while club memberships were held by only 42 percent of the working class.[44]

In the Derby survey respondents were asked to indicate the social classes with which they identified themselves. On the basis of this judgment and other criteria, the interviewer placed the respondent in the class to which he seemed to belong and, in terms of these "subjective" classes, the respondents were arranged percentagewise for club memberships as follows:[45]

| | Percentage | | |
|---|---|---|---|
| | *Upper-middle* | *Lower-middle* | *Working* |
| No memberships | 32 | 43 | 60 |
| One membership | 27 | 32 | 27 |
| Two memberships | 16 | 13 | 9 |
| Three or more memberships | 25 | 12 | 4 |

Cauter and Downham are careful to add that such a grouping is only suggestive and needs to be accepted with a degree of caution.

## Group Memberships and Social Participation

Wright and Hyman undertook an evaluation of several national and metropolitan studies of memberships in voluntary associations. In terms of

[42] Mirra Komarovsky, "The Voluntary Association of the Urban Dwellers," *American Sociological Review*, Vol. 11, No. 6, December 1946, p. 688.

[43] Warner and Associates, *op. cit.*, pp. 134, 141.

[44] T. Cauter and J. S. Downham, *The Communication of Ideas*, London, Reader's Digest Association, 1954, pp. 64-66.

[45] *Ibid.*, p. 270.

percentages they found that somewhat less than half of all adult Americans are members of such associations, and of those who are "joiners" only a small part have memberships in two or more associations. A good share of the multiple memberships are held by persons of higher social and economic status. White people are joiners more than Negroes. Jews are joiners more than Protestants and Protestants more than Catholics. Memberships are more characteristic of urban and nonfarm people than of rural farm people.[46]

The sort of people who join two or more voluntary associations tend also to be those who are interested in local and national affairs as they are likely to be active in their interests. Since such multiple joiners tend also to be persons of higher social and economic status, it would appear that multiple joining is a characteristic (if not a necessity) of higher social class people. The higher status people, because they are in better circumstances, not only need to join these voluntary associations, but it is assumed that they will or expected that they will. For them to join these organizations may burden them with more responsibility, but they also derive more power and influence from such memberships.

Some who have considered the failure of a large part, especially of the urban population, to join secondary voluntary associations have come to the conclusion that this lack of participation in organizational life bodes ill for democracy in the city. Actually, it is possible for people to participate fully as citizens in community life without being members of voluntary organizations. Moreover, one may be well adjusted to the urban way of life; he can be a most cosmopolitan person and yet he may hold membership in no voluntary association. Such nonjoiners, however, may be members of various informal groups—social, cultural, sport, hobby, and so on. Some of these primary groups may be short-lived while others may be fairly enduring, but the individual who is not a joiner of secondary groups may be able to live a full and congenial life in contact with these less formal and more intimate face-to-face groups.

## Summary

It is a safe assumption to think of social classes as having always been present in human society, although they have differed with times and places, as they still differ from one society to another. The presence of social class

46 Charles R. Wright and Herbert H. Hyman, "Voluntary Association Memberships of American Adults: Evidence from National Sample Surveys," *American Sociological Review*, Vol. 23, No. 3, June 1958, pp. 284-294.

systems in all societies may be good evidence of their utility in maintaining social order. Class systems in preindustrial societies were more rigid than those in our industrial urban society.

Most people in our society recognize the existence of classes and are able to name specific classes, although their ideas of what class is differ widely. Ideas about class are sometimes colored by views concerning what social class should be. People agree generally about the existence of an upper class and a lower class, but views about the in-between classes are divergent. People are also uncertain about identifying themselves with one or another social class.

On the American frontier the idea of social classes was repugnant, perhaps reflective of the humble origin of most Old World immigrants. The frontier in the United States attracted few persons from the European upper classes. As stable community life was established on the frontier, a type of class system developed with it. This class system, with a character of its own, was based largely on wealth and achievement through work. Later the American social class system to some extent imitated European systems in matters of culture and manners, although it retained its association with wealth and achievement through work.

For practical purposes, in order to determine a population with a class system, certain identifying criteria must be accepted, such as education, income, or occupation. Of these, occupation is most widely accepted, and to a large extent it embraces such criteria as education and income. By such a classification it is possible to rank the population of a city or nation into several categories ranging from common laborers up to managers and proprietors.

Any class system, especially at the upper levels, tends to be exclusive, but in the industrial communities the lines between classes fail to prevent people from entering upper classes from lower levels as they advance to higher occupations and incomes. As technological change and expanding economies open new possibilities, more people are enabled to become socially mobile, and class systems must continually be redefined. Also, particularly in cities, as more people enjoy civil rights through the widening of the base of citizenship, new concepts of equality develop, and these tend to neutralize the lack of equality in class systems.

The mobile urban environment necessitates continuous adaptation on the part of the individual, even though he remains in the same social class and retains membership in the same social, economic, and other groups. But the urban man is stimulated by the situation to achieve greater prestige and gain higher status. This is less characteristic of the rural environment. Urban society must be dynamic, since it can live only as it is able to stimulate greater productivity on the part of rural society. The efforts of urban man to advance himself are reflected in the changing nature of urban class systems.

The urban individual needs to be associated with groups of various kinds. These groups as organizations with collective purposes tend to be identified by their leading personalities, their elite. The elite become the symbols of the groups they lead and speak for. They are the inner circle of dominant and sometimes dominating persons in the organization. A group gets status in the community largely on the basis of the influence and effectiveness of its elite. The elite, in turn, as a group draws its strength from the character of the group it represents—the strength in unity and numbers or in unity and wealth. As there are types and social levels of groups, so there are types and social levels of elites. Taken together, the elites form a core group in community life.

For less advanced rural communities, life and work controls are in the hands of primary groups, perhaps a council of elders representing the kinship groups. Each kinship group may include lesser primary groups, forming a class pyramid. A system of primary group organization is present in the urban community, although the groups are more spontaneous and voluntary. They afford the urban man a type of emotional security, and through their informal devices of praise or censure he tends to be kept in discipline. However, his wider, more practical and purposeful interests are served through membership in formal or secondardy groups. In his efforts to advance himself, the urban man may utilize both his primary group contacts and his secondary group memberships; the two tend to be interrelated. If he is able to move upward into a higher social class or is forced back to a lower one, his primary group contacts and his secondary group memberships may change accordingly.

For different reasons, many individuals in urban society, while making and retaining primary group contacts, do not concern themselves with attaining membership in secondary groups. In larger cities such nonparticipation tends to be greater, especially for low-income groups, than in smaller cities. However, as far as private organizations are concerned, these nonparticipating individuals may participate fully in their civil status by voting, paying taxes, giving moral support to the law, and so on.

CHAPTER 11

# Urbanism and the Family

IN EVERY NONINDUSTRIAL and nonurban society the family holds a dominant position in community life and is securely entrenched in social tradition and authority. Without exception in every country, the expanding influence of industrial urbanism has been disturbing to the traditional status of the family. In the midst of technical and economic change there has been social change, and this seems inevitable; but when changes take place in the family they are sometimes difficult to understand and equally difficult to accept. "The family as a social institution is changing," wrote Ogburn, "as are other institutions. These changes differ somewhat in countries according to their degree of industrialization, of their urbanization and of their isolation."[1] Such change is in process, and seems to be gaining momentum.

The idea seems to be current that if the family changes, it can only be for the worse, and that such change endangers society. According to this view, man may change his work and his tools, his income and his level of living, the type of his house and transport, his leisure activities and his politics; but the family must go on without change.

In this chapter some of the changes affecting the urban family in different societies will be considered, as will some of the consequences of such change. We will also take account of changes affecting the family before our time, and consider some of the trends and prospects of family change.

## Background of Family Change

Describing the joint family of ancient India, Turner mentions that all authority and property were vested in the patriarch whose power over all mem-

[1] William F. Ogburn, "Why the Family is Changing," see Nels Anderson, Editor, *Recherches sur la Famille*, Tübingen, J. C. B. Mohr (Paul Siebeck), 1956, p. 1.

bers of his immediate family and their offspring was absolute. The individual then and for generations since existed only "as a member of a legal and religious group, never alone."[2] Of the Roman father, Wallbank and Taylor observed, "There could be no disobedience to his will . . . ," and further:

> What education there was, was centered in the home. The father imparted what amounted to the three R's and, more important, instilled in his children those virtues every Roman was supposed to possess, manliness and self-control. . . . The discipline that molded the young Romans was perhaps too severe by our standards, but it produced citizens who were loyal, serious, dutiful and courageous.[3]

In terms of authority, the Roman family was a miniature government headed by the patriarch father. In terms of size, it included a mixture of relatives plus slaves, a small community. This is a pattern that can be found in many cultures today. That type of family was antecedent in Europe to the family of feudal times, particularly the upper class family. Feudalism in local regions was a network of families of the upper level with its institutionalized forms and values, living in juxtaposition with the mass of people whose family life and structure was much less institutionalized. In the course of centuries, the upper class feudal family came to be the fairly generalized model in the West.

A great amount of change in the family must have transpired between Roman times and feudalism, particularly in the Christian countries of the West. Feudalism placed women on a plane to be adored and served, serenaded for their charm and beauty, and elevated to a position of ornamental idleness.

Feudalism was more than merely a family system. In the words of Burns, it was "a system of overlordship and vassalage,"[4] a way of governing and holding property. It was a way of managing the labor force, an economic system, and an organization for mutual defense. But it was organized in relation to a family system: The noble with his family and servants lived in the castle, while the people working on his lands had their community in the shadow of the castle. The vassals on his estate were a family responsibility of the noble, in that he was usually accountable for them to the king.

The idealized family of the West with its basis in romantic love, drawing

[2] Ralph Turner, *The Great Cultural Traditions, The Foundations of Civilization,* New York, McGraw-Hill, 1941, Vol. I, p. 384.

[3] T. W. Wallbank and A. M. Taylor, *Civilization, Past and Present,* Chicago, Scott, Foresman, 1950, Vol. I, p. 158.

[4] Edward McNall Burns, *Western Civilizations,* New York, W. W. Norton, 1949, p. 252.

its inspiration largely from legends about knights and ladies, began in a very rural setting where the housewife in the castle could be a lady because the noble's family could exploit the labor of a goodly supply of peasants. This ideal, according to Hansson, was not taken over by the nobles of Sweden until about the seventeenth century.

> At this time bourgeois women were still busy in their kitchens and could well take over their husband's duties when they went away to distant markets, sea ports and fishing places. The peasant woman of the 17th century must have stood at the side of her husband as a driver of the plough team and as a hard-working farm hand in all busy and important situations.[5]

Later the feudal upper-level family ideal was imitated in Sweden by the upper classes just as it came to be imitated still later by the middle classes. But changes of that kind were very different from those which have taken place in recent times because of industrial urban influences. Through these new influences the family is deprived of traditional functions, and changes in family behavior and structure result.

## Survivals of the Feudal Family

The aristocratic type of famliy has survived in two very different forms. In some European countries the titled family is still conspicuous as an archaic remnant of feudal times. The British aristocracy has been diluted by frequent invasions of families from industry or commerce, families without long and honored histories, but the old tradition of detachment from business and useful labor lives on; or service is performed, in the words of Soule, with an attitude of *noblesse oblige:*

> The nation is full of great factories and banks, just as a house may have its kitchen and its laundry, but they are accepted as a matter of course rather than glorified as the justification of national existence. Wealth exists for use or for power. The process of making it is still felt by many to be a bit untidy, to be kept out of sight when guests are in the house, certainly not to be displayed and boasted of.[6]

Russell, however, reminds us that this leisure class has contributed "nearly the whole of what we call civilization. It cultivated the arts and discovered the sciences; it wrote the books, invented the philosophies, and refined

[5] Börje Hansson, "Dimensions of Primary Group Structure in Sweden." See Nels Anderson, Ed., *Recherches sur la Famille,* Tübingen, J. C. B. Mohr (Paul Siebeck), 1956, pp. 115-156.

[6] George Soule, *Time for Living,* New York, Viking Press, 1955, pp. 9-10.

social relations. Even the liberation of the oppressed has usually been inaugurated from above. Without the leisure class mankind would never have emerged from barbarism."[7]

Another survival of the feudal family, and one that has not contributed much to cultural advancement, is found in certain new countries where industrial urbanism has not developed to any great extent. An example is the plantation-owning family of Brazil which has changed little since that country was colonized. Pierson reports three types of family in Brazil: the slave family, the lower class family, and the land-owning family. The land-owning family is the "extended" type which, in addition to the three generations, may include other relatives such as cousins and married cousins. It is a patriarchal family and authority descends through the oldest son. Pierson describes it as follows:

> By way of ceremonies on the occasion of birthdays, baptisms, graduations, weddings and similar events, the kinship group tended to absorb the greater part of the social activities of its members. By way of ties of affection many slaves, especially the house servants, the *crias* (often orphans taken into the household to be reared), and the master's favorite concubines, and illegitimate children, as well as hired servants came to gain at least peripheral status in the large plantation family.[8]

In the feudal tradition, the women of the landed class are in a protected status, and one who fails piously to safeguard her purity is cast out. Once married, a woman is expected to be the dutiful wife and mother, remaining at home except when she visits relatives or attends church functions. She is head of the household servants and supervises the making of clothing, preparing food, and distilling wine, and she is responsible for the religious teaching of the children. If widowed, she may in some cases assume the authority over the señorial household. Willems described the cluster of values pertaining to the role of the female in such a family as a "virginity complex."[9]

The male in this land-owning family enjoys full sexual freedom, and he may boast about his philandering. This cluster of values Willems designated the "virility complex,"[10] and he observed that this socially entrenched double standard seems not to be a barrier to consensus in marriage. It tends to keep the sexes in separate roles, a characteristic of feudalism and other old family

[7] Bertrand Russell, *In Praise of Idleness and Other Essays*, London, George Allen and Unwin, 1948, p. 26.

[8] Donald Pierson, "The Family in Brazil," *Marriage and Family Living*, Vol. 16, November 1954, p. 108.

[9] Emilio Willems, "The Structure of the Brazilian Family," *Social Forces*, Vol. 31, No. 4, May 1953, p. 340.

[10] *Ibid.*, p. 341.

systems. Romantic love in the Brazilian family, however, has an important place. Parents carefully select the suitors for their daughters, and try to guide their sons to proper choices, thus guarding against marriage into families of lower class.

The wife in this family does not appear if her husband has guests, unless other women are present, but this rule and others are weakening under the influence of modernization. The Brazilian upper class family is being affected in other ways, as noted by Pierson:

> Especially in those parts of the country where urbanization and industrialization have tended to disrupt the previous social organization, the dominant position of the husband and father has declined and leadership has become more dependent upon personal abilities than upon family position. A range of types can now be observed: patriarchal, partially patriarchal, and (increasingly now) egalitarian. The peripheral elements have tended to separate from the familial system and attain autonomy in essentially monogamous relationships. . . . The nuclear family is increasing in importance; and concomitantly there has been a rise in individualistic behavior.[11]

Pierson observed further that certain of the economic and other functions of the old-style upper class family are vanishing. In the city, for instance, the chaperone is tending to disappear. In the rural regions the old feudalistic family is still extant, but even there it is changing, although change is slower than in the city.

## Family Change in Other Regions

With the exception of Japan, the population in most countries of the Far East is predominantly rural. In Japan, "Westernization" has been moving at a rapid rate and family change trends are fairly comparable to those observed in the United States. In other Eastern countries where well over 90 percent of the population is rural or semirural, the family model has been very little disturbed by industrialism or urbanism. While not much is known about changes in the Chinese family, there is evidence that the Communist government has a stern program of family reorientation which aims to raise the status of both women and children. Traditional norms are being pushed aside and, according to Ni:

> . . . the Chinese family is in great process of disorganization and reorganization. Good illustrations are that fathers who used to have complete authority over

[11] Pierson, *op. cit.*, p. 312.

> their children are now frequently accused by the latter of treason or action considered as undesirable. . . .[12]

Husain, on the basis of interviews with factory workers in Decca, East Pakistan, concluded that one of the most sacred tenets of family life in that country, family size, is being re-examined.

> People in this province are generally opposed to birth control, as birth is considered to be due to the will of God, and limitation of birth is regarded as against religion. A number of factory workers and rickshaw drivers seem to hold different views. However, as one goes through the case studies of factory workers and rickshaw drivers made, one gets the unmistakable impression that, instead of taking an increase of population for granted, quite a few are probably thinking in terms of family planning, although they do not seem to be aware of the methods which would make it possible.[13]

Even though there is great overcrowding on the land and greater overcrowding in the underindustrialized cities in both the Far East and the Near East, the old institutional family is for the most part being affected very little. It is, however, being affected to some degree. Nahas, for example, mentions the problem of divorce in Egypt. Under Mohammedan law divorce is a man's right, but as increasing numbers of women seek divorce it is coming to be regarded as a social problem.

> Divorce is easy among the Moslems; Islam allows it and the husband is not obliged to go to court for it. It is his religious and legal right. The wife can ask for a divorce in the divine courts. She may be granted it if she has good reasons. This, however, is rather scarce. The outsider might think that divorce cases would be common, but this is not so. However, in the last few years, the Egyptian families have been suffering a comparatively high rate of divorce. This may be largely attributed to hard economic conditions and to some incompatibility between men and women, which started some years ago and which is exaggerated by some women's movements. These unintentionally unbalance family relations, by overemphasizing the independence of women.[14]

While a professor in Baghdad, Nahas in 1950 circulated several thousand questionnaires to middle and upper class persons in Iraq, and received a considerable response. It was sent to married and unmarried persons of both

[12] Ernst Ni, "The Family in China," *Marriage and Family Living*, Vol. 16, November 1954, p. 317.

[13] A. F. A. Husain, "Human and Social Impact of Technological Change in East Pakistan, Decca," *Social Implications of Industrialization and Urbanization* (Five Studies in Asia), Calcutta, Unesco Research Center, 1956, p. 119.

[14] M. Kamal Nahas, "The Family in the Arab World," *Marriage and Family Living*, Vol. 16, November 1954, p. 299.

sexes. Of the married he asked about happiness in wedded life and of those not yet married he asked their views on marriage. Of the married, about 28 percent of both males and females said they were unhappy in their marriage, while 24 percent of the men and 22 percent of the women expressed the wish to be separated from their spouses. They do not separate because of the social taboos, which may explain the 8 divorces per 100 marriages in Iraq compared with 28 per 100 marriages in the more urban Egypt. There is hardly any labor market for women in Iraq. If they divorce, they must return to their parental homes, often not a happy prospect.[15]

Another example of disturbance to the institutionalized family is found in Clément's observations of the frustrations faced by the African native family coming in contact with industrial urban life. Older brothers in the town find that younger brothers no longer obey them. Fathers in the villages find that sons who go to the town become disobedient. The wife, so obedient to her husband in the village, may earn her own money in the town and asserts her independence. Children of native families in town run the streets fending for themselves.[16]

Clément illustrates these difficulties as they apply to the bride price, a serious matter in African families. Family members in the village insist that the payment be made even though the marriage is consummated in the city between partners who have been some years away from the village.

> The role of the marriage payment is complex. It would seem to constitute proof of the family's consent, and a certificate of the alliance. Three facts are adduced to support this hypothesis: the fact that the sum is handed over in public; the fact that the husband's group clubs together to collect it; and the fact that it is divided out, maybe quite unequally, among the members of the wife's family. . . .
>
> The marriage payment would also seem to compensate for the sacrifice by the woman's family of certain rights over her and over the children of the marriage, and for the transfer of these rights to the husband and his family.[17]

The bride price for natives who have migrated to town, even though members of the village group cluster together after migrating, is a source of much disagreement. Subtle and even defiant efforts are made to avoid it, but the pressure from tribal brothers is also present in the city. Moreover, if the

[15] M. Kamal Nahas, "Married Life in Iraq." See Nels Anderson, Ed., *Recherches sur la Famille,* Tübingen, J. C. B. Mohr (Paul Siebeck), 1956, pp. 183-212.

[16] Pierre Clément, "Social Patterns of Urban Life," *Social Implications of Industrialization and Urbanization in Africa South of the Sahara,* Paris, UNESCO, 1956, 371-372.

[17] *Ibid.,* p. 384.

husband does not pay the bride price his wife would fear to visit the village lest her relatives prevent her return to the city until the obligation is met.

The following is the story of an incident that disturbed an Iraqi family very much. It is a problem that could not arise in an American family.[18]

> The D. family is proud of its history, running back generations. Branches of the family are in business in different cities of the Middle East. They form, in fact, a family business network which reaches to relatives in the United States and Europe.
>
> The family house is in what the stranger might call a poor part of the city, which is not unusual, for houses of the upper and upper-middle class may be found in slum-like areas. Where a house is surrounded by walls, it is a little island more than a part of an area. This house and the wall about it, like most structures in Basrah, is made of tamped earth. Beside the house and within the wall, the family has a small garden with plants but no grass. It is difficult here to grow grass.
>
> Father, mother, five daughters and a son make up the D. family. There are two trusted Arab servants, and an Arab driver for the family car. At this time (1945) few families in Basrah, however rich, possessed automobiles. The girls ranged from Alicia, age 22, to Helena who was eight. The one son and principal heir was about six, a serious boy who often delighted his father with sage remarks, as when he declared "It's not good for girls to read books." Boys, he maintained, should read many books.
>
> At one time during the war, when it appeared that the pro-German forces would get control of Iraq, Papa D., at great expense, sent his family to Bombay where they stayed about two years. Here the girls attended English private schools. What was not possible in Basrah, the girls could leave the house without being chaperoned. They could visit the stores and movies. This degree of freedom, in keeping with Basrah mores, had to be curtailed when they returned home.
>
> The two older girls while in Bombay read much romantic literature, and they brought books back with them, which were circulated among girl friends within their social group. One of these girls was the 18-year-old daughter of the junior partner in Papa D.'s firm. Papa D., however, never took him into full confidence, using one or the other of his two older daughters as his secretary, and all documents he kept locked in a huge safe.
>
> One morning the daughter of the junior partner came crying to the older D. girls. Her father in the middle of the previous night had awakened her and said that he had arranged for her to marry a past-middle-aged merchant from Iran. This man, it seems, had grieved for years because his wife had borne him no

[18] This writer had been invited to the home of the D. family numerous times. He was the unwilling listener to disputes between the mother and the older daughters, each side trying to enlist his support. He corresponded with the family after leaving Basrah. It was through these letters that he learned the happy end of the story for Alicia who now has a family of her own.

children, and now she had consented to his taking a second wife, and he decided on a young wife. The junior partner and the merchant came to an agreement over their cups of tea. The bride price would be 4,000 dinars, then about $20,000.

The two girls hurried to the office in the family car. They cornered the junior partner, rushed at him, beating his chest with their fists, screaming, "Why do you do such a thing? You sell your daughter like a sheep. Know you not this is a new time?"

Overcome by such violent behavior for well-bred girls, the junior partner answered again and again, "What else can I do? The war is over and the price will go down." Papa D. looked on, but said nothing. Mama D., when she heard about it, was scandalized.

However, the girl did become the polygamous wife of the merchant. Some months later she left him and returned home, finding some good and sufficient reason, so her father would not lose all of the bride price.

Mama D. was worried about the girls getting such ideas. She was also worried because Alicia was getting past the marriage age. Both girls were old enough for marriage, but Alicia's case was urgent. And both girls, Alicia especially, were contesting Mama D.'s efforts, claiming that they should select their own husbands. They wanted to be courted. Mama D. called the whole idea vulgar. Romance is all right in books, but in life one must be practical. Alicia argued that she, too, was practical, and she would remind Mama D. of marriages by the old method of selection that were failures.

Mama D.'s answer that she got a good husband by the old method, so final as far as she was concerned, did not settle the issue, not for Alicia. Papa D., the businessman, took no part in this family dispute, but one day when Alicia had a toothache, he proposed that she go to Baghdad to see a dentist. He would not permit her to go to the Basrah dentists. The idea suited Mama D., who went along to visit relatives in Baghdad.

For some reason that did not make sense to Mama D., several visits had to be made to the dentist. She worried about the children and returned to Basrah, Alicia remaining several days longer. Baghdad is a freer city. She could go about with friends. She met a young man of good family. He took her to a movie. They walked in the garden. They decided that their parents should come together.

On returning home, Alicia one day at the office told her secret to her father. He listened but said nothing, nor was he angry. Some days later he announced to the family that he had to go to Baghdad, and Mama D. should go along for a rest. When they returned to Basrah Mama D. was very happy. They had met a nice young man of good family. Alicia would like him. They had come to an agreement with the parents of the young man. Alicia acquiesced.

Romance had its day in the case of Alicia, but not without a little trickery. The traditional ideal, so important to the mother, had been respected.

## Functions of the Family

As far as the early institutional family was concerned, it would be difficult to think of any function concerning life and work that was not a responsibility of the family. However much responsibility may have been lost in the intervening period, the family is almost universally recognized as having a monopoly on the reproduction and maintenance of children, and the function of socialization, the rearing of children.[19] By implication, these functions also include the very basic economic function of consumption.

The functions of the family, according to Merrill, may be presented briefly in the following categories:

1. Procreation (biological function)
2. Economic (production and consumption)
3. Education and training
4. Social status
5. Religious guidance
6. Recreation
7. Protection
8. Affection

Except for the procreative and economic functions, these are deemed to be mostly social in character. This means that to whatever degree the family performs these functions, particularly the social, it participates in community life.[20] Hill and associates in their study of Eddyville tried to determine what the families there, as expressed by words and conduct, seemed to think their functions were. Eddyvillers seem to hold that the family must:

1. Preserve the traditional Eddyville way of life;
2. Preserve and continue the family name and line;
3. Maintain the family prestige;
4. Rear children to be acceptable men and women;
5. Maintain a good home; and
6. Keep good relations with other townspeople.[21]

Eddyville is being industrialized, and hence not all Eddyville families achieve these goals because, in this changing community, family life is also changing. The same would hold for any community going through such an experience. Brennan and his associates found family change taking place in the coal areas of southwest Wales where some of the labor force formerly

[19] John W. Bennett and Melvin M. Tumin, *Social Life, Structure and Function,* New York, Alfred A. Knopf, 1952, p. 546.

[20] Francis E. Merrill, Ed., *Fundamentals of Social Science,* New York, Appleton, Century, Crofts, 1946, p. 57.

[21] Reuben Hill and Associates, *Eddyville's Families,* Chapel Hill, University of North Carolina, Institute for Research in Social Science, 1953, p. 175. Mimeographed.

worked in heavy industry. Light industries are new locating in the area. Formerly the father was the person of high status in the family for he was the chief breadwinner. Today the daughter may be working in the same factory and earning as much as her father. She may even have a higher level job. The younger generation has money and with that comes a greater sense of independence; the old family life is disturbed.

> If the new industries are able to provide steady employment, family earnings, and even individual men's earnings may be higher than in the older industries. The newer industries also offer many other advantages in the way of cleaner work, more congenial surroundings and less irksome hours. Nevertheless, it seems quite clear that they, or rather the decline of the old industries, have brought about a lowering of men's status in the community.[22]

Where the urbanizing changes take place they also affect the status of women. Thus we find in towns as well as cities that:

(1) Daughters are now able to leave the parental home to seek employment, and their leaving need not involve a family crisis.

(2) Courtship and marriage have become more of an individual and less of a family affair than formerly, which means that daughters as well as sons are more free to select their marriage partners. However, in upper class families there is a tendency still for relatives to hold some degree of control.

(3) Daughters as well as sons have more freedom today than formerly to make decisions regarding their education and regarding the choice of an occupation.

## Transfer of the Economic Functions

Hansson in his study of rural life in Sweden since 1800 found that the economic functions of the farm family had changed markedly. The status of the wife and mother a century ago rested primarily on her economic role in the household. Housework and rearing children were incidental activities. Children growing up were left to themselves.[23] The story of how the factory has taken work away from the urban housewife is too well known to need any review here. However, although later and at a slower rate, the factory encroaches also on the rural housewife.

An example of such encroachment on the rural family is reported by

[22] Tom Brennan, E. W. Cooney, and H. Pollins, *Social Change in Southwest Wales,* London, Watts, 1954, p. 4.
[23] Hansson, *op. cit.,* pp. 139-142.

K

Saal with respect to Dutch farmers in the butter- and cheese-making areas of Holland. Formerly the status of the wife was largely determined by her skill in making butter and cheese. Within recent decades the farmers formed cooperatives for making and marketing dairy products. Small factories were established for doing the work. Women became less identified with the work of their husbands, and more identified with consumption than with production. Their status became less economic and more social. This was a type of advancement, but not for family solidarity.[24] The urban women, deprived of their functions as bakers, preservers of food, makers of clothing, and weavers of cloth, found employment in the factories where such work was done on a mass-production basis. Rural women had less opportunity to enter the factory, but younger rural women, educated for white-collar work, now migrate in large numbers to the city.

"Perhaps," wrote Waller and Hill, "no family in history has been so puny in home production and so conspicuously a consuming group as the American middle class family. It is likely that future studies of the family will view it increasingly as a consuming unit."[25] But it is not the only consuming unit that it once was, for it is possible today, especially in large urban places, for persons to live alone, or for two or more persons to live together in a "primary individual" or "one-person household." These are also consuming units, and they are detached from a family structure.

Thus the importance of the family as an economic unit has diminished, but it does not mean the dissolution of the family. There is involved here a process of family reorientation to work and community life. It is an orientation that may be painful to those who wish to maintain the traditional ideal, one that had its origin and gained its strength in another type of economic setting.

## Other Functions Lost to the Family

The loss of economic functions unavoidably leads to the loss of other functions once centered in the family. When the home was the chief work place, it was to be expected that whatever training needed by the individual for entering his work life was given by the family. It is beyond the capacity of the family, however, to train the child to become a factory worker, or to give him the education he needs to become an accountant in an office. Education

[24] C. D. Saal, "Some Aspects of the Farmer's Family in North and East Groningen." See Nels Anderson, Ed., *Recherches sur la Famille*, Göttingen, Vandenhoeck, Ruprecht, 1957, Vol. I, p. 309.

[25] Willard Waller and Reuben Hill, *The Family*, New York, Dryden Press, 1951, p. 14.

and training have moved out of the home, but their most important aspect, socialization, which starts the child on the way to becoming a social being, remains.

Thus have come schools of low and high level, specialized schools of many kinds. Education becomes more and more institutionalized outside of the family. Now the school becomes an ally of the family in rearing the child but, since it is an agency under law, it has authority over the family; the family cannot refuse to send the child to school. The school may perform other functions, too, such as providing midday meals, giving children medical treatment, supervising their play, helping backward children, etc. The school may also have clinics to train mothers about the care of children. In most industrial countries the school assumes the responsibility for giving vocational guidance.

The preindustrial and preurban family spent its free time—what little there was of it—mainly at home; free time was mostly the privilege of the upper classes. Holidays, such as they were, were institutionalized at the community level. Leisure activity was either a family or a community matter, hardly ever an individual matter. Even in urban communities until about three generations ago much of whatever leisure the individual had was occupied with family-centered activity. As we shall see later in the chapter dealing with leisure, this pattern has changed. Not only is there more leisure than ever before, especially for urban people, but the use of leisure becomes more an individual and less a family matter.[26]

Two types of development have served to bring about this change. One is the tendency for all forms of leisure interest to become commercialized, and the other is the tendency more and more for people in the modern community to join secondary organizations. For almost every type of leisure interest there is some type of secondary organization. Family members join such organizations as individuals: The father may belong to the rod and gun club, the mother joins clubs for women, and the children have their organizations. All such organizations uphold the family, but they take the individual family members out of the home for their leisure.

Except in minor respects, the home does not function as the center for family leisure activity. This does not mean the family is or can be less the center for mutual interest and affection, or that some backward-looking move-

[26] Certain opposing trends, however, should be noted as, for example, the tendency for television to keep children "at home." See John W. Riley, Jr., *et al.*, "Some Observations on the Social Effects of Television," *Public Opinion Quarterly*, 1949, pp. 223-234.

ment is needed to restore things as they were. It means mainly that the guidance role of the family with respect to the free time of its members must find other ways and means of being effective. The new ways give promise of being based on mutual interest and guidance, rather than on authority and supervision.

## Families and Individuals

It becomes increasingly clear that an important aspect of the modern family problem concerns the changing status of the individual. The day-to-day modern life in the urban environment not only recognizes the individual, but can hardly do otherwise. According to Ogburn, "This daily life may differ as to where the members of the family work or spend their time, or as to the times they go to work, or their methods of recreation."[27] In extremely rural societies and in somewhat primitive societies, the work unit is often the family. Sofer found that when African villagers came to the industrial town of Jinja to work in factories, they were often distressed because it was not always possible for relatives to work side by side. Brothers had to work on different jobs and they were paid separately. Relatives had to go to different places for their work, and each worked on a different kind of a job.[28]

The prospect of being a self-sufficient individual, so deeply rooted an ideal in the urban culture, may and often does give rise to problems between parents and their children. Nieminen speaks of family strain in Finland, a country that has become rapidly industrial and urban within three decades:

> This problem is connected with the democratization of the family and with the development of the family from one of a single economic unit to one in which the older children constitute their own economic units. [Here he speaks of rural families. Author]
>
> Perhaps the greatest conflicts in this regard are concerned with the work of the children, and the wages they receive. In many places in the rural districts, the parents consider it natural that the children, in accordance with old customs, work for their home without pay, or for less pay than must be paid to strangers. The young on their part deem themselves to be in a worse position than others of their own age group, who in cities have perhaps gained a relatively independent position by using their labour and abilities wholly for their own

[27] Ogburn, *op. cit.*, p. 6.

[28] Rhona Sofer, "Adjustment Problems of Africans in an Early Phase of Industrialization at Jinja, Uganda," *Social Implications of Industrialization and Urbanization in Africa South of the Sahara*, Paris, UNESCO, 1956, p. 617.

> benefit, and deciding about the use of their own income. Being wholly under the guardianship of their parents, the young complain that the parents do not appreciate their requirements nor the changes which progress in this respect has brought since the time when their parents were themselves young.[29]

Working as individuals, being trained in the schools to think of themselves as self-sufficient individuals, voting as individuals and paying taxes as individuals, are bound to have their effect on the relations between the individual and the family, especially if parents cling to what Stoetzel calls "l'image traditionelle de la famille." This image he notes concerns both the structure and conduct of family life: courtship, relations between the spouses and between parents and children, authority in the family, division of tasks between family members, relations of the family to the community and the respective statuses of the generations.[30] Any or all of these elements of the traditional image (sometimes called the model) may be affected by the changing behavior of the individual.

The individual in the modern community is in a situation that makes certain demands upon him, demands that grow out of immediate requirements and hardly at all out of traditional ones. The relationships he must establish must be suitable to the immediate situation. The traditional image of the family tends to meet present realities in terms of old forms, which the individual is prompted from time to time to challenge or ignore. Perhaps at times his conduct is irrational, for he may be in a disorganized relation to his family and not yet fully adjusted to life outside the family.

As against this tendency toward individualism, Burgess and Locke use the term, "familism," and they regard the farm family as being its prototype. The characteristics are a feeling of unity on the part of family members, complete integration of family members in work, the assumption that the product of all work is family property, willingness of members to support one another against outsiders, and a common concern for the perpetuation of the family.[31] The individual thus functions only for the family group and has no interests contrary to those of the family. Individualism tends to challenge familism.

[29] Armas Nieminen, "Effects of Social Change on the Relationships between Parents and Children and on the Development of Growing Children in Finland." See Nels Anderson, Ed., *Recherches sur la Famille,* Göttingen, Vandenhoeck and Ruprecht, 1957, Vol. I, pp. 274-275.

[30] Jean Stoetzel, "Les Changements dans les Fonctions Familiales." See Robert Prigent, *Renouveau des Idées sur la Famille,* Paris, Presses Universitaires de France, 1954, p. 347.

[31] Ernest W. Burgess and Harvey J. Locke, *The Family from Institution to Companionship,* New York, American Book, 1953, p. 60.

## Structure of the Family

The family is a group of persons "united by the ties of marriage, blood, or adoption; constituting a single household; interacting and communicating with each other in their respective social roles as husband and wife, mother and father, son and daughter, brother and sister; and creating and maintaining a common culture."[32] We need not consider the forms of the family, but we do need to mention that there are types of families. If we ask who belongs to a family and how many people may or do belong, we would find that answers differ with times and places. The normal form in family history is the extended family: grandparents, their married children and their grandchildren, even great-grandchildren; perhaps also cousins, uncles, and aunts.[33]

The extended family, or kinship group, grew to whatever size the land supply, the ways of getting a living, and the needs for mutual security permitted. Whether many lived in the same household, or whether households clustered, was also determined by such conditioning factors. The tendency, however, has been for the extended family to diminish in size as it has come under the influence of the industrial urban civilization. Goodfellow wrote of the Bantu:

> Bantu family life is essentially polygamous, patriarchal, and on the basis of a group larger than the mere initial group of father, mother and children. To deal first with the last characteristic, it is rare indeed, if not unknown, to find a father and mother with their children living alone in any part of Bantu Africa. Wherever this is found it will be the clearest sign that these people are Christianized and have abandoned their old ways of life. Under changing circumstances the size of the group living together tends to decline, and nowadays it may be common to have four primary families living in one homestead, as in Pondoland, whereas formerly twenty would have been the normal number.[34]

Changing economic conditions may, if favorable, attract subfamilies to move away or, if unfavorable, may cause fringe members to be pushed out of the extended family. In the advanced countries the extended family is rarely found, even in the more rural areas. Nor did the American frontier encourage

32 *Ibid.*, pp. 7-8.

33 The institution of polygamy, which has survived fairly well in some rural or primitive situations, meets with increasing difficulty, particularly as women have greater access to the labor market and by degree acquire a sense of independence.

34 D. M. Goodfellow, *Principles of Economic Sociology*, London, G. Routledge and Sons, 1939, p. 141.

the extended family because there was always a place where the newly married pair could begin their own home. The separate household became the American model. Sometimes in the cities in times of a housing shortage the young couple may be forced to live with the parents of the wife or the husband, or parents may be forced to live with their married children, but such arrangements are not "built into" the culture, and they frequently result in friction.

Ogburn and Nimkoff tried to learn something about the changing size of American households, but information prior to recent census enumerations is very scanty. They estimated that in 1850 there were 555 persons per 100 households as against 351 per 100 households in 1950.[35] The number of persons per 100 households in 1800 they estimated at 579. Of these, 177 were parents (including widowed household heads), 343 were children, and 59 were lodgers, servants, relatives, etc. Their figures for 1940 were 377 persons per 100 households, including 177 parents, 149 children, and 51 lodgers, servants, relatives, and others. In 140 years there was a decline of 202 persons per 100 households, but 194 of this decrease were children and only eight were other persons; the number of parents, 177, remained the same.[36]

What Ogburn and Nimkoff call the biological family is sometimes called the primary family or the nuclear family. Sometimes, however, there are academic disputes about the use of terms for identifying that unit composed of parents and their children. If persons other than these are included—grandparents, aunts, uncles, etc.—the unit is a household. Both the family and the household tend to diminish in size. The household also tends to be smaller in urban than in rural places. Baumert and Hünniger compared household size for Darmstadt, Germany (about 115,000 inhabitants), with villages around Darmstadt and also with Frankfurt am Main, the nearest large city. Table 7 shows how these households were structured with respect to family types. The chief difference is between the villages and the two cities for three- and four-generation families; 15 percent of the village households included grandparents and grandchildren as against 7 percent of the households in Darmstadt and 6 percent of the households in Frankfurt. The percentage of married pairs with no children was lower in the villages (21 percent) than in either Darmstadt (29 percent) or Frankfurt (32 percent). The percentage for married pairs with children was equal in the three groups (47 percent), and there was very

[35] W. F. Ogburn and M. F. Nimkoff, *Technology and the Changing Family*, Boston, Houghton, Mifflin, 1955, p. 99.

[36] *Ibid.*, p. 110.

Table 7. Family types, Darmstadt and four nearby villages, and Frankfurt, 1949*

<table>
<tr><th rowspan="2">Family type</th><th colspan="2">Darmstadt</th><th colspan="2">Villages</th><th colspan="2">Frankfurt</th></tr>
<tr><th>Number</th><th>Percent</th><th>Number</th><th>Percent</th><th>Number</th><th>Percent</th></tr>
<tr><td>Married pair, no children</td><td>139</td><td>29</td><td>110</td><td>21</td><td>938</td><td>32</td></tr>
<tr><td>Married pair and children</td><td>221</td><td>47</td><td>249</td><td>47</td><td>1331</td><td>47</td></tr>
<tr><td>Mother only and children</td><td>60</td><td>13</td><td>54</td><td>10</td><td>313</td><td>11</td></tr>
<tr><td>Father only and children</td><td>7</td><td>2</td><td>6</td><td>2</td><td>51</td><td>2</td></tr>
<tr><td>Married pair, one or more parents</td><td>6</td><td>1</td><td>15</td><td>3</td><td>28</td><td>1</td></tr>
<tr><td>One or more grandparents, parents and children</td><td>21</td><td>4</td><td>56</td><td>11</td><td rowspan="2">196</td><td rowspan="2">6</td></tr>
<tr><td>One or more grandparents, one parent and children</td><td>13</td><td>3</td><td>17</td><td>3</td></tr>
<tr><td>Four-generation families</td><td>—</td><td>—</td><td>4</td><td>1</td><td>—</td><td>—</td></tr>
<tr><td>Other family combinations</td><td>3</td><td>1</td><td>7</td><td>2</td><td>28</td><td>1</td></tr>
<tr><td>Total</td><td>470</td><td>100</td><td>518</td><td>100</td><td>2885</td><td>100</td></tr>
</table>

* Gerhard Baumert and Edith Hünniger, *Deutsche Familien nach dem Krieg*, Darmstadt, E. Roether, 1954, p. 47.

little difference for single parents with children. The general impression is that larger households are fewer in the city.[37]

## The Changing of Family Size

Shrinkage, or the decline in family size, from the large kinship group, including a cluster of subfamily units in the same household, to the nuclear family of parents and their children, involves a variety of separations. That such numbers clustered in the same household was generally made possible and necessary because of the nature of family work. Industrial and urban influences have tended to disrupt the old work arrangements, the signal for fringe members to be separated from the larger unit. Other situational influences make it possible, even necessary, for subfamily units to separate. Still other influences separate young unmarried adults from the larger family. They may go to the city for employment or migrate to another country. In the census

[37] Prabhu in his study of city workers of rural origin found they were burdened with dependents from the villages and feelings of responsibility to their kin in the villages. Average number of dependents per worker in the city was 3.6 and in the villages 4.2. Extended families in the villages expected members in the city to send aid, it being regarded a duty. P. N. Prabhu, "Social Aspects of Urbanization of Industrial Workers Migrating from Rural Areas to the City of Bombay," *Social Implications of Industrialization and Urbanization* (Five Studies in Asia), Calcutta, Unesco Research Center, 1956, p. 69.

they are called "primary individuals" in contrast to "primary families" of parents and children. In 1950 primary individuals comprised but 1.4 percent of the farm population in the United States, but nearly 4 percent of the population in the large city. Thus the nuclear family remains, the final familial unit.

The nuclear family is the procreation unit, but it exercises the procreative function with great irregularity, having few children or many. Here we are confronted with another type of family shrinkage, not by people going away from the family but by people not being born to it. In some cases the married pair are not able to have children, but probably more frequently they refrain from having children, or they limit the number of births. This is a situation which sometimes becomes controversial. Couples, as Zimmerman noted, may favor having three or four children, but may limit themselves to one child or two. Some feel that reform is needed. Zimmerman answers:

> Any reform of the family will require a great many reforms of an institutional nature, along with the re-creation of the feeling that familism is the way of life. People living in a state of constant institutional insecurity simply will not do what we wish them to do in regard to family life, even though they believe it is right.[38]

Family limitation as against positive family planning is mainly a problem of the industrial urban countries in the West where governments have tried without success to discourage and punish family limitation practices. In Great Britian, a low birth rate country, a Royal Commission on Population was asked to study the family-size problem among other aspects of population. The Commission found no simple answer.

> Our review of the position of the family shows that in all classes of society, except the very wealthiest, married couples with young children to support are at an economic disadvantage as compared with childless couples; and parents with a family of several young children are at a disadvantage as compared with those with only one or two. In nearly every income range this disadvantage is substantial. Even where the income is high enough to rule out any question of actual want, the support of a fair-sized family will nonetheless entail very large sacrifices of comforts and amenities and may make it more difficult to give the children as good a start in life as the children of small families.
>
> The most important of the non-monetary handicaps of parenthood are felt by mothers who have shared little, if at all, in the great growth of leisure in modern times. The worsening of the position of parents relative to leisure is in part an incidental result of great social advances. In the process of social ad-

[38] Carle C. Zimmerman, *The Family of Tomorrow*, New York, Harper & Brothers, 1949, p. 237.

vance, until recently, the family has been overlooked, or given only a minor place in social policy.[39]

As external conditions made the extended family possible and necessary in former times, so the modern Western family is shaped in the direction of smallness by external conditions; the decision making is largely in the hands of the married pair. This debate, which is not within our area, concerns finding ways and means (external pressures) to influence the married pair in making "right" decisions. But there are different groups with different opinions about what that influence should be. In Hertzler's view, the decision should be left to the people and there should be open and free discussion (which some groups oppose). "Convictions must be developed at the individual, family, village and local level." He holds also for a strategy of popular education which would start with building "effective minority beliefs and actions."[40]

## Factors for Family Break-up

We have mentioned the factors which diminish the extended family, leaving only the monogamous nuclear family, and we have considered the diminishing size of the nuclear family. But there are other factors that operate to disintegrate the nuclear family. One parent or both parents may die, or one parent may disappear, or the parents may separate legally through divorce. Families may also partially disintegrate through the moving away of the children as they marry, but often before they marry. All of these changes, step by step, can be associated with the expanding influences of industrial urbanism.

An example of how the family as an institution can be disturbed by the urban situation is seen in a report by Busia of a study made in the twin cities, Sekondi-Takoradi, Ghana (formerly Gold Coast), a transport port and commercial city of about 44,000 including about 600 non-Africans. About 50 tribes are represented in the population, tribes that have their locale in the hinterland. Members from these tribes tend to form groups in the city, and thus a variety of family types and traditions are present: different ideas about marriage, bride payment, inheritance, etc. One factor making for friction is the sex ratio, 14,810 males and 9,707 females between the ages of 16 and 45 in 1948. Women here find themselves in an advantaged position. In addition

[39] *Report of the Royal Commission on Population,* London, Stationary House, 1949, p. 227.

[40] J. O. Hertzler, *The Crisis of World Population,* Lincoln, University of Nebraska Press, 1956, p. 261.

to tribal differences, there are tribal class differences, all of which are being challenged by a different class system in the city, one that places a higher worth on the individual. Moreover, the immigrant groups, more numerous than the labor market can serve, are badly housed. Families crowd into any type of shelter available.

About 50 percent of the marriages are intertribal, which means that the marriage partners have somewhat different cultural backgrounds, perhaps language differences, and each may be bound by different tribal laws. That may be one source of family conflict. Native law permits polygamy but the civil law does not (or did not under British rule). Still another source of conflict concerns "Europeanization," generally resisted by the older generation but favored by younger people. This gives rise to different forms of conflict between parents and children. Often parents are confused and helpless as their children roam the streets and countryside in gangs.[41]

The Busia report pictures a situation in which people are trying in a single step to move from tribalism to urbanism. People faced with change in the modern city have the advantage of generations of community living, much of it in urban communities. But that experience is lacking in Sekondi-Takoradi. All that ever happened to the Western family, from the extended household to the nuclear family, to the family practicing control and to the broken family, is present in the experience of a single generation in Sekondi-Takoradi.

Balandier found a high degree of family disorganization in Poto-Poto, a part of Brazzaville in French Equatorial Africa. Here, again, village families settling in the city are caught between the demands of rural (and tribal) tradition and the immediate realities. Here, again, as in Sekondi-Takoradi, African men far outnumber women, and women sometimes prefer transient matings to the long-term obligations of marriage. Women have their own mutual-aid associations, as if in league against masculine domination. In the meanwhile family stability is placed in jeopardy.[42]

All the factors making for family break-up are found in all of the Western countries, although not to such a high degree as in the African cities mentioned above. American students of family problems remind us that these different

[41] K. A. Busia, "Social Survey of Sekondi-Takoradi," *Social Implications of Industrialization and Urbanization in Africa South of the Sahara*, Paris, UNESCO, 1956, pp. 74-86. This is a summary of Busia's report.

[42] Georges Balandier, "Sociological Survey of the African Town of Brazzaville," *Social Implications of Industrialization and Urbanization in Africa South of the Sahara*, Paris, UNESCO, 1956, pp. 106-109. This is a summary of a preliminary report on a research still in process.

types of family break-up are more prevalent now than some decades ago, and that, whether now or several decades ago, they are more prevalent in urban than in rural communities. Divorce is the most commonly cited index of family break-up, since, because they involve legal action, divorces can be counted; but even this measure is a faulty one.

### *Increasing Divorce Rates*

In our courts divorce is usually handled as a criminal matter, in that one party is considered innocent and injured, and the other must be punished.[43] "Divorce," say Ogburn and Nimkoff, "reflects marital unhappiness, but not all marital unhappiness results in divorce." They note that in religious communities where divorce is not permitted the unhappy couples may separate. Separations would, in fact, be a better index of marital unhappiness than divorce, but there are no statistics on them.[44] Divorce in a country may be made easier or more difficult, but this is not reflected in the statistics. But in spite of all the shortcomings of divorce statistics, it is evident that the number increases. In the United States divorces per 100 marriages rose from 7.9 in 1900 to 23.05 in 1950, or from .73 to 2.6 per 1,000 population.

Burgess and Locke present comparative figures assembled by the Metropolitan Life Insurance Company showing changes in divorce rates for the period 1910-1914 and 1950. These figures are for each 1,000 population in the specified countries. We mention several of these only: Australia rose from .1 to .8, an eightfold increase; Belgium from .2 to .6, a threefold increase for this strongly Catholic country; Canada from .05 to .4, an eightfold increase; England and Wales from .05 to .7, a fourteenfold increase; France from .4 to .8, a doubling of the rate; Sweden from .1 to 1.1, an elevenfold increase. Burgess and Locke observe that the rates of increase for divorce have been greater for countries predominantly urban than for countries with predominantly rural populations, and of the more urban countries those with homogeneity of population, as England and Wales, have lower divorce rates than countries with heterogeneity of population, as the United States.[45]

The divorce rate may also reflect various social and economic changes. In countries having only a limited labor market to which women can turn, unhappy wives are less encouraged to seek divorce than they are in the United States, for instance, where women not only have access to a labor market but are trained for special work outside the home. Also in the more urbanized

[43] Zimmerman, *op. cit.*, p. 45.
[44] Ogburn and Nimkoff, *op. cit.*, p. 214.
[45] Burgess and Locke, *op. cit.*, p. 576.

countries, especially in large urban communities, people tend to become freer from divorce taboos. Some countries have tried to take legal measures to discourage divorce. Sweden, according to Segerstedt and Weintraub, requires the engaged pair to wait three weeks, during which time their names are announced to their respective church congregations. This measure notwithstanding, the divorce rate also rises in Sweden.[46] In countries where divorce is forbidden people resort to a tolerated type of separation which also tolerates persons living together outside the legal marriage bond.

## The Changing Roles of Women

In the labor force of the United States for 1950 women comprised from 33 percent to 34 percent of all workers in urban places, the percentages being higher in cities under 250,000 and in those of 3,000,000 or more. In rural areas females comprised 21.1 percent of the nonfarm and 16 percent of the farm labor force. In preindustrial societies the proportions were quite different, especially in rural communities where at least half of the labor force was made up of female workers. Statistics are not available, but it is a safe guess that the percentage of females then in the urban labor force was smaller than today.

The median 1949 income for males in central cities of the United States was $2,808 and that for females was $1,393. For males in suburbs and fringe areas in 1949 the median income was $3,166 and for females it was $1,310.[47] Working women comprise about a third of the American urban labor force but the median income for a woman worker in the city is about 50 percent of that of the average working male. It is only 41 percent for the suburban woman. We do not concern ourselves here with the relative fairness of these incomes, merely with the fact that the American urban woman has an important place in the labor force and an equally important place in the American urban payroll.

We do not have information regarding what part of the income of female workers is used for family support, but for 1940, the Department of Labor estimated that 35 percent of the female workers were married and at least half of the female workers were in lower level jobs.[48]

[46] Torgny T. Segerstedt and Philip Weintraub, "Marriage and Divorce in Sweden," *Annals of the American Academy of Political and Social Science*, Vol. 272, November 1950, p. 188.

[47] The figures on women in the urban labor market and income are from Otis Dudley Duncan and Albert J. Reiss, Jr., *Social Characteristics of Urban and Rural Communities, 1950*, New York, John Wiley and Sons, 1956, pp. 81, 130.

[48] Janet M. Hooks, *Women's Occupations through Seven Decades*, Washington, Department of Labor, Women's Bulletin 218, 1947, pp. 4-5.

To a high degree in the United States, and to a lesser one in other industrial countries, women are increasingly being educated and trained for higher level employment. This fact may result in a higher percentage of women in the future labor market, including a higher percentage of married women. Resistance to education and training for women is, however, fairly general. Speaking of the attitude in East Pakistan, Husain observed that Muslim religious leaders "are generally opposed to the education of women beyond the acquisition of some ability to recite or read verses from the Koran." He found that factory workers favored education for women, especially women who might become clerks, teachers, or nurses.[49]

Some figures collected by Landry for France, a country that is still 46 percent agricultural, show the changing role of women in the labor market. Out of each 1,000 working women in 1866, 547 worked in agriculture compared with 363 of each 1,000 working women in 1931. Yet for each 100 men in agriculture in 1931 there were 71 women. Between 1866 and 1931 female workers in commerce and industry increased by 60 percent, while the number of women employed in domestic service decreased by half.[50]

The community may have an interest if the working woman is a mother and is unable, because of working, to care properly for her children. Baumert and Hünniger found that of 9,901 younger children in the Darmstadt, Germany, schools, 14 percent reported that their mothers worked, and of 10,909 school children in nearby villages, 10 percent reported that their mothers worked. Slightly more than a third of the Darmstadt children and slightly less than a third of the village children were without adult care when not in school during the working hours of the mother. This is not necessarily a measure of neglect, but it represents the possibility of child neglect. This situation may be found in any modern industrial community, and may concern a good share of the mothers who are employed.[51]

## *Wives with Time on Their Hands*

It is likely that in the not-too-distant future it will be regarded as normal for married women to take outside employment. Perhaps even today the married woman with a job is less of a problem to herself and to others than are some housewives who have very little real work to do, particularly those

[49] Husain, *op. cit.*, p. 123.
[50] Adolphe Landry and others, *Traité de Démographie*, Paris, Payot, 1949, p. 151.
[51] Baumert and Hünniger, *op. cit.*, p. 53.

without children, or some who are able to leave their children with relatives or servants. The process of removing functions from the family has consigned these housewives to minor roles. Perhaps they occupy themselves earnestly with finding new ways of making thc home "beautiful" and "comfortable," without adding anything substantial either to its beauty or to its comfort. Or they may occupy themselves with worry about their own looks, their dress, or their weight. The extreme examples of such housewives, victims of the family-change process, are identified by Myrdal and Klein as belonging to the "costly cults of the lily-white hands, of lavish entertaining" and of frequent changes of clothes.[52]

It is the married woman with time on her hands, if she has access to money, who may be very occupied with setting the social vogue. She is also active in social life which, as between the classes in the community, is the arena where the struggle for social prestige takes place. It may be very much linked with economic competition. Married women who have acquired occupations to which they can return, or in which they remain active even after marriage, are less likely to be numbered in the "lonely housewife" group. The proportion of women being trained for such occupations is increasing.

## Parents and Old Age

The aged in all civilizations, wrote Sumner, have been a burden, but they did not always lose status. If they were not respected for wisdom, they were feared because their ghosts might return to haunt those who misused them.[53] Even in modern societies parents tell their children, "The way you treat us is the way you will be treated by your children." In all societies the aged have been secure mainly as they have held control of property or as they have been able to exploit fear.

A study of old age by the Council of State Governments reported that between 1900 and 1950 the increase of persons 65 years of age and older had been double the rate of total population increase. These older people totaled 4.1 percent of the population in 1900 and 8.1 percent in 1950. At the same time the labor market has been excluding older workers. Between 1890 and 1950 workers 65 years of age and older dropped from 68 percent to 42 per-

[52] Alva Myrdal and Viola Klein, *Women's Two Worlds*, London, Routledge and Kegan Paul, 1956, p. 5.

[53] William G. Sumner, *Folkways*, Boston, Ginn, 1940, p. 309.

cent of the American labor force. The aged are described as a "roleless" group in our society.[54]

> Today the chief reliance of older persons for support outside of their earnings is less on their savings, property and children and more upon government through social security legislation, veterans' pensions and upon employers through public and private pension plans. Adult children per family are fewer in number and less able and willing to assume support of their parents. The smaller size of the present-day dwelling makes it difficult and sometimes impossible to offer a home to an aging parent as compared with the large house of the past. The increase in the number of years of life after 65 increases the duration of such care. Although the trend is toward public provision, many adult children still, in whole or in part, support their aging parents.[55]

With a variety of exceptions, which help to confirm the generalization, the modern family is a one-generation arrangement. It waxes up to and through the adolescence of the children and declines as the children marry and move. Families with property can and do hold together longer. In former times the strength of the family was partly in its numbers, and children were assets. The opposite is true in communities under the urban way of life. Children are a burden, even though a joy, and there is little prospect that parents will "recover" the cost of their rearing. Parenthood, Moore declared, is not a form of social insurance but "a type of sacrificial duty."

> Even the small family does not assure the position of the parents after their children reach adulthood. The parents' freedom from responsibilities is likely to be too late for effective utilization for enhancement of the parents' social status of the money, time and effort formerly spent on the young. Even more important is the lack of satisfying social participation, as the available activities substituted for those centering around the family are likely not to provide the emotional sense of belonging and being needed. The aged are in effect members of no family except their own truncated one.[56]

The problem of aging parents is in part economic, although the prospect of deep poverty is much less than formerly. Some degree of economic security is provided for the aged in most Western countries. A good share of the American aged have managed to accumulate some means, but very few of

[54] *The States and Their Older Citizens,* Chicago, Council of State Governments, 1955, p. 9.

[55] *Ibid.,* p. 7.

[56] Wilbert E. Moore, "The Aged in Industrial Societies." See Milton Derber, Ed., *The Aged and Society,* Madison, Wis., Industrial Relations Research Association, 1950, p. 38.

them, whatever their economic status, are prepared for the shock of starting life over after completing their family-rearing task. They are confronted with a future of loneliness, made worse by exclusion from the labor market.

In the United States increasing numbers of retired persons turn to a mobile life. Perhaps they buy a trailer which they attach to the automobile and drive away "to see places." As a result, across a wide southern belt from Florida to California outside almost every town is found a trailer camp. Some of these are small communities where from 1,000 to 5,000 trailers may be parked. Whether or not this is a solution to the family break-up problem, it at least illustrates a type of resourcefulness which is to be found among the aging people.[57]

## Community Services for Families

How the family tends to be guided, supervised, materially aided, and "pushed around" is too wide a subject to be discussed at length here. The problems handled by the various agencies may be intrafamily ones as well as those which make the family to some degree a community problem. Some of the sources of help and guidance are public agencies, others are private. In addition, there is, especially in the United States, a growing number of individuals who concern themselves with the family. These may be clergymen or members of the medical profession, particularly psychiatrists, or aid and guidance may be given by members of that new profession which specializes in family counseling.

Much of what may be called institutional help and guidance, whether by public or private agencies, aims to keep the family more or less in line with community expectations. As would be expected, some of these agencies are much more conservative in their views regarding a particular family problem than other agencies, and it would be unrealistic to expect otherwise. The problem family or the problem individual in the family may have still other views which, in long-term perspective, may not be without merit.

The important thought for our purposes is that the modern community, whether through public or private agencies, does exhibit an interest in the

[57] Two articles, each based on a separate study of the problems of retired couples, are found in the *American Journal of Sociology*, Vol. 59, No. 4, January 1954. One is by L. E. Michelon, "The New Leisure Class," pp. 371-378, and the other by G. C. Hoyt, "The Life of the Retired in a Trailer Park," pp. 361-370.

welfare and continuity of the family. Duvall has listed some of the services available:

1. General family service, usually involving material aid.
2. Security programs for family members: mothers, children, the blind, etc.
3. Health aid and nursing service, general health guidance.
4. Mental hygiene facilities and services.
5. Marriage and family counseling both before and after marriage.
6. Advice on and encouragement of family interest in religion.
7. Instruction in home-making skills and offering of home-management aids.
8. Encouragement of housing developments for groups of families.
9. Premarriage and postmarriage parental education.
10. Youth education for marriage and family life.[58]

Such services are found mostly in the urban communities and mostly in the Western countries. In the Western countries, of course, they differ widely. In no Western country are such services available in such variety and volume as in the United States, but even in the United States they differ widely from region to region and from city to city. They are in their purpose more characteristic of urban than rural communities.

## Family Stability and Equilibrium

"The family," says Homans, "will not go wholly to pieces." He holds that marriage is still a successful human institution, and that the family is finding a new equilibrium. There is nothing new in saying that the family is not what it used to be. It is changing, but change will not prevent it being a "successful, and perhaps more flexible instrument."[59]

Wurzbacher and Pflaum in their study of rural communities near centers of industry, for example, found that the German father-dominated authoritarian family is much less in evidence than is sometimes believed. Formerly the German father spent his free time in the *Gasthaus* but today he has become more domesticated and spends more time with his family. The trend of change is for the out-of-town family to become more like the town family and the town family to become more egalitarian. Even in the village the family tends

[58] Summarized from an article by Evelyn Millis Duvall, "Organization of Social Forces to Promote Family Stability," *Annals of the American Academy of Political and Social Sciences*, Vol. 272, November 1950, pp. 77-85.

[59] George C. Homans, *The Human Group*, London, Routledge and Kegan Paul, 1951, p. 280.

to be held together by inner bonds; it is more private than formerly, more closed, even against relatives.[60]

Stoetzel finds that the family in France is also in process of evolution. As a social institution beside other social institutions, it must adapt. The result of this change is the emergence of a new family type; Stoetzel is not ready to say whether the direction of change is favorable or not. The final test must be whether the family provides sufficient children to insure a strong and balanced population. Apparently this is not the case in France, but there are other tests involving family integration, harmony, and interpersonal relations.[61]

Bott, in connection with a study of ordinary families in London discussed in Chapter 2, found that families at different social and economic levels tend to adjust differently to the urban life. Adjustment depends in part on the length of time the family has lived in a particular area, whether or not the family has moved from one occupational level to another, whether or not relatives of the family live nearby, and other conditions. Families of the working class, who in London seem not to move as frequently as families at upper occupational levels, appear to be more stably adjusted. But the family in process of changing its status may be confronted with a complex of adjustment problems: adapting to the work, fitting into a new type of neighborhood, selecting new friends while holding or discarding old ones. Bott found that each of several different types of ordinary families was confronted with different adjustment problems. In general, however, the adjustment problems become more complex for the more upper-level families, and these families are more conscious of continuous need for adjustment to change.[62]

Burgess advances the optimistic theory that the family, while shaking off the shackles of institutionalism, is evolving into a type of partnership, an inner-control group, which he calls the "companionship" family. It is a relationship of give and take between the spouses and between parents and children, and it is a type of marriage relationship which is sufficiently flexible to withstand most of the shocks that have proved so disturbing to the institutional family. He compares the companionship family and the institutional family:

1. A union of internal attraction versus external pressures;
2. Permissive and democratic as against authoritative and autocratic control in the relations of husband and wife;

[60] Gerhard Wurzbacher and Renate Pflaum, *Das Dorf im Spannungsfeld industrieller Entwicklung*, Stuttgart, F. Enke, 1954, Chap. 4.

[61] Stoetzel, *op. cit.*, p. 369.

[62] Elizabeth Bott, "A Study of Ordinary Families." See Nels Anderson, Ed., *Recherches sur la Famille*, Tübingen, J. C. B. Mohr (Paul Siebeck), 1956, Vol. I, pp. 29-68.

3. Development of adaptive and creative rather than traditional, static and conventional designs for marriage;
4. Personal happiness and affection versus duty and respect in the marriage relation;
5. Self-expression of family members rather than their subordination to family objectives; and
6. Assimilation rather than accommodation of the differing attitudes of husband and wife.[63]

The idea of companionship relieves the wife of her former "chattel" status which inheres in almost every institutional family system and which is quite out of harmony with the liberating character of the urban way of life. If this partnership or companionship is the emerging type, it means that the family is more the private holding of its members than a collective possession of many relatives. It may not mean the end of divorce, but it may mean that more families will be internally integrated than formerly. It would not necessarily mean security for parents in their old age, but it would mean that parents would understand well in advance what the life cycle of the predominantly one-generation family is and would be emotionally prepared for changes. There would still be family problems, but with more people better informed about the nature of these problems, they should not constitute a serious community challenge.

## Summary

The evolution of the family is reflected in the evolution of society. In all phases of the development of cities the family has been under compulsion to change, and it has changed. It has appeared in one form or another and has changed from one type of inner organization to another as circumstances from time to time seem to have made necessary.

Although the forms have differed, many of the older family systems have been of the extended type—that is, including three and sometimes four generations in addition to other relatives, and perhaps servants and other fringe individuals. The feudal family, which still survives in some places, was of that type. Such a family did not take root on the American frontier where con-

[63] Ernest W. Burgess, "Companionship Marriage in the United States." See Nels Anderson, Ed., *Recherches sur la Famille*, Tübingen, J. C. B. Mohr (Paul Siebeck), 1956, Vol. I, p. 70.

ditions favored setting up a household with each marriage. The economic and other influences that have affected family life have tended to keep down the size of the extended family, leaving only the nuclear family of parents and children, with occasional other relatives.

Along with the change in size there has been progressive loss by the family of many of the former household functions. Much of the economically productive work has been taken away from the urban family, although the economic function of consumption remains. Much of the work of education has been taken over by the school, but the function of giving the child the initial social instruction remains. While leisure activity has tended to move out of the family, the function of affection remains, and the family still has the monopoly on the procreative function.

This changing of the family has other implications, involving situations of conflict between the family and members, behaving as individuals. The married pair may refrain from having children or may limit the number of children. Children in young adulthood before marriage may leave the parental home. Spouses may separate or divorce, leaving the children with one parent. Finally, parents who have reared families may find themselves alone and perhaps not needed in the families of their children. Thus the modern urban family tends to be a one-generation household. This means that modern parents must find their old-age security in social insurance, unless they have savings, since only rarely can they look to their children for support or for shelter. Rarely does the modern dwelling have "extra rooms."

Since women have lost much of their traditional household work and responsibility, many modern housewives have time on their hands. If they do not find outside employment they may turn to social activity or various non-utilitarian occupations. It appears that as more women are given more training for work there may be an increase of "lonely" housewives taking employment.

In American communities, particularly in the larger cities, may be found a large variety of private and public agencies that devote themselves to the guidance and aid of families in trouble. These bear witness to the interest of society in the importance of the family in community life. Family guidance has also become the professional work of various specialists.

In spite of all efforts to persuade or pressure families to return to traditional ways and values, the trend of change has been away from the old "institutionalism." Various students of family problems are optimistic about these trends. They declare that the modern family, where it holds together, enjoys a solidarity based on inner bonds, while the institutional family was held together by outside pressures. The unhappiness in the old-style family was hidden. It had to be endured. It is claimed that the emerging family is more egalitarian and less authoritarian; more a partnership of spouses and parents with children. It has its problems which come out in the open (and can be statistically reported). According to these views, which seem to gain wider recognition, the family is the most basic and vigorous primary group in the modern urban society.

CHAPTER 12

# How Communities Get Things Done

IN CHAPTER 9 WE CONSIDERED the community as a behaving entity, a human aggregate having collective experience and a collective memory. We noted that a community can behave with pride, or sentimentally, and that it can rally as an entity in a crisis. In this chapter we are directing our attention to the day-to-day behavior within a community. Many kinds of activities go on within the community and these make up its life. Much of this activity is to a purpose, but there are many kinds of activities and many different purposes. Goals are constantly being reached and efforts toward other goals begun; things are accomplished. We are now to consider how "things get done."

## Beginnings of Group Cooperation

Settlers sometimes moved to the frontier in an organized group. A group of families, by agreement, would start together and agree on the various types of tools, implements, livestock, and other properties they would take along. They knew approximately their destination, how long the journey would require, and how much of a load could be carried, including food needed en route. They took along seeds and plants, chickens, cows, and pigs. Effecting the organization for travel was the beginning of community organization. The journey was a test in which each member of the group acquired a status in the group. Unforeseen problems en route had to be met by the resourcefulness of the group.

Arriving at its destination, the group took possession of a tract of land. A site for the central point of the settlement was selected; decisions were made about how the different types of land might be used. The difficult task of determining which families would occupy which pieces of land was usually settled by drawing lots. Various kinds of work had to be done. Each family took the responsibility for building its house from whatever materials were available. Groups of two or three families usually helped one another in putting the first bits of land under cultivation. Tools and household utensils and equipment were borrowed back and forth. While each family was a free agent, much that was done had to be on the basis of agreement, in the interest of group welfare.

But there was some work which concerned all members of the group. Roads had to be made, perhaps bridges built. This work had to be shared by all able-bodied men. The livestock put out to graze during the day had to be herded so that the animals would not wander away. Boys were given this work. A common meeting place had to be built to provide the settlers with a building for religious service and social activities. In order to facilitate the accomplishment of such common work the group as an organization assigned persons to leadership positions and gave them authority to act.

In time the settlement perhaps became an incorporated village with certain officials elected to perform certain functions. This was civil government, and the settlers, now established farmers, paid taxes for the support of the civic organization which was responsible for the school, the police, fire protection, the public meeting house, and other properties which belonged to the group as a whole. Perhaps the first achievement of the corporation was to organize the work of putting a hard graveled road through the village thereby eliminating clouds of dust in summer and bottomless mudholes in winter. This completed, it was regarded with local pride. The same feelings resulted later when the corporation took the lead in erecting a safe bridge over the stream.

The large village in time became a small town with stores and paved streets, electric lights and a piped water supply. Installation of a sewage system was often a controversial issue between people in the town and those outside. No longer were they settlers sharing their work, but the grandchildren of settlers, and different special interests had meanwhile arisen. In time, the sewer would be completed, and the contest about it forgotten.

Different organized groups are to be found in the town today. An occasional farmer near the town belongs to certain of these groups, but for the most part the farmers have their own organizations. There are various church groups and groups organized for social service, as well as social clubs, fraternal societies, veterans' organizations, and organizations for the youth. The businessmen have their organizations and the workers have a trade-union center. The local government now has special departments: welfare, police, a health service, and a department for construction and maintenance of buildings. The department of education now includes a high school. The town has a baseball team and an orchestra, and a library is being started. There are private agencies for welfare and social service. Any one of these groups or agencies represents an effort on the part of groups of people to get things done.

The community whose evolution is reviewed above—and it is typical of many—was a center of activity from its founding. Always, as at present, it has been confronted with new problems. What it has become is a composite of many accomplishments, from the graveled street to concrete pavements, from log houses to houses of brick and concrete, from the first wooden bridge to an arched concrete structure, from the one-room school to a modern school system, from the justice of the peace and the constable to municipal courts and a trained police force. Many other changes could be mentioned, each of which exemplifies the process of getting things done.

## Our Expectations and Community Work

In a report by Beals on a Sierra Tarascan village we find an example of a community with all its equipment and activity; we see it working, playing, worshipping, engaging in politics, and in every way living its community life. Beals describes the activities by which people get things done: farming, herding, going to market, digging ditches, making roads. Much of the work is private, some is public, and some is under public supervision, but all of it "fits together" into a continuing process which engages the attention of all the population.[1]

Whatever the activities of the agglomerate in any community, the people involved know what they are about. They know the worth of the different activities. They share certain expectations with respect to each type of activity. In terms of such expectations the people in one village are able to judge a neighboring village. Whatever these criteria of measurement may be—and we know they vary with people, times, and places—they are used in comparing one place with another. Whether they are objective about it or not, Americans tend to expect in any town or city of a given size certain features or characteristics found in other towns or cities of similar size. If these are not present,

[1] Ralph L. Beals, *Cherán, A Sierra Tarascan Village,* Washington, Institute of Social Anthropology, 1946, p. 109.

such a place may be described as "backward." This thought is expressed by implication in the following item about Butte, Montana, a mining community that has produced great wealth. The writer of the book quoted noted that Butte had a population in 1940 of 53,450 but during the Great Depression the federal government on its work relief program, the WPA, spent $35,000,000 there on public works.

> Maybe there's a clue in the fact that despite the vast fortunes which have been made here, it finally remained for the WPA to pave hundreds of miles of streets, lay a vast network of sanitary sewers, develop parks and playgrounds and provide an art center for the community, while symphony orchestras, libraries and art galleries endowed by the copper kings are enjoyed in the cities to which the millionaires moved.[2]

A mining town is sometimes a "wide open" lively place, but it is usually not very community conscious. The private organizations which are active in other places working on community problems tend to be absent or few in communites occupied in large part by mines and smelters. The barrier to getting things done is more a lack of interest than a lack of means.

Washington, the national capital of the United States, is often criticized by architects. The architect who might not become disturbed about what he would find in a mining town, even one of more than 50,000 inhabitants, might have very different expectations when visiting the much more attractive national capital. But architects differ very much about what needs to be done in Washington. Gutheim, for example, called Washington a "mess," and he urged that something must be done about it.

> Nor is the "mess" limited to the parts of the city which are of first importance to the nation. Behind the imperial façades are the slums of the city which must be torn down and redeveloped. Washington suffers from the shortages of school buildings, hospitals, libraries and other public facilities that exist in all rapidly expanding cities. It is tormented by traffic jams, parking problems, and the obsolescence of thousands of buildings as the result of the impact of the automobile. Its system of mass transportation fails to serve more than half the metropolitan area. The city lacks an adequate concert hall, a convention auditorium, a well-located place for mass entertainment. These shortages are parts of the "mess" and must be mentioned in any balanced view of it.[3]

In the case of Butte, many families of means whose source of income is there have their homes in other cities; pride of community does not flower.

[2] *Copper Camp, Stories of the World's Greatest Mining Town, Butte, Montana,* American Guide Series, New York, Hastings House, 1944, p. 11.

[3] Frederick Gutheim, "The Mess in Washington," *Journal of the American Institute of Architects,* Vol. 19, No. 1, January 1953, p. 15.

As for Washington, it is a city without citizens, without a local government. One may be born in Washington but he can never vote there. The responsible government is a committee of congressmen. There are many private organizations in Washington and these in various ways try to get things done but make little headway with a population of people who have no citizenship status in the community. The local government does not belong to the people who live there. That important half of the combination needed for getting things done—the other half being private organizations—cannot function. So much that is obviously needed does not, therefore, get done.

The Hammonds, writing about England as it was a century ago, call attention to the filth that was everywhere in the cities. There were no sewers. People began digging cesspools about 1840 but these served only to saturate the earth with sewage. Streets were in poor condition, housing was of low quality and congested, and those who were most inconvenienced were unable to do anything. Few people, either of the lower or the higher classes, were greatly disturbed by the conditions of a period which the Hammonds later called an age of bleakness.[4] To the people of an English city in 1840 the scene did not have the same meaning as it did to the social historian a half-century later. Somewhere in between a change in values had taken place and old standards were replaced by new ones.

## How Things Get Started

Few New Yorkers who admire and enjoy Central Park are aware of the efforts which were made shortly after 1800 by a group of idealists to have the city set aside "that wasteland area" for public purposes. It was public land mainly occupied by shanty towns with their herds of goats. Years passed before the idea was accepted, but now the park is there, a model for other cities. Ironically, it was the inspiration for a Kansas City real estate man, Thomas H. Swope.

As reported by Garwood, there had been some agitation on the part of the newspapers for parks in Kansas City, but there was little evidence of interest by the citizens, and the authorities did nothing. But Swope was interested and he offered the city a tract of hilly land, 1,331 acres. In 1896 when the gift was made, however, the proposed "park" was somewhat remote for most town people. The city has since expanded to the park. In this case, and

[4] J. L. and Barbara Hammond, *The Bleak Age*, West Drayton, England, Pelican, 1949 (1934), p. 58.

it is true in other American cities, the action was started by a private person.[5] It frequently happens in cities that private properties are turned over to public use for hospitals, playgrounds, libraries, and so on.

In the days of walled cities fire dangers were great and everyone had to be fire-protection conscious. Each person had the duty to remind others of fire hazards. Every household was required to keep water and fire buckets available for instant use at any time. There was no public fire service, but public authority was used to set up continuous fire watches. Periodically every chimney had to be cleaned, and in many cities the chimney sweep still operates, his services being required by law, although the danger of chimney fires today is very slight. In those cities the work of fire protection is now a public responsibility, and fire protection has become the work of professional firemen.

Volunteer fire fighters are still used in most American towns and villages. Of about a million firemen in all types of communities, no more than 100,000 are employed as full-time firemen. Most of the rest are members of volunteer companies which are under public authority or are privately organized groups.[6] The establishment of professional fire-fighting departments came slowly and under pressure. For example, the volunteer companies in New York City were an efficient service at one time, but later these groups became attached to various politicians and their integrity declined. Rival companies competed for the right to put out a fire, and fights took place while the building burned. Followers of the firemen, helping carry property out of burning buildings, were often no more than organized thieves. Insurance costs went up and property owners became so concerned that they brought organized pressure which the politicians dared not resist. Thus New York City got a professional fire service. But a great amount of loss and inconvenience was sustained before group pressure took form. This is another way by which things get done and services are established.

Lighting of streets at night became necessary when it was dangerous in towns and cities for persons to be outside after dark. Some American cities followed the practice of English cities of requiring householders to place lights in front of their buildings. Or groups of householders joined in maintaining lights along their streets, perhaps as late as midnight. Such services proved to be uneven and unreliable. The idea of street lighting as a public responsi-

[5] Darrell Garwood, *Crossroads of America, The Story of Kansas City*, New York, W. W. Norton, 1948, p. 204.

[6] See *The Encyclopaedia Americana* under "Fire protection."

bility began to spread after 1800. Different methods of meeting the problem were used.

The city authorities of Birmingham, England, accepted the responsibility for street lighting about 1780. Contracts were given to private individuals for from 160 to 170 nights of lighting per year. Standard lamps burning whale oil were used. These contract services proved unsatisfactory, and even changing from one contractor to another did not meet the problem. In 1798 the Birmingham Council decided to make street lighting a public service. This began as an experiment, and it was continued. The public service was more efficient and cheaper than had been the service provided by the contractors.[7]

## Private Work and Public Work

Regardless of the kind or size of a community in which people live, many kinds of work must be done in order to serve all their physical, social, and spiritual needs. They need to work without interruption and their property must be safe. There must be facilities for their educational and cultural needs. They must be secure against undue health hazards. Community life is satisfactory as it is possible to get these many kinds of work done, whether by individuals, corporations, private groups, or public bodies.

All this effort is either private work or public work, and we are reminded by Anderson and Weidner that it is not easy always to distinguish between the two.[8] Each kind of work may supplement the other, as when in ancient cities public authority established fair weights and measures and maintained order in the market place. The private sellers in the market place, however, were taxed for the service. The traders contributed to the total work of operating the market by keeping the rules. Public work is often a service to those performing private work.

How much of all work in the community and what kinds of work in the community should be public cannot be stated in any but the most general terms. Should the water supply service of a community be public or private? It is public in most communities over the world, but in some places it is a private service under public control. In most communities education may be either a public or a private work; usually it is both. The only service on which government must have a monopoly is that of law and order. The police and

[7] Conrad Gill, *History of Birmingham*, Oxford, Oxford University Press, 1952, Vol. I, p. 168.

[8] William Anderson and Edward W. Weidner, *American City Government*, New York, Henry Holt, 1950, p. 95.

protective services must be public. A hospital, a clinic, or some other health service may be a private work, but the final authority for the integrity of its work must be public. The factory may have its strictly enforced rules about safety and the protection of workers against industrial hazards, but this private work must meet the minimum standards of public authority.

## The Utility of Public Work

Civil government in a community is essentially a creation for service, for getting things done. The same could be said even of the most elementary forms of tribal government. Modern government tends to be more diversified and specialized, having different departments or branches for performing different types of services. A good share of one's education (outside of school) is concerned with coordinating one's private activity with some public service or other, and with learning how to make use of various public services. Anderson and Weidner speak of government as a "process for doing things," and to understand this process:

> . . . we must see through and even beyond the mechanism of organism. We must actually see things in operation or in process, seeing how they function as well as how they are constructed. If we then find that some men do actually get things done by government as they wish them done, we may even say that there is an art in government. Those who know this art we call statesmen, or politicians or leaders. The basic material with which they work is the same as that of which the city or the state is composed, namely, the people themselves.[9]

While the different branches of government are utilities to be used under certain rules of fairness, and should be so used, some individuals, corporations, or groups are more skillful (or artful) in making use of these services than others. In this connection we may hear of some practices as being fair and others as being corrupt. Some people or groups may be described as having privileges which others cannot get. The existence of such conditions does not alter the force of the observation that government and its branches are utilities; their effectiveness depends on the varying abilities of people to make use of them.

While these observations apply to government at all levels from local to national, our interest is primarily in their application to local urban government. Local governments differ widely, depending on the size and character of communities. The larger the community, the more diverse and specialized

[9] *Ibid.*, pp. 247-248.

are the services; the smaller the community, the less expert and professional are the services. It is often true in the small town that anyone may be named or elected to fill most any post. In the small community various services may be grouped in a single branch, whereas in the city each would be a specialized branch. On the other hand, the demand for public services varies from one community to another. In the smaller place the demands are fewer and simpler, not requiring so much the services of experts. In all modern communities, however, even the smaller places, all public services tend to move in the direction of using trained personnel. This trend, as Lampard noted, is an accompaniment of the evolution of industrialism.[10] This means that the quality of public work must keep on a level with that of private work.

It is less than a century since most American cities were governed according to the principle that any average citizen can perform most of the duties of public office, apparently an idea that was inherited from England, the initial source of most American ideas about local government. Simon, writing the story of government in Manchester, England, reported that even after that city had been filled with hordes of factory workers, it was being governed according to the form of village government. Responsibility for government rested with the Lord of the Manor who twice a year called a one-day session of a "great court leet" to handle all problems and complaints. Each year this court appointed a mayor, two constables, a tax "layer," a tax collector, "lookers" to supervise the various markets for fish and flesh, an inspector of weights and measures, testers of beer and ale, and other officers who were appointed on the spot and required, as responsible citizens, to accept the assignments and to serve without pay. Appointments went to the leading families.[11] The franchise was a monopoly of the property-owning families, while the working population newly arrived in the city had no part in the government. They were "subjects," not "citizens."[12]

We would not find that conception of government in Manchester or any other city of England today, nor would we find it in any American city of any

[10] Eric E. Lampard, "The History of Cities in the Economically Advanced Areas," *Economic Development and Cultural Change,* Vol. III, No. 2, January 1955, p. 133.

[11] Shena D. Simon, *A Century of City Government, Manchester 1838-1938,* London, George Allen and Unwin, 1938, p. 59. Simon quotes a 1834 newspaper article by Alexis de Tocqueville who was impressed with the individual power of man in Manchester, but saw little of the "regulating power of society." There was creativeness in private matters but no show of "continuous action of government." P. 17.

[12] "Citizens are those in the body politic who share the allocation of power; subjects, those who do not." Harold D. Lasswell and Abraham Kaplan, *Power and Society,* New Haven, Yale University Press, 1950, p. 217.

size. There is good reason to believe that the trend toward specialization in public services and professionalism among public servants has had much to do with eliminating graft and corruption. Certainly, American municipal government has moved far from the conditions described by Lincoln Steffens and other "muckrakers" who were writing prior to 1910.[13]

## Citizenship Performance and Learning

It is not difficult to demonstrate the increasing efficiency of municipal government over the past century, particularly during the past four or five decades and throughout the so-called Western urbanized world. Important as that evolution has been, it perhaps only reflects something that is more important: an evolution in citizenship. The American at home is probably much less aware of this evolution than one who may be sent to an underdeveloped country in the role of an expert. Here he finds it not easy to build industrial cities with a population that has yet to pass through the learning process that has already been achieved by the urbanized peoples. The "know how" which such an expert brings to the less advanced people can have little meaning before they have risen out of illiteracy, before they have been trained in various work skills, and before they have acquired an orientation in and have become identified with the type of community in which industry is at home.[14]

Marshall has interpreted citizenship in terms of rights, and these rights are of three types: civil (free speech, liberty of the person), political (voting, sharing political power), and social (social living, sharing social heritage). He recognizes that these rights tend to be interwoven and they are enjoyed as the individual participates in the life about him.[15] The idea of citizenship is one with which the person lives. Some of it he received in the social heritage and some from his own experience, and he shares in passing on the idea and what pertains to it. We may, in short, speak of a community or a society as having a *level of citizenship*.

Moreover, a level of citizenship is indicated in terms of informed participation in all phases of life—civil, political, economic, and social. Citizenship participation in the life of the modern community, with the scenes ever chang-

[13] Harold U. Faulkner and Tyler Kepner, *America, Its History and People*, New York, Harper & Brothers, 1947, pp. 537-538.

[14] Fred H. Klopstock, "Agricultural Development in Tropical Africa," *Social Research*, Vol. 17, No. 2, June 1950, p. 172. A description of the indifference or resistance of the natives to imported "know how."

[15] T. H. Marshall, *Citizenship and Social Class*, Cambridge, Cambridge University Press, 1950, pp. 10-12.

ing, is a continuous learning process, a process in which some communities are more advanced than others. The most advanced appear to be the most industrial and urban communities.

This thought is illustrated in part by the land reform program which the Italian government has initiated since World War II. Large estates were divided into numerous small farms and new farming communities were formed. On these projects the government has established wineries, sugar factories, dairies, and canning factories. There are stores, schools, playgrounds, and other community facilities. Everything is planned in advance by the experts. The people have only to settle on the land, occupy the houses, and go to work. Government officials will supervise the projects for twelve years during which time the people will be learning to operate them as cooperatives, much as rural cooperatives are operated in Israel. After making a study of the projects and after considering the background of these Italian farmers, Nelson concluded that it would be surprising indeed if these people will be prepared to operate these large cooperative communities after only twelve years of tutelage.[16]

A rural population cannot be moved out of a traditional small-farm type of agriculture into a complex cooperative system in the expectation that they will be adjusted within a decade. The learning of new ways is hindered by the resistance of old ways and beliefs. Hence, before the new ways of agriculture and rural community living can be fully entered into, the old ways must be uprooted and forgotten. That means setting aside all folklore about the weather, seeds and plants, soil behavior, the influence of the moon, ideas about the timing of work, and so on. This learning process does and can move faster in the urban community.

## Action on Community Problems

It is often said that many of the problems of the community could be disposed of easily if we applied our knowledge to them. Beeley has made such an observation with respect to a number of community problems:

> Another problem arising from this one-sided application of technology to society is the increase in mobility, the movement and migration of peoples, and the drainage of rural areas to metropolitan regions. It is in this sociological climate of anonymousness that personal and social disorganization proliferate. The urban environment emphasizes appearances and thus substitutes the

[16] Lowry Nelson, *Land Reform in Italy*, Washington, National Planning Association, Pamphlet No. 97, 1956, pp. 33-34, 45.

> veneers of personality for the sounder traits of character. This is the chief reason why vice and crime are more prevalent in cities than in simpler societies where human relations are more intimate.
>
> The city has also had a deleterious effect upon the home and the family: late marriages, fewer children, mothers employed away from home—these factors taken together have weakened parental responsibility and forced it upon the school and the community.[17]

Such charges have been made many times by many different people, and doubtless will be made many times more. However, Beeley argues that we have the knowledge for dealing with most of these problems, and we should be using that knowledge. We should be doing something. That determination to do something about social problems is deeply rooted in American community life. In no other country, except perhaps England, have so many "movements" been started or so many agencies established for the purpose of doing something about community problems of one kind or other. These are found in every community and they increase in number and variety with community size. In any American city of a million inhabitants can be found hundreds of such organized groups. If the problem is divorce and the broken home, perhaps fifty agencies will step forward saying they are interested and trying to do something about it. If the problem is juvenile delinquency, another group of agencies will step forward. Another group of agencies may be trying to abolish commercialized vice, another group is interested in housing, and so on.

These many kinds of agencies do have the knowledge, but different agencies have different viewpoints in regard to defining a problem or agreeing about procedure. Some organized groups, acting in good faith, are opposing other groups, also acting in good faith. This does not mean that they are mutually frustrating in their efforts; quite the contrary. The work goes on and these problems are in many respects abated by these efforts. Not all organized groups, of course, are concerned with social problems; other groups deal with problems in economic, political, cultural, religious, and technical areas.

These organized groups, diverse though their purposes may be, have two characteristics in common: Each is trying within the community to get something done, something important to itself, and each in some way tries to influence the work of civil government in the locality. Each group tries to get government to perform some work beneficial to its members, or it may try to

[17] Arthur L. Beeley, "The Unused Knowledge of Human Behavior," Commencement Address, University of Utah, August 26, 1955, p. 5.

prevent government from doing work which it holds to be damaging to its members.

Some agencies or groups in the urban community may be called socially oriented. A welfare agency, whose purpose is to serve others, is an example. Other organized groups are essentially self-oriented but are willing to serve social purposes "up to a point": A businessman's group may support a relief program for the unemployable but may resist a work relief program for the able-bodied unemployed.

In the following section we will examine some of the public services that are found in an urban community. In doing so, we will consider the relations between the public and private services in the community and will also consider the types of community groups that may have a special interest in one type of service or other.

## Public Aid to Private Work

Government represents the solid background of civil authority under which private work may be safely performed. Says Hawley, it "holds the police power through which it exercises many regulatory functions."[18] Man-to-man relations, whether in the labor market or in trade, in the office or the shop, are carried out under the protective aegis of civil authority. True, civil authority does not always operate ideally, and the little man when cheated or exploited by the big man does not always get redress, or he may fear to invoke the law, but in general, and in the long run, government does perform this function. In some respects this may be called the stabilizing role of government.

Government also performs public work which is of immediate use to people in their private work. It may dredge the harbor and build docks which enable larger ships to arrive; thus business is aided and more workers are employed. It builds and maintains roads and streets, or bridges over which goods and services are delivered, and it insures safe travel. It keeps the community clean so that the populace is protected against disease from unsanitary conditions. It inspects work places against the hazards to life and limb. It maintains surveillance over construction so that buildings may be safe for work or other uses. It extends its good offices during labor disputes to maintain industrial peace.

No less important, and in some respects more important than these regular services, are the emergency work services which only public authority

[18] Amos H. Hawley, *Human Ecology*, New York, Ronald Press, 1950, p. 229.

can provide in a crisis. In a time of disaster public authority is paramount. It can on such occasions order and execute any type of work, and does. In times of depression local government may provide work for the unemployed, although this type of public work may be controversial. The local employers' association and the taxpayers' association may oppose such public service, while trade unions and welfare organizations would take an opposite view.

## Public Utilities

A public utility may be any industry that supplies electricity, gas, water, transportation, or communication within a political area. In our discussion, this area is the municipality. But there can be other utilities: a public slaughter house, for example, a community grain elevator, a warehouse, or a public scale. The utility is a service that is recognized as indispensable to the community and "affected by a public interest," and it may be either public or private. If it is privately owned, or publicly owned and privately operated, the quality of its service and service costs are normally regulated by public authority. Some utilities, such as transportation, may be partly private and partly public.

Utilities that are public in some communities may be in private hands in other communities. In European cities streetcars, buses, and other common carriers are operated as a public service, but this is not true of all cities in the United States. In many American cities the debate between those who favor privately owned and operated utilities and the defenders of publicly owned and operated utilities is not finally resolved. On this issue the different organized groups in the community usually take sides. Private utility companies may sometimes spend much money in "lining up" various local groups to oppose the idea of their kind of service becoming a public one. Usually a utility becomes public if (1) it is not work that private enterprise is able to finance or operate at a profit, or (2) it is not being satisfactorily operated as a private service. In some cases private enterprise will ask government to meet the cost of creating a utility but will demand that its operation be in private hands.

Water supply and sewage disposal are types of service which are almost everywhere a public work. It would be almost impossible for private enterprise to provide these services at a profit, yet there have been and still are private water supply systems in the United States. Per capita water consumption in American cities is higher than in other countries, ranging from 60 gallons per capita per day in small towns to around 200 gallons in large cities (246 gallons per capita per day in Chicago, 1947-1948). The amount used depends largely on two factors: industrial needs and street-cleaning needs. Because of this high consumption, cities in some eastern states are being confronted with shortages, and some type of interstate or federal regulation may one day be necessary. How much per capita consumption is enough is a question that draws many answers, but it is pertinent to note that water consumption increases with increasing urbanization.

Earlier we mentioned the expectations people tend to have regarding communities. Not infrequently the negative or favorable judgment of a place or a country will include the quality of the utilities. The effectiveness of the utilities are reflected in the conveniences and comforts that may be enjoyed. But that is not all; effective utilities are needed for industry, and their absence is a hindrance to work.[19]

## Educational Services

One phase of education, that which is primarily associated with the schools, is concerned with the formal imparting of knowledge. Another phase, usually identified as training, is concerned primarily with the imparting of skills. For our purposes the two phases can be included under the term education. In all Western countries education may be either a public or a private service and in most communities it is both, although the responsibility for setting the minimum standards is almost always a public function.

As community life becomes more complex the demands on education become greater, and special demands are increasingly made. The individual today needs to devote more time for a similar education than did his father and still more than was necessary for his grandfather. His education is more costly than was that of his father and still costlier than in his grandfather's time. Teachers need to be better prepared, more equipment is needed, and more administrative organization is required today than ever before. Even though children in the urban community, percentagewise, are fewer than a few decades ago, the organization needed for their education is much greater.

[19] The development of utilities and public works in Africa south of the Sahara is an urgent need. The natural resources in that vast region cannot be exploited because of distances. Railroads and roads are needed. Coal is scarce and electricity must be developed so that factories can start, towns can be built, and transportation provided. This must be public work, since the cost is too great for private enterprise. See *Investments in Overseas Territories*, Paris, OEEC, 1951, pp. 28-33; also see *Water Supply and Sewage Disposal*, Paris, OEEC, 1953, a report by a group of experts on these two utilities in American cities and a comparison with European cities.

This evolution in the United States has not been without difficulty, but the American experience has been fairly uneventful when compared with the evolution of education in some other countries. Elementary education in England could not advance for years because of its control by the dominant church, a control that was partially broken in 1870. Trevelyan states:

> By the religious compromise of 1870, England was enabled to obtain, better late than never, a system of universal primary education without which she might soon have fallen into the rear among modern nations. Between 1870 and 1890 the average school attendance rose from one and a quarter million to four and a half millions, while the money spent on each child was doubled.[20]

In the beginning of industrialization the urban community could get on very well if only a minority of the population could read and write. In the most advanced urban communities today, however, it is almost necessary for lower-level workers to have at least a common school education, for supervisors to have a secondary school education, and for persons in higher positions to have at least some college education. For particular kinds of work special training, and as the work a person does changes, additional special training may be needed. Dubin emphasizes that each type of work has its own language, and language changes much as tools and workways do.[21]

It is in this area of service that efforts to establish industrial communities in underdeveloped countries meet serious obstacles. The people are illiterate and without skills, and educational facilities are lacking. Xydias reports that in the Belgian Congo about 900,000 native children are in the schools, over 90 percent in the first three or four grades, and of these well over 90 percent are in church schools where religious education is given primary emphasis. Nearly half of the religious schools receive no state subsidy, but those that receive subsidies are being pressed to raise their standards. High costs prevent public education from expanding, but until education improves industrialization and urban development will lag. More advanced schools are only beginning.[22]

In the United States education is a community responsibility, which explains the differences between the school systems from one community to another; often the better schools are in the richer communities. This uneven

[20] G. M. Trevelyan, *English Social History*, London, Reprint Society, 1948, p. 586.

[21] Robert Dubin, *The World of Work*, Englewood Cliffs, N. J., Prentice-Hall, 1958, pp. 340-341.

[22] Nelly Xydias, "Labor Conditions, Attitudes, Training in Stanleyville, Belgian Congo," *Social Implications of Industrialization and Urbanization in Africa South of the Sahara*, Paris, UNESCO, 1956, pp. 319-333.

quality would not be found, for example, in a country like France where all education is directed and standardized by the Ministry of Education in Paris. In the American community the school belongs to the people, and what education should be or should not be is everybody's concern. While business organizations may try to lower the cost, civic groups of different kinds may try to expand the work of the schools. The trade unions usually take an active interest. All local associations become disturbed if for any reason the schools come under criticism. This interest leads to a competitive effort between communities to keep the local schools up to standard. The schools are looked to for a wide variety of services of which the following are typical:

1. Surveillance of the health of children by school clinics;
2. Instruction for sick or crippled children in the homes;
3. Nurseries and day schools for the children of working mothers;
4. Training for citizenship and the study of local problems;
5. Special classes for backward children or very bright children;
6. Guidance of wayward children and advice to their parents;
7. Vocational guidance to children in selecting occupations;
8. Technical training in preparation for life work;
9. Domestic skills and home-making courses for future housewives;
10. Guidance for leisure, playground supervision, school sport;
11. Classes for the guidance of mothers with small children;
12. Free or low-cost lunches for children during school hours.

## Health Services

As civil authority has the final responsibility for education, so it has for the health of the community, and always in the history of cities this authority has been invoked in crises. The effectiveness of such service depends on the availability of knowledge regarding the health hazards being faced. Conditions in a community which are recognized by all as health hazards today may not have been so recognized a century ago. The following is a description of living conditions in Manchester, England, about 1840:

> The greatest evil that arose from the crowding together of houses before there was any system of main drainage was that of the disposal of refuse, personal as well as household. Privy middens and ashpits, which was the accepted method, even in the houses of the well-to-do, were far too few. In the working class parts of the town, one privy for twelve buildings inhabited by seventy-eight people was common, in fact about a third of the whole population of Man-

chester lived under these conditions. The reason was that there were so few of these conveniences that they had to be put up outside the houses, and land being so valuable for building, the owners did not want to waste any that could not bring rent. There had been no regulations to force the builder to provide a back yard, a privy or ashpit for each house, and, in fact, builders had evolved a highly ingenious way of making two houses out of a plot of land that would normally hold one. . . . These privies and ashpits, as can well be imagined, soon became so full that the contents overflowed the seat into the road. . . .[23]

The quality of health service has gained with the advance of technology and science; in fact, few services in the community are so quickly responsive to scientific and technological advance. Much of the credit for this alertness must be given to the various private organizations, found especially in the larger cities. Some of these are engaged in research, while others are occupied with the treatment of either physical or mental health problems. They and the organized medical groups are the conscience of the community in health matters. The local public health service is also one of many functions. It keeps morbidity and death statistics, carries on research, performs laboratory services, and enforces health laws. Its work is seldom noticed unless an emergency condition develops.

## Sanitation and Community Appearance

Keeping the community healthy is closely related to keeping it clean and presentable. The man who sweeps the streets and the inspector who looks for rubbish-strewn places are among the functionaries in this service, but engineers, architects, doctors, and others are also engaged in activities related to sanitation and beauty. Cole mentions beauty as one of the essentials of urban design and he quotes Julian Huxley, "Beauty is part of the necessary emotional cement of society."[24] Beauty, or the appearances of the community, is the opposite of ugliness and disorder. With this in mind, neighborhood groups bring pressure to have the mosquito-breeding swamp made into a public park. Thus, as Tunnard observes, beauty is a reality.[25]

Sanitation means providing light and fresh air in buildings, ridding the city of unnecessary smoke fumes and odors, and it may mean putting pressure

[23] Simon, *op. cit.*, p. 286. Regarding health in American cities prior to 1900, see Robert DeForest and Lawrence Veiller, *The Tenement House Problem*, New York, Macmillan, 1903.

[24] William E. Cole, *Urban Society*, Boston, Houghton Mifflin, 1958, p. 569.

[25] Christopher Tunnard, *The City of Man*, New York, Charles Scribner's Sons, 1953, p. 347.

on householders to clean their back yards. Pittsburgh for years was called the "smoky city," not a pleasing identification for people living there. But something was done about it. Few industrial cities today are more free of smoke.

Community problems of sanitation and appearances are not confined to the city alone; some of the more disturbing ones are outside. Factories beyond the city limits may pollute the waters with industrial waste. Perhaps outside the cities are slums and shanty towns not serviced by sewers. Deshmukh reported on the colonies of homeless families outside Bombay and other cities of India. These families were found living ". . . along the railroad lines, banks of drainage streams, wastelands adjoining the dirty localities of the towns, open spaces earmarked for dumping the town refuse, etc.," and these places were usually without drainage, clean running water, or latrines.[26] Sanitation authorities are helpless because of the overwhelming number of people without means of support. That such conditions do not exist outside Western industrial cities is due, aside from the economic reasons largely to a long history of sanitation effort. Much cleaning up and beautifying has been necessary.

Sanitation is a shared responsibility of different departments of local government. Enforcement is often an unpopular aspect of this work because it involves different types of cost to private individuals. The work is better accomplished if the diligent public official gets support from the organized groups in the city, or if these groups bring pressure to bear on local government.

## Safety and Protection

Although the modern urban community becomes healthier and cleaner, it is still not free from hazards. Mechanisms come increasingly into use and hazards created by mechanical defects or misuse multiply. Technological change brings new mechanisms, and more people must learn to use them. Until they learn, children must be protected against them. The individual must not only beware of the mechanisms operated by himself but he must be on guard against mechanisms operated by others.

Hazards at work, on the street, or at home are avoided only as people learn to use the mechanisms. It is probably no less of a task than that faced by

[26] M. B. Deshmukh, "A Study of Floating Population, Delhi," *The Social Implications of Industrialization and Urbanization* (Five Studies in Asia), Calcutta, Unesco Research Center, 1956, p. 186. Deshmukh also reports that when these people apply to the authorities for medical care they are often refused because "if you are treated all of you will remain in the city and the congestion will increase." Here the problems of health and sanitation are one. P. 214.

primitive man in learning to avoid the dangers of wild beasts. It is now as then a kind of learning for which many persons must be responsible. However, too much is at stake to permit safety to be left to private volition. In many respects, as in using automobiles or elevators in buildings, safety is enforced by public authority.

In addition to the hazards inherent in mechanisms, chemicals, and other things that man uses, hazards may relate to the misconduct of persons. A man's valuables may be stolen from his home, or he may be robbed by force. Inconsiderate persons may intentionally inconvenience or harm others. Such persons may commit other crimes of the most serious nature, even murder. It is a basic responsibility of the organized community to protect its members against such dangres, and this is the function of the police service. Another protective function is to safeguard property against the hazards of fire. The fire department may spend more time teaching the community to be fire-protection conscious than it does in actually fighting fires.

Safety and protection are types of work that are attended to by the public services without much notice from the public. It is only when these services lose efficiency that the public is aroused.

## Services for Leisure and Culture

The subject of leisure in the industrial urban society will be examined more fully in Chapter 14, but something needs to be said here, and we need also to take notice of various cultural activities in the community. The problem of leisure is more of an urban than a rural matter, both because it has developed in relation to industrial employment as working hours have become shorter, and because the use of leisure for urban people more than rural people involves the spending of money, and hence leisure becomes business.

Moreover, in the urban community may be found a large variety of private organizations that occupy themselves in some way with leisure problems. They provide leisure activity for different groups, they train recreational leaders, they are organized with reference to their own leisure (dancing, sport, hobbies, etc.), or they are concerned with the social or moral problems arising in connection with leisure activity.

Most private organizations with an interest in recreation and the services for leisure expect some type of action on the part of the civil authority, although they may have different views about what that service should be. As Butler points out, local government in the United States is assuming responsibility to an increasing degree.[27] In other countries where the private organizations are less active, the initiative tends to be more with the public officials, both local and national. The services that civil authority may provide would include:

(1) Regulatory services to uphold local social values and to safeguard public morals. Certain types of entertainment my be banned or held within acceptable limits. Sports may be kept under scrutiny to prevent corrupt practices.

(2) Guidance and training may be provided as in the supervision of playgrounds, swimming places, camp grounds. Programs for special groups, such as mothers, may be sponsored.

(3) Facilities for leisure may be provided, including sport fields, playgrounds, swimming pools, dancing pavilions, parks, promenades, beaches, concert halls. The variety is great and the initial cost may be high.

When we speak of cultural activities in this connection, we need to keep in mind that the term is used in its popular meaning. It is not, quoting Herskovitz, "the man-made part of the environment," although it is that to a degree. The popular term is associated with social refinement and "implies the ability of a person who has 'culture' to manipulate certain aspects of our civilization that bring prestige."[28] In this sense, culture embraces the manners, the vanities, and the cultivated graces in select social groups. Once assumed to be a monopoly of the leisure class, these qualities are now more widely diffused.

Cultural activities in many Old World cities are major public functions. In Germany, Cologne is proud of its new municipal opera house. The people of Frankfurt took up collections and a great fund-raising drive was sponsored to get money to rebuild the bombed theater. With public funds some cities support opera companies, theater troupes, and orchestras. Culture in this sense is not so much of a public activity in American cities, although in some cities it is supported by private groups. However, in the American cities may be found more activity on the part of amateur theater groups than in most European cities, and perhaps more art activity among the hobby groups.

## The Competence of Communities

We have already noted in this chapter, and also in the previous ones, that in the urban community the individual is usually able to function better if he

[27] George D. Butler, *Introduction to Community Recreation*, New York, McGraw-Hill, 1949, p. 50.

[28] Melville J. Herskovitz, *Man and His Works*, New York, Alfred A. Knopf, 1952, p. 18.

associates in groups with other persons having like interests and purposes, and that he may belong to a variety of groups, as they relate to his work, his leisure, his social aspirations, and so on. These organized groups collectively try to achieve objectives important to their members. Some may exist to render services to the community (welfare, health, education, moral reform). Others may serve social objectives to some extent, so long as such services do not conflict with their special interests and purposes.

We have also noted in this chapter that most organized groups in the urban community try to influence government to get specific services performed, to effect changes in services being performed, or in some way try to control the distribution of public benefits. Since every public service involves a distribution of benefits, it is to be expected that different groups will promote their own interests and points of view.

The public services that were considered in the preceding pages are illustrative of the thought that getting things done in the urban community is in most cases a joint action on the part of public agencies and various private groups. The community, as MacIver observed, is "a system of ordered relations" including a network of groups, "presided over by a central social agency"[29] which is the civil authority. It is probably more correct to see the civil authority as holding an in-between position in this relationship. On matters concerning which there is general agreement in the community, public action is likely to be firm and direct.

But on most issues, there is "criss-cross and clash," and the civil authority may waver, finally taking a course which may be one of compromise. "In the conduct of government, compromise is proper as well as necessary, and the art of compromise is one that all but fools and the least successful dictators recognize."[30] Compromise may mean tardy public action or only limited action, which may be disturbing to some. For example, Sinclair speaks of his own London as having a standing record of "incompetence and mal-administration." He describes that city as a place where "nothing is done until it is too late, and where things are not always done then."[31] Many Britons are not that impatient about London.

Earlier in this chapter we indicated that persons visiting another city usually arrive with certain expectations. What one really finds in going from

[29] Robert M. MacIver, *The Web of Government*, New York, Macmillan, 1947, p. 22.
[30] *Ibid.*, p. 289.
[31] Robert Sinclair, *The Big City, A Human Study of London*, London, Reynal and Hitchcock, 1938, p. 11.

one city to another, whatever his expectations, is that cities differ in their competence as well as their interest in getting things done. Tradition-rooted London may seem slower perhaps than such a relatively traditionless city as Chicago, but London does get its work done, and apparently not less well. This does not indicate a competence of a lower order, but merely one of a different kind, as, for example, the difference between San Francisco and Los Angeles. Civil authority behaves differently from one city to another and in any city from one time to another—for example, when reform administrations are elected.

Here we must mention a pertinent point; namely, that the citizenry may behave differently from one city to another, as from one period to another. Citizen groups may have the best of motives, as many public officials may have, but they can also make grave blunders. Citizen groups about 1900 were captivated by the "city beautiful" idea, which will be described in the final chapter. And there was the "garden city" movement, which also turned out to be very unrealistic in our kind of urbanized society. Often the very best schemes of one generation come to be regarded as costly mistakes in the next generation.

## Popular Vigilance and Public Service

It would not be incorrect to speak of the civil authority, comprising the different levels of public officials and other public workers, as also an organized group in the urban community. In the big city it may be a cluster of groups. These groups have a professional interest in their work and are often the most articulate in expressing community consciousness and pride and in stimulating citizen participation in the many kinds of work that must be done. They in turn, as they are blamed or praised by citizen groups, are under pressure to improve.[32]

Civil authority, whether consciously organized or not, does acquire the characteristics of an organized group. It may, whether under leadership within or from without, evolve into a powerful and even a corrupt organization, a "machine." It may be dominated from the outside by some individual who has been able to gather to himself much power. Or the outside control may be in the hands of a small group of politicians joined with a few business or indus-

[32] On the relationship of citizenship to participation in community life, see Bert F. Hoselitz, "The City, the Factory and Economic Growth," *The American Economic Review*, Vol. 45, No. 2, May 1955, pp. 166-184.

trial leaders who stand to gain advantages. Such concentrations of power leading to corrupt public administration were much more prevalent in American cities in 1900 than today. Their existence led to various reform efforts through which certain organized private groups gained experience in using their influence for community service. Not all of the groups that now take active interest in community problems can be called reformistic. Some may be business organizations that do not benefit from the various corrupt combinations mentioned above.

We may say that corruption and public administration scandal in some American cities served a good purpose in that they aroused community conscience. This influence, among others, helped to render active and articulate a great many organizations which, if they existed previously, were only mildly active. Today the corrupt, as well as the inefficient, civic authority is more apt to be "cried out against" than a half-century ago. The so-called pressure groups have learned much about the ways and means of using influence, not only with respect to the shortcomings of civil authority, but against other groups. This way of behaving, so necessary in managing the affairs of large communities, is generally designated as popular vigilance.[33]

Vigilance is everybody's business, like the duty of voting. In the large community, however, while it may be individually exercised, in its most effective forms it is exercised by groups. It is a service to the honest public official who is interested in getting his particular kind of work done well. And it is the only community defense against civil authority becoming indifferent or corrupt.

## Summary

In order for people to live together in a community they must resort to different arrangements for accomplishing purposes which are important to them—either to all of them or to parts of the whole. The purposes of a com-

[33] In Great Britain, as in some other countries, cities are under national administrative control, but local civil authority acts with great independence. Local authorities are so interrelated with the private groups and prominent citizens that Parliament rarely intercedes. Interference or intercession would be resisted by organized groups as well as the local civil authority. Says Miller, "Everybody knows that areas and functions in local government are out of alignment; everybody knows that something ought to be done; nobody does anything to speak of, because it is recognized that local government authorities are a series of vested interests of a particular awkward kind." Thus each city goes on getting things done in its own way. Bruce Miller, "Citadels of Power," *The Twentieth Century*, Vol. 162, No. 968 (special number on "Who Governs Britain"), October 1957, p. 330.

munity full of people are of many kinds, and the volume and variety of purposes increase progressively with the size of the place. These purposes may relate to work or to leisure, to health and safety, or to a variety of normal needs. They are the general needs of the total community or special needs of certain of its component parts, but they are all of concern to all of the people sharing a limited environment, and sharing in different ways its limited resources.

In this chapter we have considered somewhat the role of organized secondary groups, as well as other groups, but our attention has been on the uses such groups make of government in the community. We have described government as an instrumentality for getting things done. The incorporated municipality is a public service agency having within itself a number of branches or departments for performing particular types of services. The municipality is also a legal authority, responsible at the same time to local people and to a higher government for maintaining order. In large measure, its order-maintaining fuctions are realized through the provision of services.

The services provided by any local public authority, while including some that are found in all communities, also include those which are peculiar to that community. Or a similar service provided in different communities may have peculiar characteristics in each of them. This evidences the extent to which local public service can be adapted to particular local needs which for social, economic, or other local reasons are thought to be necessary. Community wishes are made articulate through local groups or through other means.

We took note that the most meaningful evidence of citizenship is found in ways that people, as individuals or in groups, participate in different community efforts to get things done. In most respects these efforts are an outgrowth of a relationship between the citizenry and the civil authority. Citizenship in this sense is the product of a learning process which is more advanced in some cities and less advanced in others. It is more advanced in American communities and in most European communities than in the less urbanized and less industrialized places—for example, the so-called underdeveloped countries.

A number of types of public services were described in the chapter and it appeared that each of these may have a different value for different organized groups or sections of the population. Sections of the community favoring more of a particular type of service, or initiation of some new type of service, may be equally unfavorable to some other type of public service. Community life with respect to getting things done is in many ways a competitive relationship.

The civil authority, while dominant, typically stands on neutral ground in these contest relationships and usually follows some compromise course, perhaps changing to another course later when the issue is not in debate.

Civil authority is itself a type of organized group with its own continuing interests. It may, however, assume a tightly organized form under political leadership within or from without, with results not beneficial to the community. Or the civil authority may come under the control of various combinations of economic and political interests. American cities have lived through a period of municipal corruption due to such control. It seems that with the present growing vigilance of organized groups—not so vigilant in the days of corrupt urban politics—the likelihood of such political control of local civil authorities diminishes. Such vigilance evidences a rising quality of citizenship and the promise of an increasing social usefulness of the instruments of civil authority in getting things done.

CHAPTER 13

# Work and Its Place under Urbanism

THIS CHAPTER IS CONCERNED with work, particularly as it relates to urban people. As we will note, man's application to work, because he lives by it, permeates all other interests of his life; and this is true whether he lives in the city or the country. Necessarily, the chapter will deal with work as seen at close range, but we need to keep in mind that there is also a wider perspective, one which applies more to urban than to rural work. Urban work, which is best described as industrial, is interrelated not only in local terms but in networks that extend into wider areas. Because it is interrelated, it is also highly interdependent, and this interdependence tends to be as wide-reaching as industrial urbanism itself.

In the previous chapter we recognized the close relationship in the community between private work and public work. This is but one aspect of a network of interrelationships through which most of the productive activity in modern society is linked. This phenomenon is new in the history of labor; man's work is global, the extreme opposite from work in a primitive society sustained by a subsistence economy. The confirming evidence for this observation is seen everywhere. We need only to ask whence came all the items that make up the family dinner—all the equipment needed to set the table and all the equipment and utensils needed to prepare the food. Or we might look at the work of the family head, and ask how it is related to work done by others, and so on. Or we might ask about the destination of all the products, and services, that are put on the market by any industrial city.

## Work in Town and Country

Seen in such global dimensions, all work within the influence reach of industrial urbanism tends to be urban-oriented, and in various respects urban-dominated. The farmer must read the city paper to learn the price of eggs, butter, meat, grains, and other land products. Yet, there are differences between rural work and urban work; in fact, the social implications of work are different for rural and urban workers.

It is sometimes said that rural work is oriented to the natural environment while the work of the city man is oriented to a man-made environment. This, however, is not a sufficient distinction. Rural work is also more tradition-oriented, and is often enmeshed in a network of folk beliefs or is limited variously by well-established taboos, which is seldom true of urban work. The tools that the urban man uses are mere artifacts which are quickly set aside when better ones can be found, and his working organization is rationalized and may be changed at any time if that seems advisable.

Rural work is much more embedded into all parts of the culture. And when we speak of rural work we must include a wide variety of people, from the modern farmer to somewhat primitive peoples, who also send products into the world markets. For most of these, work is not something measured by the clock a few hours daily; it is a good part of daily living. They apply themselves as the work needs require, and these needs tend to be situationally defined.

A very great difference between rural and urban work is also found when we consider the expectations of people with respect to rewards. The urban worker, since he associates work with the sale of his time, tends to be quite concerned with getting more money for the time he works. If he cannot get more money per hour or per week doing the same work, he strives to find work that will net more money. The rural worker is rarely disturbed by such aspirations. The rewards for his labor tend to be fixed by the established elements in his environment: the relatively fixed amount of land, the relatively fixed methods of work, and the equally fixed social attitudes in the community. However, while such limiting influences are present in rural places generally, and even more so in tribal communities, they were not so fixed on the American land frontier, and they are not so conspicuously present in the more advanced of American rural communities. But even when these exceptions are taken into account, we must recognize that the idea of the individual trying to advance himself through his work is more of an urban than a rural trait.

Moreover, since for most urban people work is often an activity which does not involve their deeper personal interests, their attitudes toward it tend to reflect economic considerations more than others. Unlike the rural man, urban workers are less interested in and identified with the product of their labor. We must recognize there are exceptions to this observation, but the exceptions concern a minority of urban workers, those who have some chance to participate creatively in the production of goods and services.

## Work and a Man's Identity

We will return to these rural-urban comparisons, as they relate to work, at other points in this chapter. First, we must mention that these observations regarding the identification of the urban man with his work pertain primarily to the modern city of industry and commerce. They did not apply with equal force to towns and cities of preindustrial times. In the days of the guilds men were very much known for the work they did, and a great number of family names originated from this source. Carpenter in English is Zimmermann in German and Charpentier in French. Miner in English is Bergmann in German and Mineur in French. It might be interesting to speculate on the origin of such names as Wood, Stone, Clay, Land, Waters, Meadows, Plant, even Mudd. We further have such occupational names as Tinner, Tanner, Baker, Butcher, Miller, Shoemaker, Taylor, Needlemaker, Powdermaker, and many others. Lipson calls attention to many names from the textile field: Fuller, Webster, Weaver, Spinner, and others, and he notes how from this industry came such folk expressions as "spin a yarn," "weavers of long tales," "web of sophistry," and "thread of the discourse."[1] We would not expect such names to originate in the contemporary industrial urban environment, although that may happen in underdeveloped countries where industrialism is only now getting started.

A man may be identified with his occupation in other ways, however. Edwards wrote that "The most nearly dominant single influence in a man's life is probably his occupation." He elaborated this thought:

> More than anything else, perhaps, a man's occupation determines his course and his contribution in life. And when life's span is ended, quite likely there is no other single set of facts that tell so much the kind of a man he was and the part he played in life as well as a detailed and chronological statement of the occupation or occupations he pursued. Indeed, there is no other single characteristic that tells so much about a man and his status—social, intellectual and economic—as does his occupation. A man's occupation not only tells of each work day, but it indicates with some degree of accuracy, his manner of life,

[1] E. Lipson, *The Growth of English Society*, London, Adam and Charles Black, 1949, p. 53.

> the kind of clothes he will wear, the kind of a house he will live in, and, even to some extent, the kind of food he will eat.[2]

But the urban man may have a status which stems from his occupation although even his children may not know what his occupation is. They think of him as working for a corporation of one name or another. It is his income that determines the kind of clothes he wears, the kind of car he drives, the sort of neighborhood he lives in, and all these together help to identify him with a certain intellectual level, even with a social class. If he lives in a neighborhood, more is likely to be known about him and his work; it all depends on the amount of mobility and anonymity there is in the neighborhood.

Formerly, and in some European countries still, it was almost impossible for the worker in his community to shake free from the occupation with which he was identified. If it was an honored occupation, people would salute him, but, whether honored or not, the occupation was associated with fairly specific expectations regarding the worker's way of life. His occupation defined his role, and he usually accepted that role. His family belonged in that same "niche," which was in fact a class definition. If the father belonged to one of the hand-work occupations, it was assumed that his children would also be workers, and they were so received and educated in the schools. Pipping, perhaps thinking of his own Finland, wrote, "Children are 'listed' in the same class as their parents. Occupations, too, are not seldom inherited." But this, he noted, is much less true than formerly.[3] Pipping recognizes that this observation applies mainly to occupations least subject to change.

Caplow observes that the status of the individual in the village is defined in part by his occupation, for here the individual is also identified with his family history and activities outside his occupation. However:

> The metropolitan community is driven by its own complexity to associate each status with a particular economic function. The urban dweller tends to define his own relationship to his fellows in functional terms, since other means of identification with the community are attenuated. Then, too, the separation of home from work place and the necessity of casual interaction with many unrelated people requires various shorthand methods of recognizing others, of which occupational designations are the most convenient, after sex, age and race.[4]

[2] Alba Edwards, "Comparative Occupation Statistics for the United States, 1870-1940," *Population*, Washington, U.S. Census Bureau, 1943, Preface.

[3] Hugo E. Pipping, *Standard of Living, The concept and its place in Economics*, Kobenhaven, Ejner Munksgaards Forlag, 1953, p. 103.

[4] Theodore Caplow, *The Sociology of Work*, Minneapolis, University of Minnesota Press, 1954, p. 30.

## Identification of Places by Their Work

Butte, Montana, calls itself the world's biggest mining town, and any visitor to that city will very soon be made aware of its identification with a particular kind of work. Not only can he see the mines and the coming and going of workers, but he hears about mining from conversation. Butte is psychologically as well as occupationally identified with mining. In somewhat the same sense, Fort Worth, Texas, is identified with ranching and the cattle industry. Other economic interests are also present, but the main preoccupations of this city have been associated with that complex of activities symbolized by the ten-gallon hat. Gary, Indiana, was established less than half a century ago to be a city of iron and steel, and it has been that almost to the exclusion of other elements which lend completeness to a community.

A town or city that acquires a character which is identified with a particular kind of work tends to think and behave in relation to that group of occupations essential to its work. Akron, the "rubber town," is a good example of occupational identification:

> There is about the city the turbulent and impatient manner of a miracle town, with all the symptoms of a community that has grown fast and does not want to stop growing. In shops and restaurants and along Main Street, people move fast, eat fast, talk fast. It is as if the boom that rubber brought the town thirty years ago has subsided in the factories only, and still agitates the people.
>
> Neither booster pamphlets nor fact books are needed to inform the visitor that Akron is the rubber manufacturing center of the world; it is proclaimed in Akron's air, especially on hot summer days. It is written on the clothing and the hands and faces of the workers, who, unless they have scrubbed and brushed vigorously, bear a thin coating of soapstone. And rubber dominates their talk; they discuss the big industry expansion, the threat of decentralization, strikes, A.F. of L., C.I.O. and the future.[5]

It is only the exceptional city, however, that tends to be identified with one or another special kind of work, and the larger the city the less is it likely to have a special occupational character. Some larger cities that are associated with particular types of work are occupied with other industries as well. For example, Milwaukee and Munich, Germany, are widely known as brewery cities. But the visitor to either city is not made aware of this special occupation as he is made aware of the rubber industry in Akron. Paris is called the world's fashion center, but this is a very small part of the many activities that take place there.

[5] *The Ohio Guide,* American Guide Series, New York, Oxford University Press, 1940, p. 167.

## The Dynamic Quality of Modern Work

It would be difficult to overemphasize the ever-present element of change with respect to man's work under urbanism. This is not solely an urban phenomenon, since the influences for work change radiate outward and affect various kinds of work that are timed with or geared to urban work. If industry itself moves into such outer areas, it means, according to Salz, "a qualitative change of the economy, a frequently fundamental alteration of existing skills, and the introduction of new techniques in, and organization of, productive work."[6] Industry entering a new area may, in fact, not use any of the existing occupations, but certain skills learned in connection with such occupations may be transferred to new kinds of work.

People who previously knew nothing about industry may soon learn that the new work is not as stable as their old occupations were. The special tasks that they have learned to do well may change on short notice. If new machines are introduced and a new production arrangement instituted, they will need to be taught to do new kinds of work. Perhaps some skill used on the old task can be transferred, but it will need to be blended with other skills. This process is repeated with each change in the production process.

It could be that modern industry, some industries at least, will meet with less long-term difficulty in the underdeveloped countries than has been met in the more advanced countries over the past century. In these countries certain essential occupations were well established, using hand tools and a limited amount of machinery. The skills were the monopoly of organized groups, the guilds. Work secrets were kept within these groups and the ways of work were not confronted with the need to change, or the demand for change was at least very minor until the advent of powered industry. With their symbolism and their rituals for initiation and work performance, the guilds are likened by Caplow to a form of caste system.[7]

While eventually in some European countries the guilds, because of their organized resistance to industrial advancement, were outlawed, the social conservatism, which had become institutionalized in several respects, lived on. One of these institutionalized practices concerns the idea that the foreman in a factory not only should have served an apprenticeship in a trade or handicraft,

[6] Beate R. Salz, "The Human Element in Industrialization; A Hypothetical case study of Ecuadorean Indians," *Economic Development and Cultural Change*, Vol. IV, No. 1, Part 2, October 1955, p. 1.

[7] Caplow, *op. cit.*, p. 15.

but must also have become a master workman. Much of his training is in hand-work skills which usually have little to do with his work as a factory foreman. In the United States the apprenticeship system exists almost entirely in such hand-work occupations as the building trades. It does not exist in the industrial plants, although most industries have their "job training" systems. But in many European industrial plants the apprenticeship system still exists, the idea being deeply rooted that the youth must spend four years of learning, even to be a factory worker. While the idea is now weakening, the presence of these remnants of the guild system illustrates how stolidly traditional work ways stand against the entry of new work ways.[8]

Work in the industrial urban community is first of all nontraditional, but we must recognize that this is true in varying degrees. It may be more traditional for self-employed skilled workers than for industrial skilled workers. Managers may be more tradition-oriented than the engineers and technicians, but least traditional of all are the tasks of the great mass of workers: machine operatives, laborers, clerks, and others who handle mechanisms. Secondly, this work is functional; that is, it is concerned with immediate tasks, and is, therefore, practical work related to other practical work. Finally, it is sensitively responsive to changing needs.

## The American Ideology of Work

We can scarcely consider the subject of modern work without recognizing that the American record is unique. Many of the jokes told about Americans by foreigners hinge on their passion for work, their resourcefulness, and their proclivity for bragging about their work achievements. This is not accidental, but the outcome of a combination of circumstances, a development that must be seen in relation to the frontier. Any frontier is conducive to social change, both for the people who lived there before and for those who move in. On the American frontier the original people were pushed aside,[9] and a mixture of

[8] William Bascom, "Urbanization among the Yoruba," *American Journal of Sociology*, Vol. 60, No. 5, March 1955, p. 449. Bascom reports that Africans in the Yoruba cities still practice the ancient crafts of carving, weaving, and metal work, each craft having its mysteries. But the guilds of specialists are in various ways coming under Western influence. The tight family or group domination over these crafts is beginning to weaken.

[9] The American frontier was largely one of land resources. We can think of Kuwait on the Persian Gulf as one of resources below the surface. The land is desert. Here a quiet, very old pearl-fishing town has been occupied by American oil people. The local people are not pushed aside, but get half of the gain. Don Cook (*Herald Tribune* [Paris Edition], October 16, 1957) reports that 250 oil wells yield $300 million annually to the

people moved in. Some are content to say that the American capacity for doing things is due to the isolation of the frontier which was followed by the self-isolation ("splendid isolation") policy of the government. Other new countries were similarly isolated, but did not develop in the same way.

Several other factors in addition to isolation must be included to explain the American work ideology as it relates to the frontier. One of these is the mixture of peoples of different origins who, on the American frontier more than any other, stimulated one another in a culture-blending process. The new culture was the product of unguided mixing, and it differs greatly from the results in the French settlements in Canada where guided isolation has led to resistance to all change. To this isolation and heterogeneity of first settlers must be added the factor of social origin. The migrants who went to the American frontier came from the lower classes of Europe and they knew nothing about the ways of life of the cultured leisure classes. They knew only one way to higher social status and that was through work. The people who had other ideas about work, those who knew leisure and the cultural life, did not go to this frontier.

Still one other factor needs to be mentioned. We may call it the religious outlook that prevailed over most of the frontier. It is sometimes called the Protestant Ethic. The censor for right and wrong was an inner one, and the responsibility thus rested with the individual. Although it was an outlook which permitted a wide range of individual choice, the community code of conduct was strict. One must "work out his own salvation," and much that was considered right and moral hinged on work. Work was the chief protection against sin—for idleness opened the doors to sin—and hence work was the basic virtue. This was the prevailing ideal of the frontier, but it was an ideal which tolerated a high degree of individualism.

On the frontier, individualism was very much in demand. One had to be

---

group of sheiks, one family, who have ruled for 200 years. They have ruled by the simple Arab standards of "reasonable social conscience and outlook" with a sheik at the head of each department and with all decision making for the 10,000 population in the hands of the one family that has always had absolute power. The city now has more wealth per capita, more automobiles per capita, and a higher living standard than almost any city in the world. Most of the elements of a frontier boom town are present, but in fantastic contrast to that of the American frontier, and here the wealth came without work. In the view of this writer, a good approach to the study of the American frontier, by the method of extreme types, could be made in a careful comparison with this boom town which has suddenly moved from one way of life and work to another. Two books on background are H. R. P. Dickson, *Kuwait and her Neighbors*, New York, Macmillan, 1956, and Zahra Freeth, *Kuwait Was my Home*, London, George Allen and Unwin, 1956. Dickson was many years a British top official in Kuwait and Mrs. Freeth is his daughter.

able to stand alone, to be inventive in new situations, and, as Webb observed, to create new tools and new workways as necessary.[10] A man on the frontier took pride in his work, how much he could do, how fast he could do it, or how ingeniously he got a particular job done. And on the frontier he could enjoy a feeling of independence with respect to his work. De Tocqueville found that such attitudes of individualism and independence were general. The servant felt equal to his master. The relationship between servant and master was one of "covenant," beyond which they were equal citizens.[11]

Work was the measure of the man on the frontier, and originality paid dividends on work well invested. The situation, or the illusions it offered, never ceased to hold promise. Those attitudes were not set aside when Americans turned from the land frontier to building towns and establishing industry. They had become integrated into the American way of life, whether urban or rural, although as recognized by Gist and Halbert these attitudes may be differently expressed in country and city.[12]

## Social Implications of Work

As the American is described by foreigners, he is not only hard working but practical and resourceful in his work. He is impatient to advance himself and to accumulate money, he is always in a hurry, and he is not able to give himself to leisure without feelings of guilt, unless he can use leisure for self-improvement. He is likely to be generous with his money and if he acquires a fortune much of it after his death will be given to educational or some other social use. These and other qualities which make up the American stereotype found in the minds of most foreigners tend to cluster around *work*. If we think of Americans on the frontier and Americans up to the 1900-1910 decade we not only find these traits, but they were proclaimed in most of the biographies of "self-made" men.

As we have noted previously, these attitudes toward work, whether expressed or implied, were strongly supported by religious convictions: man must work out his salvation, man's faith can only be judged by his works, man must "earn his bread by the sweat of his brow." Even if the religious injunction was muted, the frontier itself inspired devotion to work, a striving for success,

[10] Walter Preston Webb, *The Great Frontier,* Boston, Houghton Mifflin, 1952, p. 13.

[11] Alexis de Tocqueville, *Democracy in America,* New York, Alfred A. Knopf, 1946, Vol. II, p. 181.

[12] Noel P. Gist and L. A. Halbert, *Urban Society,* New York, Thomas Y. Crowell, 1948, pp. 327-328.

and something like reverence for resourcefulness and creativity in matters of work. It is this complex of attitudes, whether it may be called a rudimentary ideology or not, that has been identified by some as the Protestant Ethic.

Whyte in *The Organization Man* has a chapter on "The Decline of the Protestant Ethic" in which he holds that this ideological tool that served so well in the days of land occupancy and initial expansion has lost its utility in a society dominated by great corporations, and while it was somewhat of a myth, it is now being replaced by another myth, the Social Ethic with its ideal of service to others rather than to self. "So I argue of the Social Ethic; call it mythology, if you will, but it is becoming the dominant one."[13]

Whatever their ideologies may be, or have been, people in the industrial urban countries who impose work upon themselves—one of Mumford's apt terms[14]—tend also to link their work with most other values. But occasionally other peoples who are not industrial and urban tend also to behave in about the same way. Such an example is provided by Redfield in his report on life in a Yucatan village:

> But it is a practical ethic that requires no promise or hope of salvation to support it. The people are not concerned with salvation. The villager is closer to Benjamin Franklin than he is to Wesley. He is concerned with earthly affairs, he sees the gods as rewarding man for his prudence and piety with good fortune on the earth and in this life. Piety and practical good sense are closely entwined; good husbandry is virtue, virtue hardly distinguishable from the virtues of domestic life and religious piety. To make the land produce and the cattle multiply is to be respected by man and to be blessed by God and the gods.[15]

Miller and Form observed that most of man's waking hours are occupied with work, and the competitive interests concerning work are the "source of much of his happiness and despair." And further:

> Almost every aspect of work is social. The production, servicing and handling of goods is essentially a cooperative or group process. In addition, the nature of our industrial economy makes all workers dependent on one another for the goods or services they buy or sell.
>
> In a narrower sense, work is sociological activity for it is often done in the

[13] William H. Whyte, Jr., *The Organization Man*, New York, Simon and Schuster, 1956, pp. 9, 14ff. "The young men speak of the 'plateau.' If they were to find this haven they would prove that the Social Ethic is personally fulfilling. For the goal of the plateau is in complete consonance with it; one's ambition is not a personal thing that craves achievement for achievement's sake or an ego that demands self-expression. It is an ambition directed outward, to the satisfaction of making others happy." P. 157.

[14] Lewis Mumford, *The Condition of Man*, New York, Harcourt, Brace, 1944, pp. 4-6.

[15] Robert Redfield, *A Village that Chose Progress*, Chicago, University of Chicago Press, 1950, p. 158.

> presence of others. Work plants bring many people together; and when people assemble they inevitably interact. Anything done in a social atmosphere becomes defined as a social activity. Thus the behavior or operations of work are soon evaluated for their prestige or social importance. The motives for working cannot be assigned only to economic needs, for men may continue to work even though they have no need for material goods.[16]

The writers just quoted see work as social in two ways: First, it is social in a very indirect way as people in one industry are benefited by the labor of people in other industries, through the exchange of goods and services in the market; and, second, it is social as workers on the same job establish interpersonal relationships. We must add that work is social in at least two other ways. The individual may acquire high or low prestige as he is known to be employed at one type of occupation or another. As we shall note later, occupations have different prestige value. And, an individual may have high or low social status even if people do not know what work he does. It depends upon how much money he can earn from his occupation and the skill with which he uses this money to gain social recognition. In other words, the type of work an individual does, coupled with the income derived, plus the necessary personality traits, enable him to be accepted into one social class or another.

## The Prestige of Occupations

In their study of German rural villages, Wurzbacher and Pflaum asked the villagers to rate a selected number of occupations. The list included 17 occupations, all of which were familiar in the villages. There was some difference of opinion—for instance, older persons tended to rate pastors and teachers higher than did the younger people—but these differences were minor. The ratings from highest prestige to lowest prestige were:

1. Physician
2. Public official
3. Employer-manager
4. Druggist
5. Teacher (upper)
6. Priest, Pastor
7. Teacher (lower)
8. Farmer (big farm)
9. Skilled worker*
10. Secretary
11. Agent, trader
12. Storekeeper
13. Skilled worker*
14. Farmer (small farm)
15. Conductor (bus)
16. Sales girl
17. Unskilled worker

* Skilled worker No. 9 is permitted to work independently, is often self-employed. Skilled worker No. 13 may work for No. 9.

[16] Delbert C. Miller and William H. Form, *Industrial Sociology*, New York, Harper & Brothers, 1951, p. 115.

The public official in these small places is often a postal or railroad employee above the hand-worker level. As a member of the *Beamte* (official) level he enjoys a higher social status than a clerical person of the *Angestellte* (employee) level, who in turn enjoys higher social status than does the hand worker, or *Arbeiter,* although he may earn less.[17]

Gadourek has assembled similar information about the status ratings of different occupations in a tulip-growing village in Holland:

1. Physician
2. Clergyman
3. Teacher
4. Exporter (of tulip bulbs)
5. Civil servant
6. Artisan*
7. Bulb grower
8. Storekeeper
9. Traveler (a salesman)
10. Foreman
11. Skilled industrial worker
12. Office clerk
13. Agricultural laborer
14. Horticultural worker
15. Unskilled industrial worker

* The artisan is a self-employed skilled worker.

Note that the clergyman is given the second place in the very religious Dutch community, but sixth place in the German ratings. The rating for the teacher is higher than in German villages, while the civil servant (public official) is rated lower.[18]

The idea of social prestige being attached to occupations was found by Xydias to be developing among African children in the Belgian Congo, in the industrial city of Stanleyville. These children, if not born in the tribal villages, were born of parents who migrated from such villages to the city. Of the boys in the primary grades, 51 percent wished to become skilled workers and 24 percent to become clerical workers. Of the primary school girls, 54 percent would seek employment in some needlework occupation, while 21 percent

[17] Gerhard Wurzbacher and Renate Pflaum, *Das Dorf im Spannungsfeld Industrieller Entwicklung,* Stuttgart, F. Enke, 1954, p. 33. In most European countries some kind of distinction is made between officials, white collar workers and hand workers. In Germany the *Beamte* is an official in the employ of some level of government (including railroad, post, etc.). He is a low, middle, or high *Beamte* and he receives not a wage or salary but a *standesgemasser Unterhalt* (support measured by his *Stand* or class). He has life security. One may not "insult" him. The *Angestellte* is a salaried worker, usually white collar and in private employment, but some are in public employment. The *Angestellte* has higher social status than the *Arbeiter* but lower than the *Beamte.* He may earn less than some in the *Arbeiter* class, but he may also earn more than some *Beamte.* Distinctions between the three classes are traditional and in many ways crystallized in the system of education, in work relations, in labor law, even in public housing developments. Such distinctions are found in other European countries as well.

[18] Ivan Gadourek, *A Dutch Village,* Leiden, H. E. Stenfert Krosse, n.v., 1956, pp. 277-278. The village studied by Gadourek included more than 3,000 population.

aspired to be teachers. It appears that these children are assuming industrial urban (to them, "European") attitudes toward occupations.[19]

Clément, associated with the same Stanleyville project, secured the prestige ratings of 12 occupations common to that city. He interviewed male workers only, and with these results:

1. Clerk
2. Teacher
3. Mechanic
4. Shopkeeper
5. Driver (auto)
6. Tailor
7. Shoemaker
8. Small farmer
9. Head salesman
10. Labor foreman
11. Boy (house servant)
12. Unskilled worker

They are, of course, occupations that in a European-dominated African city are within reach of the African worker. Clément found that these occupations tend to fall into three (African-conceived) social levels: Group I includes Teacher, Clerk, and Mechanic; Group II includes Shopkeeper, Driver, Tailor, and Shoemaker; while the remaining five occupations are in Group III. He found that workers in these respective occupations tend to select friends in the same groups, or try to make friends in higher groups.[20]

Caplow made the following tabulation of factors that figure in shaping attitudes toward different occupations:

1. White collar or clerical work is held to be superior to manual work (may lead easier to advancement).
2. Self-employment is generally held to be superior to being employed by others.
3. Occupations on which the work is clean are held in higher esteem than dirty ones.
4. Occupations with important business firms are more desired than those with small firms, but this seems not to hold for agricultural employment.
5. Personal service is regarded as degrading. It is considered more desirable to be employed with an enterprise than to be in the service of an individual.[21]

The prestige of occupations may be the result of social or other factors which are not related to the work itself. In a small industrial city which does

[19] Nelly Xydias, "Labor Conditions, Attitudes, Training," *Social Implications of Industrialization and Urbanization in Africa South of the Sahara*, Paris, UNESCO 1956, p. 357.

[20] Pierre Clément, "Social Patterns of Urban Life," *Social Implications of Industrialization and Urbanization in Africa South of the Sahara*, Paris, UNESCO, 1956, pp. 461-468.

[21] Caplow, *op. cit.*, pp. 42-43.

not experience much mobility, the top positions of a factory or other enterprise may be monopolized by certain families. The holders of such positions have a social standing which remains regardless of their occupation. They do not climb up occupationally to become managers or top technicians, although they may prepare themselves in schools. Persons in lower occupations cannot climb up to these pre-empted positions.[22]

Warner and Low observed that in "Yankee City" the traditional "skill hierarchy" of occupations by which workers had their status both in the factory and in the community is continually being broken by the technological changes which management must introduce. With each new machine or process or product new jobs are created and old jobs are set aside. New skill hierarchies are not given time to take form.[23] Workers must find other approaches to gaining social prestige in relation to their work.

In their study of an Egyptian factory, Harbison and Ibrahim found that concepts of social class, quite unrelated to the work, tend to keep workers, supervisors, and top management separated into three groups. Supervisors in their middle position have superior attitudes toward the workers, while they are themselves held in low esteem by top management. One does not rise from worker to supervisor or from supervisor to top management. This does not mean a lack of social prestige connected with occupation, which is also present, but the prestige is overshadowed by social class concepts which are related to occupation level.[24]

## Neutral Quality of Urban Work

Much is often said of the dullness of work in the urban community, of routine jobs that do not challenge the interest of the worker. Tilgher wrote that people toil on jobs that have no apparent connection with their inner lives, repeating the same movements over and over "as if they were machines and not human beings."[25] We cannot gainsay the complaint that much of the

[22] See F. van Heek ("The Method of Extreme Types as a Tool for the Study of Causes of Social Mobility," *Transactions of the Second World Congress of Sociology*, London, International Sociological Association, 1953, Vol. II, p. 393) for an example of a city in Holland where management positions are a monopoly of a limited group of families.

[23] W. Lloyd Warner and J. O. Low, *The Social System of a Modern Factory*, New Haven, Yale University Press, 1951, Chapt. V.

[24] F. H. Harbison and I. A. Ibrahim, "Some Labor Problems of Industrialization in Egypt," *Annals of the American Academy for the Political and Social Science*, Philadelphia, Vol. 305, May 1956, pp. 114-124.

[25] Adriano Tilgher, *Work, What it has Meant to Men through the Ages*, New York, Harcourt, Brace, 1930, p. 149.

work in the industrial urban environment is deadening and dull, but we may very well question whether the proportion of such work is as great as is sometimes supposed. However, there is one quality which most urban work has, the quality of neutralism or impersonalism, which may negatively affect the interest of workers in their jobs.

The thought is expressed by Saltz who describes industrialism as being neutral and efficiently operated only as it maintains an impersonal relation between the production process and people who work.[26] The management of urban work is neutral and impersonal, much as law may be so described. The idea is expressed by Drucker in these words:

> In fact, the worker no longer produces, even in the plant; he works. But the product is not being turned out by any one worker or any one group of workers. It is being turned out by the plant. It is a collective product. The individual worker usually is not even capable of defining his own contribution to the productive organization and to the product. Often he cannot even point to a part or a process and say: this is *my* work.[27]

This neutral relationship, Drucker adds, is not a characteristic of manufacturing alone; it holds for banks, insurance companies, department stores, the railroad shipping office, and so on. The work is something separate from the personality of the worker. This applies even in the front office of an enterprise where top management sits. Thus, Drucker holds, "Social status, social prestige and social power cannot attach to the individual's work. They can only attach to the individual's job."[28] A job is a relationship in which the worker sells his time for a wage or a salary. The Drucker thesis is useful if it is not overworked, because a good share of urban people who hold jobs take an interest in their work even though their relation to the product is an impersonal one. The situation, however, is not one that encourages the individual to identify himself with his work in the way a farmer can.

Urban work is generally recognized as impersonalizing and neutralizing with respect to the personal interests of those who work. While such an observation is apparently correct for most urban people, it must be seen in relation to the urban situation as a whole, which it reflects. The isolating of the person in the city is evident in other respects as well, two of which need to be mentioned. The first of these is the impersonal flow of activity, the on-going of urban life. When the urbanite rises in the morning he knows that

[26] Saltz, *op. cit.*, p. 5.
[27] Peter F. Drucker, *The New Society*, New York, Harper & Brothers, 1950, p. 5.
[28] *Ibid.*, p. 7.

the milk and his newspaper have been delivered. He knows that his eggs and bacon arrive at the stores daily, that his streetcar or bus will leave a certain point at a certain time, that when he arrives at his work all his fellow workers will be there, although they must come from different parts of the city. His telephone will be working and people will be calling him now and then, so that what is happening in his place of work is linked with work in other places. Not only does he know, but others know where he will be and with whom he will be at different times during the day.

The urban man functions in a wide variety of networks, and what he does must gear into what other people do, and what he and they do gears in many ways with what many others are doing. All these activities, guided by no single person, are interlaced into a coordinated activity network which is paced by a single coordinating mechanism, the clock. Even the urban man's wife, although not employed outside her home, must also pace her activity by the clock, and so his children who have no other tasks than going to school. In other words, urban people, whether they work or not, live in a situation of round-the-clock schedules from which there is no escape except the annual vacation. For some, certainly not the majority, this day-in and day-out regularity may be a strain, even though they do not recognize it.

The second type of urban depersonalization is the anonymity of the crowd. Whereas the regularity of schedule keeping and racing against the clock enmeshes the individual in a vast flow of activity, on the other hand he may be detached in the impersonal crowd and, as some tell us, is here immersed in loneliness. While he may walk among thousands, socially the individual in the crowd is sometimes said to disappear from notice and control. And there are some who feel that this is pathological. Hallenbeck describes this isolation:

> These appearances of anonymity have deeper roots, which have serious implications for the personalities and behavior of city people. When individuals are isolated from intimate continuous associations with others and have to make their decisions and take their actions largely on their own responsibility, they do so outside the most powerful of social controls—what other people think of them, and the gossip of the neighborhood. Many break under the stress of this responsibility, and others use the circumstances as a protection for nefarious practices and immoral self-indulgence.[29]

For the individual to lose himself, or to be lost, in the crowd exposes him to anonymity, perhaps to loneliness. He is not known and perhaps not

[29] Wilbur C. Hallenbeck, *American Urban Communities,* New York, Harper & Brothers, 1951, p. 42.

noticed. His contacts with people are impersonal. He may not be floating, as might seem, and instead of being lonely, he may be very preoccupied. Anonymity may be for him an escape into privacy. It is true that urban anonymity provides an ideal hiding place, but one must be urbane indeed to make use of it. Doubtless there are people in the city who break under the strain of being alone in the mass society, but they are relatively few.

Actually most people in this mass society are not detached from fairly close contacts. They have their families and their groups. If they do not have families, they are not without friends. Usually the urban man or woman has a greater variety of contacts and acquaintances—and we must emphasize variety —than would be possible in the small town.

## Classifying Occupations

Perhaps no aspect of urban life is more fluid than the pattern of occupations at which people are employed, and it in turn is conducive to change and fluidity in all aspects of urban life. If we try to classify occupations we find there are different ways of doing so and each method might produce a different type of pyramid. And there are different purposes for which occupational classifications might be made. Whatever methods are used and whatever the purposes of classification, the work must continue from year to year because of changes in jobs and occupations. Industry uses labor to perform different kinds of work needed in the production of goods and services. A particular industry is a complex of mechanisms geared to one another as a single mechanism. How labor is used and the kinds of labor used depends in large part on the needs of the enterprise, and these needs change.

Florence classified factory workers on the basis of the uniform or adaptive character of their work. Workers in a uniform process include all those who work at tasks involving little or no individual choice. They do not design the task or think it through, or change it, unless instructed. They include about 80 percent of the labor force in the factory, although this 80 percent is divided in a variety of task-type groups. Some are brain workers (clerks), hand workers (assemblers), or machine workers, and there are others. The remaining 20 percent of the work force of the plant are identified with what Florence calls the adaptable processes. They are at the upper levels of the occupational hierarchy.[30]

Actually the hierarchy or pyramid of occupations (jobs) in a factory or

[30] E. Sargant Florence, *Labour*, London, Hutchinson House, 1950, p. 66.

other modern enterprise is not a fixed arrangement. We may say that the pyramid continues to be present but the component parts are in constant change, and the jobs which are found at the different levels continue to change. Two trends are generally present in this change. One is *simplification,* which aims at economy and efficiency by dividing various complex work processes which use labor of higher skill into a number of simple operations which use labor of lower skill. Simplification tends to lower the status of jobs and to reduce the prestige of workers.[31] The complementary trend to simplification is *specialization,* which does not include the mass of workers, but the few who hold key positions for planning, designing, inspecting, supervising, and managing. This is the group with responsibility, but the responsibility is usually precisely divided by function and closely integrated.

Not only does the hierarchy of occupations change in an enterprise, it also changes for cities as well as for countries. These changes affect the arrangement of occupations within the pyramid and they may also affect the general shape of the hierarchy or pyramid. Major changes may greatly influence the social structure within the community. This is illustrated by the changes in the British labor force between 1931 and 1951, here seen in percentage of increase or decrease:

| | (in percent) |
|---|---|
| Increase in the labor force | 19 |
| Increase in electric, gas, and water service | 62 |
| Increase in public administration | 74 |
| Increase in engineering, shipbuilding, and electrical manufacture | 123 |
| Increase in vehicle manufacture | 136 |
| Decrease in agricultural labor | 6 |
| Decrease in clothing and textiles | 14 |
| Decrease in mining and quarrying | 20 |
| Decrease in domestic service | 68 |

While the British labor force during the decades 1931-1951 increased 19 percent, the American labor force increased more than 55 percent. In many ways the amount of increase or decrease in the labor force affects industrial as well as occupation changes. Each of the changes indicated in the figures for Great Britain reflects a variety of other changes. What, for example, could be more disturbing to household organization than the decline of domes-

[31] Elliott Jaques (*The Changing Culture of a Factory,* London, Tavistock Publications, 1951, p. 250) identifies status with the work or job and prestige he links with the person. The status of the job and the prestige of the worker are both social evaluations.

tic service by 68 percent? The increase of 74 percent in public administration, an increase nearly four times that of the labor force, reflects changes in government. Certain of the changes in industrial employment are due to technological change and changes in the market for goods, but each in different ways affects the hierarchy of occupations at home.

According to the studies of Duncan and Reiss, "There is a declining gradient in the socio-economic status of the labor force from the largest urbanized areas to the rural groups."[32] The labor market of very large cities uses more workers who have special training. The percentage of higher-trained persons in the labor force increases with the size of the place from village to big city, while the percentage of untrained labor decreases. Thus, as a place grows from town to city to big city, there will be changes in the shape of the pyramid of occupations. In the course of urban growth there may be, and generally are, inner changes, because some types of enterprise may expand very much and others very little. These industrial changes, in diminishing or enlarging the labor market, or in using different occupations in changing number, may affect urban growth. Whether a place may grow or decline, the occupation pyramid changes.

## Marginal Groups in the Labor Force

A look at the labor market is sufficient to remind us that there is great variety among the people offering their services and great variety in the demand for services. Whereas the services offered are determined in different ways by the demand, the demand in various ways is influenced by the types and amounts of labor offered. In addition to the give and take between the supply and demand, other factors add to variety and change in the supply of workers. The most obvious of these are sex and age, but there may be differences in the capability of workers and differences which may be socially defined. All these differences tend to be present to varying degree from time to time, and they are present in both the rural and urban labor force.

During the lifetime of the individual he moves from one category of the labor force to another. He finds himself in one group as he enters the labor force as an inexperienced young worker, a member of a marginal group. As he gains experience and training he moves into another group, the qualified central body of the labor force. From here he ultimately moves out in one

[32] Otis D. Duncan and Albert J. Reiss, Jr., *Social Characteristics of Urban and Rural Communities, 1950*, New York, John Wiley & Sons, 1956, p. 96.

direction or another. The great majority move into that marginal group of the aging and aged. Others become chronically ill or permanently injured and move into the marginal group of partially employables.

Another marginal group in the industrial urban labor force includes the women, and we are reminded by Myrdal and Klein that women are in two roles, one being the natural role of bearing and rearing children, the other being the work role. All preindustrial societies had solved the problem of utilizing the labor of women in a double role arrangement that comprised a satisfactory division of labor, but this goal is being achieved only slowly in the modern society.[33] Not all the barriers are industrial; there is the widely held belief that women belong in the home, even if they have no function there except an ornamental one. However, since women have lost most of their traditional work to industry, the idea becomes more acceptable that women may and even should seek employment outside the home.

In most industrial countries, the United States included, women initially entered the industrial labor market as operatives in the textile mills, and from that they have entered other industrial fields. Hooks notes that in the United States between 1870 and 1940 women workers increased in the following occupations from:

903 to 1,862,154 for clerical workers,
893 to 475,685 for bookkeepers and accountants,
9,027 to 853,670 for sales personnel,
1,154 to 362,837 for registered nurses.[34]

These are a few of the extreme examples. The implications are to no small degree revolutionary. In most of the industrial urban countries today from 30 percent to 35 percent of the adult women are in the labor force. However, a good share of the employed women are in the younger age groups and for many of these employment is an interim experience until they enter marriage. Some who marry may leave and later return to the labor market. Others remain in the labor market after marriage, often more because of necessity than desire to continue working. But there is another group of women who actively desire to hold outside employment and they continue in the "two roles," those of worker and wife or mother.

Zweig, following a study of the subject, concludes that women are suit-

[33] Alva Myrdal and Viola Klein, *Women's Two Worlds,* London, Routledge and Kegan Paul, 1956.

[34] Janet M. Hooks, *Women's Occupations through Seven Decades,* Washington, Department of Labor, Women's Bulletin No. 218, 1947, pp. 208-223.

able for a great variety of jobs in modern industry. He found that many industrial managers prefer women workers because they are "more patient and long-suffering and placid" than are male workers and because "Frustration in industry is much more frequent among men."[35] According to this view, trends in the personnel organization of industry are such that we may expect an increasing use of female labor.

A somewhat opposite view is given by Dubin who reminds us that most industrial work performed by women is of the semiskilled type and that the result of the trend of industry in the direction of the "automatic factory," or automation, is to eliminate both unskilled and semiskilled labor. More and more the industrial workers will be specialists, and relatively few women become specialists. "Science and engineering are traditionally male specializations." He concludes:

> These limitations on future employment opportunities may be very threatening to women. There may not be sufficient alternative employment opportunities for them. This problem may be worked out, in part, by shifting our traditional preference for men in science and engineering. Women could then more easily gain entrance into schools in these disciplines preparatory to becoming automation specialists.[36]

Dubin then considers the likelihood of women retreating to the home, if job opportunities in the labor market disappear. One difficulty with his thesis, however, is that he has not taken into account wide areas of work. A larger part of the work in the modern community is outside the manufacturing field; moreover, the ideal of automation, if realized, would not include all of manufacturing.

While the problem of children in the labor market is a diminishing one in the urban community, since more and more time of childhood and youth is needed for education, youth remains a marginal group in the working population. The problems involved are less concerned with children than with later teen-agers entering the labor force. For these the problem assumes one aspect in times of full employment and still another in times of unemployment. During the Great Depression in the United States, special programs for youth were established—the work camps of the Civilian Conservation Corps (CCC), for example.[37]

[35] Ferdynand Zweig, *Women's Life and Labour*, London, V. Gollancz, 1952, p. 87.

[36] Robert Dubin, *The World of Work*, Englewood Cliffs, N. J., Prentice-Hall, 1958, p. 202.

[37] Howard M. Bell in a depression time study of youth in Maryland found that 40 to 46 percent were still unemployed a year after leaving school. *Youth Tell Their Story*, American Council on Education, 1938.

The marginal group of youth is a temporary one for those in it, since most members of this group move into the adult class after four or five years. Ferguson and Gunnison, in their study of teenage boys in Glasgow, found that 76 percent of these had their first "permanent" job before being out of school three years, but "one-fourth of the boys, three years after leaving school, still find themselves, at age 17, in jobs in which they have no interest."[38] Unlike that of rural life, much of the work given to urban youth is dull and constitutes a negative introduction to the labor force, although there are, of course, exceptions. Hollingshead made this observation:

> That the job is a means to an end—pleasure—is a potent contributory factor to the poor adjustment these youngsters make. The pleasures the job will enable them to purchase in their off hours are more important to many than what they do during working hours. From the stories they tell it is apparent that thoughts about these pleasures, flitting through their dreams while they work, often hinder the efficient execution of their duties. But memories of past pleasures and dreams of those to come form a pleasant fantasy world which helps the time pass when one is forced to spend long hours alone cleaning house, cultivating corn, polishing cars, washing milk bottles, or one of the half-hundred other monotonous, always menial, and often hard jobs these youngsters work at from 8 to 12 hours a day, six days a week.[39]

Whether his introduction to the labor force is easy or hard, the youth has the illusion of better days ahead. Ultimately, however, he will move into the marginal group of the aged, and this may become a dreadful prospect after he passes the age of 40, unless he has a profession or property. Pre-industrial social systems, by one method or another, afforded security to the aged, but the modern labor market is dominated by economic, and now also social, rules. Those who work must retire, usually at a stated age, but retirement, states Tibbitts, "is a relatively new phenomenon in our society, and the challenge of a new way of life for most Americans." And he reminds us that between 1870 and 1953 persons of 65 years and over increased from 3.5 percent to 8.4 percent of the population.[40]

In the less developed countries, meaning the less industrial ones, as Moore has demonstrated, the percentage of persons 65 and older in the population is less than in the more advanced countries. People are able in

[38] T. Ferguson and J. Gunnison, *The Young Wage-earner*, New York, Oxford University Press, 1951, p. 98.

[39] August B. Hollingshead, *Elmtown's Youth*, New York, John Wiley & Sons, 1949, p. 377.

[40] Clark Tibbitts, "Retirement Problems in American Society," *American Journal of Sociology*, Vol. 59, No. 4, January 1954, pp. 301, 302.

the more industrial countries to live to more advanced ages. He found that the ratio of children under 15 years to adults is higher in the less advanced than in the more advanced countries. The populations are younger because more children are born and adults die earlier.[41] The advanced industrial countries face the prospect of having an increasing percentage of old people. If ways are not found of utilizing them for more years in the labor force, these older people must be supported more years by the economy. But the increasing number of old persons has various social implications as well, for the thinking and behavior of an older population may be very different from the behavior and thinking of a younger population.

## Work Change and Social Change

It is well known that changes in the work people do may in different ways affect their social relationships. Such work change takes place in many areas, of which the following are examples:

(1) *New materials to work with.* Steel may replace wood, plastics may replace wood and steel, cement may displace brick, etc. In agriculture new crops may replace old ones.

(2) *New products to make.* New materials may give rise to new products, although in competitive situations new products are being developed from familiar materials.

(3) *New machines and tools are made.* As new materials or new products may require changes in work, they may also stimulate the making of new machines and tools which, in turn, may facilitate the making of new products and may change the occupations of people.

(4) *New work organization may develop.* With technological changes in tools and machines and with the making of new products, new methods of organizing the work in industry may be needed. This involves a realignment of the occupations being utilized.

(5) *Market changes come about.* Whether markets expand or decline, the work organization must be adjusted to meet the need, and employment contracts or expands.

These different types of change affect the work people do, how they do their work, how much work they have, and how much they will earn. Workers may have to set aside old lines of work and learn new ones. The individual

[41] Wilbert E. Moore, "The Aged in Industrial Societies." See Milton Derber, Ed., *The Aged and Society,* Madison, Wis., Industrial Relations Research Association, 1950, p. 25.

is affected in both his work and nonwork activities, and both family life and community life are affected. In fact, community life, in consequence, is in constant change, and some writers like Brownell, feel that this is undesirable:

> The textural change in the community during the last fifty years or so is in many ways an indication of social decay. It lacks human coherence. The social fabrics binding one man to another are fewer and often weaker. Though he is connected in a vast net in which millions of men gyrate and heave anonymously, his relations with any other one person are fewer, farther between, and usually feebler. This is a fundamental loss in human life that no number of partial contacts, anonymous relations, or specialized services can replace. The community in this process is lost and that is irreplaceable.[42]

Brownell is protesting that the rural type of community life is not surviving in the city. It may be asked, however, if this survival is necessary, since the urban man finds other ways of integrating himself into community life. That new type of integration is in large part oriented to the individual's work and, as we shall see in the next chapter, to his urban type of leisure.

## Work Change and Job Security

The urban community lives by selling its work. From the rancher it buys hides for leather and to the rancher it sells cowboy boots. It buys corn from the farmer, processes this corn, and sells packages of cornstarch to the farmer. If the urban community would live better it must do more work and it must produce with increasing efficiency. The fear is ever present that work will be less and work efficiency may decline. Florence observed that British workers have such a fear of unemployment that they can hardly cooperate in plans calling for work changes that would increase production. Not even the new system of unemployment insurance relieves them of this entrenched fear. Their old work habits also motivate resistance to change.[43] It is this fear of insecurity that draws workers together into trade unions. As Dankart observed, workers join together to achieve collectively what they cannot achieve by individual action, but he adds that trade unions also serve a social purpose for the workers.[44]

Beveridge, writing about this sense of insecurity associated with industrial

[42] Baker Brownell, *The Human Community,* New York, Harper & Brothers, 1950, p. 26.

[43] Florence, *op. cit.,* p. 22.

[44] Clyde E. Dankart, *Contemporary Unionism in the United States,* Englewood Cliffs, N. J., Prentice-Hall, 1948, p. 5.

change, concluded that workers come to acquire feelings of inadequacy if exposed to insecurity beyond certain limits. This is largely due to the lack of control they have over the processes of change.[45]

Actually the security problem is not only beyond the control of workers, it is beyond the control of employers as well. It is even beyond the control of organized labor and organized employers combined. Nor is it a problem that can be met by the local community, even big cities, no matter how well organized they may be. It is a problem that must be approached on the national level under the authority of government with all private organizations cooperating. In its widest aspects, work security is a global problem which is fully met only when all countries bound together in the wider markets for the exchange of goods and services enjoy full and effective employment.

## Summary

Work in many ways is social as well as economic, although the social implications may at times be indirect and vague. This is less true of rural work, since the work of the rural man tends to cut across all other life interests. The work situation and the conditions of work for the urban man, on the other hand, separate the activity of his labor from other interests. There are other differences between rural and urban work. Urban work is less oriented to the cycles of nature and more oriented to the man-made environment, more regulated by schedules and the clock and more interrelated.

But the modern urban man, the American in particular, is motivated by the desire for advancement and he is very conscious of work as being a means to that end. Particularly in the highly competitive American community, attitudes toward work are highly rational. Here there is the least resistance from the preindustrial hand-work traditions, but there are the relatively recent traditions of the frontier which emphasize creativeness and individual advancement.

Although it is less true than formerly that individuals and families tend to be identified by the work they do, a man's occupation may still be for him a source of prestige. Although it may be decreasingly true for many industrial workers, large sections of the urban population identify themselves with occupational levels, if not to specific occupations.

Places, too, may be known for the prinicpal work that is done there, but this observation does not apply to all larger cities. It is apparently more true

[45] William H. Beveridge, *Unemployment, A Problem of Industry*, London, Longmans, Green, 1930, p. 111.

of smaller cities and towns. What is important about larger places is that they are centers of dynamic effort where the ways of work are in constant change, stimulated by competition and facilitated by technological ingenuity. The result is that occupations as such are highly unstable. Changes in machines, materials, products, and markets may develop new occupations (jobs calling for particular skills) and eliminate other occupations.

Urban work is often described as neutral or impersonal. Since it is economic activity, it can hardly be otherwise. It cannot be socially oriented as much rural work generally is, which means that the social interests of the urban man are in a sphere apart from his job or occupation, except for the few who have interest-absorbing work. Industrial urban work is largely impersonal, and it serves an impersonal market. This may be a passing phase in modern work and other ways of identifying work with the worker may develop.

We noted that each community has its own unique labor force and that the labor forces in neighboring communities may differ widely. This fact is a reminder of the global interconnectedness of most modern work. We would, of course, be surprised if the work performed in two neighboring communities living on a subsistence economy were not about the same. The modern industrial urban community, however, is not living on a subsistence but on an exchange money economy. Its goods and services are sold not only in the immediate hinterland but in far-flung markets. The community that sends goods to all places also receives the products of labor from all places. The result is a global interdependence of work being done and of people performing this work.

We can describe this global-connectedness in economic terms, and we can recognize its political implications. But the social implications are equally important, for these bring us face to face with ways of living which both influence and reflect the work people do to sustain and to advance themselves.

CHAPTER 14

# Leisure: By-product of Urbanism

We can understand the life of a community in part if we know about the work people do and how they behave with respect to their work. We use the term "in part" because what to many in the urban community is of greatest interest does not concern their work time as much as their free time. This observation is increasingly applicable when we compare the life of communities from the least urban to the most urban. When we ask what people do when they are not working or going to work or coming from work, we are confronted with a vast and lively mixture of answers. Not only do urbanites engage in thousands of different activities when not working, but we must recognize that they are likely to enter these activities with more zest than most of them exhibit when occupied with their work. Admitting various exceptions, we must conclude that urban life for most urban people appears to be concentrated on leisure interests or in getting the money needed to enjoy leisure.

Not only Americans, but all industrial urban peoples are getting more leisure "thrust upon them" than they are prepared by their traditions to use. The gift of leisure gives concern to many while it inspires visions of cultural progress for others. The purpose of this chapter is to review some aspects of leisure as they relate to life in the entire urban community.

## Work Time and Leisure Time

Before coming to a definition of leisure, we need to consider leisure in relation to work, as both terms have to do with time. Lindeman observed that one does not have leisure if he has not obeyed "the injunction to work." One earns leisure as he earns his wages. The loafer has free time but he does not have leisure.[1] Work and leisure alike are ways of using time, but they are related uses of time. Man works so he may live, and if he is not effective all of his time must be used in finding his subsistence. If he is effective in his efforts, he may be able to live with less work. Thus, in Lindeman's terms, man has earned some free time. Modern man has been effective with the aid of technology and science to get more return for his labor.

It is not necessary here to review the struggle during the past century for the shorter work day and the shorter work week. That is too well known. Once the seven-day week was fairly universal, whereas in the most modern community the five-day week or the five-and-a-half-day week is becoming increasingly general. The 12-hour work day is a thing of the past for almost all workers and that is fairly true for the 10-hour day. It means that during the past several decades most workers have gained up to 40 hours of free time per week that was not available to most of our grandfathers.

As the hours of work have shortened, due to the increasing efficiency of work, the earnings of workers have increased. More goods are available than ever before and at prices which most people can afford. In other words, while work demands have decreased, the level of living for most people has been rising. Furthermore, most modern workers have vacations with pay, something not dreamed of a century ago. To this must be added the thought that in most countries workers are insured against their old age, so that people are relieved to some extent of saving and slaving for their later years. We have moved far from what the Hammonds called the Bleak Age in the mid-nineteenth century industrial cities of England:

> For no problem today is more urgent than the problem that arises with the growth of leisure and the spread of common enjoyment. At first sight it may seem paradoxical to suggest that the way to examine the problems of an age of leisure is to study the experience of an age without leisure. Yet a little reflection will show that there is a vital connection between the two.
>
> In the society described in the following pages, the lot of the great mass of mankind was supposed to be the routine of eating, drinking, working and sleeping. Leisure was the privilege of the few. The few therefore were educated to enjoy and use leisure; the mass of the population were educated for work, and for work that did not demand any considerable intelligence. Thus, for the general population, reason and feeling were left at the very lowest level. Standards of taste and culture were in the keeping of a small class which had

[1] Eduard C. Lindeman, "Recreational Planning." See George B. Galloway, Editor, *Planning for America*, New York, Henry Holt, 1941, p. 448.

> inherited with leisure, the aesthetic and literary sensibilities that had been developed by generations accustomed to an atmosphere of ease and elegance.[2]

Leisure, as we will consider in a moment, is time man has been able to salvage in the process of work. It is time which is unobligated and it is a problem only when there are not traditional provisions in the culture for making use of it (obligating it). The leisure class cited in the above quotation had acquired traditional uses for such time.

## Defining Leisure and Recreation

How much leisure is identified with work is indicated by the ever-present necessity of defining leisure in some opposite relation to work. Thus we hear such expressions as "work time and nonwork time." To Dumazedier,

> Leisure is a pursuit to which each man can devote himself according to his inclination, outside the demands of his work, his family and society, for relaxation, diversion or personal enrichment.[3]

Soule calls work a market phenomenon to which the individual must adapt, but which he can escape through leisure, which is "time one has apart from work. It may be employed in rest or complete indolence. It may, however, be employed in play."[4] Leisure is time which one may use as he wishes and such time need not yield a product. Actually this turns out not to be true. Although it is time not obligated for work, other obligations may be equally exacting. Thus the mother may be grateful for a few hours of leisure "so I can get something done." The worker may worry before his weekend comes how he is going "to get done what must be done" and still have some time for himself. The clergyman wonders that people don't have more time for church work. People have obligations to organizations and family that usually exceed their free time.

Recreation is a term more used by Americans than other people and they even use it synonymously with leisure. The following definition by Butler reflects the viewpoint of the National Recreation Association:

> Recreation has been variously defined as experience engaged in either alone or with others for its own sake and for the gratification in the doing; as an expres-

[2] J. L. and Barbara Hammond, *The Bleak Age*, West Drayton, England, Pelican, 1949 (1934), p. 5.

[3] Joffre Dumazedier, "Loisir et Pédagogie," *International Review of Education*, Vol. I, No. 2, 1955, p. 103.

[4] George Soule, *Time for Living*, New York, Viking Press, 1955.

> sion of the inner nature of man; as the satisfaction of basic human appetites; as a form of leisure-time experience in which physical, mental or spiritual satisfaction comes to an individual for participation in certain forms of activity. Expressed in terms of activities, recreation has been defined as an activity which is not consciously performed for the sake of any reward beyond itself, to which we give ourselves in our leisure time, which offers man an outlet for his mastery, or in which man engages because of inner desire and not because of outer compulsion.[5]

Sanders observed that recreation is the way people spend their leisure time, but some make the word *re-creation* and apply it "to what they and the community generally might describe as 'constructive uses' of leisure, giving it the idea of creativity, refreshment and renewal of energy for the work-a-day world."[6] He reminds us of the wide area of activities and interests covered by the term. It includes all sorts of activities engaged in by individuals or groups in all types of situations.

## Recreation for a Purpose

Nash holds that "leisure must be made to contribute to man's advancement, to aid him in his pursuit of happiness, and to give him a sense of worthwhileness. Otherwise, it will be a liability and lay the basis for his destruction." Man, he continues, "must grow through doing, achieving and creating."[7] The idea that people should engage in recreational activities that are constructive is found in much of the modern literature on the subject.[8] Leisure is regarded as a gift, something that man should use to enrich his personality, to gain knowledge; in short, to improve himself.

The point of view that recreation should have a purpose is not uniquely American, although it may be more vocal among Americans. Russell made the observation, apparently having in mind middle and upper classes, that American men work hard even when it is no longer necessary. They become indignant at the idea of leisure for wage earners, and they expect their sons to work as they have. "Oddly enough, while they wish their sons to work so hard as to have no time to be civilized, they do not mind their wives and daughters having no work at all." He adds:

[5] George D. Butler, *Introduction to Community Recreation*, New York, McGraw-Hill, 1949, p. 8.

[6] Irwin T. Sanders, *The Community*, New York, Ronald Press, 1958, p. 322.

[7] Jay B. Nash, *Philosophy of Recreation and Leisure*, St. Louis, C. V. Mosby, 1953, p. 63.

[8] For example, see Martin H. and Esther S. Neumeyer, *Leisure and Recreation*, New York, A. S. Barnes, 1949, pp. 13-22.

The wise use of leisure, it must be conceded, is a product of civilization and education. A man who has worked long hours all his life will be bored if he becomes suddenly idle. But without a considerable amount of leisure a man is cut off from many of the best things.[9]

Many Americans also feel that recreation should have a purpose so that man can get what Russell called "the best things." But here they think differently from Russell, to whom the best things are cultural values. They would use recreation to solve social problems by some systematic method, but they would also expect the individual to secure tangible values from his recreation. Thus recreation becomes for many what Steiner called "a kind of cult aiming at physical, mental and moral efficiency."[10]

Recreation programs, whether promoted by private organizations or public agencies, are very apt to be motivated by institutional objectives. Whatever these objectives are, they are symbolized by the professional recreational leaders. This must be recognized as a characteristic of Western society, and it is especially characteristic of the United States. How it appears to a non-Western person is illustrated by the following words expressed by an Indian woman who was a member of one of the round-table groups at the Corning Glass Conference:

I am wondering why leisure is a problem at all. Surely nowhere else in the world do people fuss about what to do with their spare time. I think it is rather sad that some kind of guilt has to be built up in this particular society so that people feel that they should be productive in their spare time. Production rolls with such a speed that now you feel you are useless unless you are making a bookcase or turning out a car or whatever you do. I think it is unfortunate that respect is gone for the man who simply sits in his rocking chair and thinks, if he happens to feel like it, or does nothing if he happens to feel like it.[11]

It must be added that these attitudes of puritanic seriousness about the use of leisure are changing, and perhaps nowhere faster than in the cities of the United States. Describing the thinking of The Organization Man, Whyte notes there is now an urge for more leisure and "more of the good life."

To the organization man this makes abundant sense, and he is as sensitive to the bogey of overwork and ulcers as his forebears were to the bogey of slothfulness. But he is split. He believes in leisure, but so does he believe in the

[9] Bertrand Russell, *In Praise of Idleness and Other Essays,* London, George Allen and Unwin, 1948, p. 19.

[10] Jesse F. Steiner, "Recreation and Leisure Time Activities," in *Recent Social Trends in the United States,* New York, McGraw-Hill, 1934, Vol. II, p. 913.

[11] Statement by Santha Rama Rau. See Eugene Staley, Ed., *Creating an Industrial Civilization,* New York, Harper & Brothers, 1952, p. 52.

> Puritan insistence on hard, self-denying work—and there are, alas, only twenty-four hours in the day. How, then, to be "broad gauge"? The "broad gauge" model we hear so much about these days is the man who keeps his work separate from leisure and the rest of his life. Any organization man who managed to accomplish this feat wouldn't get very far. He still works hard, in short, but now he has to feel somewhat guilty about it.[12]

Whyte is not thinking about the rank-and-file workers, but the man who is in the lower and middle ranks of management. In substance he is saying that however much lip service these leaders in American corporate life may give to the worth of leisure, they are so deeply rutted in the American work tradition that they cannot partake of leisure without feeling guilty about it, but they feel equally guilty if they overwork, and many must overwork.

## Leisure and the Preindustrial Society

Much that is said or written about leisure today relates to the industrial urban community, but a century ago even in the urban community the subject was, relatively speaking, of little interest. In 1850 the idea of workers having vacations would have provoked laughter, even among factory hands. Few of the private organizations that made business of leisure existed then; almost all that exist now are types of urban enterprise. In those days the American rural man had his Sunday for church going and visiting, and he looked forward to certain holidays: the Fourth of July, Thanksgiving Day, Christmas, and New Year's Day. Until recent times he did not heed Labor Day, that being mainly a city holiday. But the holidays just named also belonged to the city man. The rural pastimes inherited from frontier living included such community "get-togethers" as corn huskings, rabbit hunts, barn raisings, and the like, where work was mixed with fun making. Such early rural leisure-time activities have since been supplemented by various urban importations: baseball, basketball, the movies, radio, and now television.[13]

[12] William H. Whyte, Jr., *The Organization Man*, New York, Simon and Schuster, 1956, p. 18. On the origin of this Puritan urge in England, see Roy Lewis and Angus Maud, *The English Middle Classes*, New York, Alfred A. Knopf, 1950, p. 39. They note, "The Puritans set themselves to conquer and frown down the indolence of the working people, and indeed of people of all degrees, which had been noticed and commented by foreign visitors."

[13] William E. Cole may be right in the following expression about the need for pastimes by rural people. Some will disagree. "Although we would not underrate the importance of recreation for rural folk, rural farm people do lead physically active lives, whereas city people do not. City people, on the other hand, need physically active forms of recreation to compensate for the more sedentary nature of their lives and work. Though the rural world is not without its strains and tensions, they do appear to be more prevalent in cities." *Urban Society*, Boston, Houghton Mifflin, 1958, p. 333.

Many other American rural pastimes of the earlier decades were linked with the work of the people. Steer roping and broncho busting are very much just that. On holidays the timbermen had their wood-cutting contests and the miners their drilling contests, while sheepmen before clipping machines were invented had shearing contests as dairymen had milking contests before the advent of milking machines. In other words, the rural man made his fun and he was able now and then to find time for it; furthermore, these activities did not constitute an effort to get his mind off his work. The urban man deliberately turns to occupations unrelated to his work.[14]

The peasants of India are a most rural people. Moore reports that many of these peasants are very much burdened with debt "incurred by ceremonial expenditure." While people face food shortages, they still place high value on traditional social events and some spend as high as 10 percent of their income on such functions; it is the thing to do. Moore notes that various leaders are urging that something be done to diminish such activity, and he quotes from the autobiography of Pandit Nehru, who is aware of the waste aspects of such ceremonies, especially among the poor:

> The poor are also extravagant, even at the cost of burdensome debts, but it is the height of absurdity to say, as some people do, that their poverty is due to social customs. It is often forgotten that the life of the poor is terribly dull and monotonous and an occasional marriage celebration, bringing with it some feasting and singing, comes to them as an oasis in a desert of soulless toil, a refuge from domesticity and the prosaic business of life.[15]

Whether among American farmers or the peasants of other countries or the agricultural people of the underdeveloped countries, one finds that the burden of work does not wholly exclude some type and degree of leisure activity. It may, however, look different from leisure activities as we know them. For that reason Salz concluded that leisure is unknown to the Ecuadorean Indians. He found that these Indians had no concept of leisure and no problem of leisure. He therefore used the term "non-work" to describe what people did for fun or for social reasons, apart from work-a-day activity. He noted, however, that these same Indians have their family celebrations, for which they spend more money on liquor than they can afford, and they have their frequent fiestas. The Indian saves for these occasions, or he may

[14] At Essen, a steel city in Germany's Ruhr district, the companies sponsor occasional "hobby exhibits," which are occasions of great interest. At the 1957 exhibit someone pointed out that not one of the hundreds of hobbies presented had any direct relation to the work in the plants where the hobby fans are employed.

[15] Quoted by Frank A. Moore, "A Note on Rural Debt and Control of Ceremonial Expenditure in India," *Economic Development and Cultural Change,* Vol. II, No. 5, June 1954, p. 410.

go in debt. Salz also reported that Indians who go to the towns very soon fall into the leisure-time practices that prevail there.[16]

Writing about the Indians of Guatemala, Tax also mentioned the many social, religious, and ordinary fun activities for which they spend their money —much of it for liquor—and which he estimated took up 11 percent of local expenditures. These activities include parties among friends, baptisms, marriages, and various community festivities. "Indians speak of an eleven-bottle ceremony, a twenty-two bottle ceremony and so on. The social pressures are great to do right by the Saints being honored and it is difficult to reduce the traditional amount."[17]

Everywhere, however, the urban way of leisure activity encroaches upon these rural ways, and it gives rise to some serious problems because the new pastimes often require facilities that are not present in rural and underdeveloped regions. This was observed by Saville with respect to rural England where the lack of cinemas, dance halls, sport fields, and educational and cultural possibilities (also the lack of talent) turn the people cityward:

> We have to recognize the acceptance of what may be called urban values by our people, whether they live in town or country. Such values include a regular visit to the cinema or a reasonable proximity to a dance hall or to one or other of the multitudinous social and cultural activities that occupy the nation's leisure time.[18]

## Characteristics of Urban Leisure

This world-wide phenomenon, the leisure side of urbanism as a way of life, has characteristics which in many respects are not dissimilar from the work side of the urban way of life. Work is money-earning activity, while leisure is enjoyed mainly as one is able to spend money. (This thought will be developed further under the next section.) That spending money on leisure activity is mainly a means to various social ends is a theme that was developed by Veblen who saw money as the medium in the struggle both for social prestige and self-esteem.

16 Beate R. Salz, "The Human Element in Industrialization, a Hypothetical Case Study of Ecuadorean Indians," *Economic Development and Cultural Change*, Vol. IV, No. 1, Pt. 2, October 1955, p. 102.

17 Sol Tax, "Changing Consumption in Indian Guatemala," *Economic Development and Cultural Change*, Vol. V, No. 2, January 1957, p. 157.

18 John Saville, *Rural Depopulation in England and Wales, 1851-1951*, London, Routledge and Kegan Paul, 1957, p. 36.

> Those members of the community who fall short of this somewhat indefinite, normal degree of prowess or of property suffer in the esteem of their fellowmen; and consequently they suffer also in their own esteem, since the usual basis for self-respect is the respect accorded by one's neighbors. Only individuals with an aberrant temperament can in the long run retain their self-esteem in the face of the disesteem of their fellows.[19]

Urban leisure activity may be competitive in other ways than social; in sports, for instance, competition is of the essence. Any leisure activity that becomes popular stands a good chance of becoming professional, and to become professional means to become commercialized. One after another, leisure activities in the city have become commercialized for mass consumption. Thus horse racing, boxing, baseball, basketball, football, like the movies, radio, and television, have become leisure industries, each in many respects highly competitive.

Another characteristic of urban leisure is that people enjoy it increasingly as individuals. The individual becomes more the unit and the family less so. This, too, is in keeping with other aspects of the urban way of life. It is in keeping with the individuating nature of the urban labor market.

Moreover, urban leisure tends to be activity which is detached from the work interests of the individual. As urban work has taken the individual from his home, so urban leisure tends to detach him from his work. The groups and organizations he joins in relation to his leisure are often very different from those he joins in relation to his work. It is recognized, however, that the urban man may join social organizations in order that he may, through proper contacts, advance himself in his work; and to advance in his work may give him more prestige in such organizations.

Urban leisure-time recreation activities are more likely to be controversial than rural activities. If something new emerges in the way of leisure activity, it appears first in the city. But new ways of enjoying old activities may also be controversial. New dances, for instance, are apt to be protested by certain sections of the population. In the Christian community dancing was once called indecent, but later after dancing was tolerated the label was put on the various new dances. Sunday baseball has been a controversial subject in many American communities, but it began in the big cities and is now becoming very general. Sunday movies were also an issue and here, again, the taboos began relaxing first in the large urban communities.

[19] Thorstein Veblen, *The Theory of the Leisure Class*, New York, The Modern Library, 1931 (1899), p. 30.

Still another characteristic of leisure activity in the urban community is variety. Every type of leisure activity can be found in the city, whereas that is not possible in the small community. It is this variety that contributes much to the individuating influence of leisure.

## Leisure Activity as Consumption

As indicated above, leisure is usually associated with the spending of money. It would be difficult to find any leisure enjoyment that does not involve cost. There are many levels of money cost even for the same types of leisure: different priced seats at the theater and different priced skates and toys, and for any sport one would join there are different grades and prices for the essential equipment. If one wants to join a rowing club, for example, he will find that memberships vary from the poorest to the most exclusive club. Those who like to pass their leisure in camping will find that the equipment needed (or what people think they need) can be had at many price levels. Whatever game one would play or whatever amusement or entertainment he might turn to, he is a consumer of goods or services, and he must pay.

We don't know what part of the national income is spent for leisure. Not even the budget-honest individual can make such an accounting of his own spending. Bendiner, after considering the many ways that money is spent for enjoyment, concluded that all estimates in this area are "contradictory at best and extravagantly unscientific." He offered his own estimate, about 10 percent of the national income, which for 1956 would have been well over 30 billion dollars. He then discussed the fact that the cost of leisure is one thing, but what we get out of it is more important.[20]

But the cost of leisure is not a small item; in fact it concerns the economic life of various industries. Saunders and Parker estimated that for 1953 more than 30 billion dollars were spent on the enjoyment of leisure by Americans. And this estimate excluded such items as food, automobiles, home entertaining, and liquor.[21] This estimate was made for a year of full employment and high wages when skilled workers were buying motor boats and joining golf clubs, as only managers did twenty years earlier.

[20] Robert Bendiner, "Could You Stand a Four-day Week?" *The Reporter,* Vol. 17, No. 2, August 8, 1957, p. 13.

[21] Dero A. Saunders and Sanford S. Parker, "$30 Billion for Fun," *Fortune,* Vol. 49, No. 6, June 1954. They defined leisure as activity or expenditure "that is undertaken by choice, not of necessity, and is pursued for its own sake, not merely to avoid some larger expenditure."

Steiner estimated the American cost of leisure for 1930, a year of deep unemployment. According to his estimate, the leisure bill for that time was about 10.2 billion dollars. Among the items he excluded were home visiting, liquor, and gambling.[22] The Steiner estimate for 1930 was about a third the estimate of Saunders and Parker for 1953. There is no sure answer to any question we might ask about how much of the difference was due to depression conditions in 1930 and boom conditions in 1953.

In his history of the Great Depression, Wecter mentioned that between 1929 and 1933 the sales of amusement and sporting goods dropped by half a billion dollars. Between 1930 and 1934 federal taxes paid by athletic clubs, country clubs, and golf and tennis clubs fell by a quarter-billion dollars. Golf clubs lost a million members. Many clubs ceased to operate. On the other hand, sales increased on all types of home games. Card playing at home increased. The nation spent 10 million dollars on bridge lessons in 1931. Hobby activities increased.[23] These facts illustrate how all areas of leisure activity had to be modified with the thinning of the nation's pocketbooks.

## The Ambiguity of Leisure Spending

If there is any doubt about how or to what extent leisure interests and activities enter into all phases of modern living, particularly urban living, there is no better approach to the answer than to examine in detail the spending habits of people. The answer comes only in part from the clearly recognizable leisure expenditures: toys for children, sports equipment for youth, memberships in clubs, phonograph records, equipment for hunting, fishing, or camping. Much more is told about the relation of leisure to other areas of modern living by the marginal expenditures.

The family buys an automobile; what part of the cost may be charged to leisure and what part to general utility, and what part may be charged to display? The same question could be asked about the clothing people buy, the sports clothes, evening clothes, and so on. How much leisure enjoyment is derived from the furnishings in one's house, the hundred things that are placed on display there to be admired when visitors come? The dinner for guests, and the casual or formal party are home meetings that involve the spending of money, but we cannot definitely say to what extent the expenditure is one

[22] Steiner, *op. cit.*, p. 949.
[23] Dixon Wecter, *The Age of the Great Depression, 1929-1941*, New York, Macmillan, 1948, p. 219.

for leisure as against social duty. The line between social activity and leisure activity is not always clear.

Toynbee expressed the view that too much leisure for so many who received it suddenly without being prepared results in a "cultural proletarianization of life," and whereas leisure has always been misused by some, it is now being misused by many. What the many need is more cultural maturity.[24] In this sense, man must be trained for leisure much as he is trained for work, and the two types of training overlap at points. One may well ask if training and education in the pursuit of culture is part of leisure. Perhaps the answer is that it depends on the zest and interest of the person involved.

We need a wide and flexible definition of what leisure is. Denney and Riesman ventured the view that we must keep in mind *play*, a capacity for enjoying the useless; *recreation*, what some people think that play should be; *off-time sociability*, *intensively private pleasures*, and *intensively public pleasures*—the radio, movies, and so on. They arrived at this definition:

> Our discussion is concerned with all of these within the context of the larger issues of the use of leisure in modern industrial life. We employ the term "leisure" not only to refer to the time away from the work place, or the work hours and from home obligations, but also to refer to certain demands which are made on us not to squander unimaginatively the resources which industrialization has opened up to us. Leisure in that sense implies a quality of life which we seek to capture or recapture as an element in all of our attitudes.[25]

Leisure interests and activities become very much interwoven with other activities and interests, and the nearer they come to the deeper interests of the individual the less they can be separated out and counted. Leisure in this more intimate sense becomes as much a cluster of attitudes as it is a kind of identifiable behavior.

## Distribution of Leisure Activity

In thinking of leisure in general terms and considering the different sections of the population, the age groups, the sexes, and the classes, we find different kinds of activities and interests. Before the advent of mass commercial entertainment, class differences in types of leisure activity were well recog-

[24] Arnold J. Toynbee, *A Study of History*, Oxford, Oxford University Press, 1955, Vol. IX, p. 610.

[25] Reuel Denney and David Riesman, "Leisure in Industrial America." See Eugene Staley, Ed., *Creating an Industrial Civilization*, New York, Harper & Brothers, 1952, p. 246.

nized. Horse racing was the sport of kings, as was fox hunting. Only the well-to-do played golf. The opera was largely a monopoly of high society, although music lovers of the lower classes might get seats in the gallery. Class characteristics in leisure pastimes are still very much present, but White reminds us that there are age-level differences within classes, that the class ideas about leisure become more fixed with maturity.[26]

For leisure enjoyment we find several "generations," and people, as they grow from childhood to old age, move from one to another. Toys and playgrounds are the province of little children who must continually be observed by adults. Beyond this is another age when children are on their own. The interests of this group are different from those of the young children and yet not the same as those of older teen-agers. The next "generation" is between teen-age and marriage, at the age when the sexes begin associating. Then come the newly married with small children, and here there is one level of recreation for the children and another for the parents. The picture changes again when the children are older, and there is still another change when the children leave home for marriage. Not only is urban leisure activity different for each age level, but there are different types of organizations with which the different age groups associate.

These "generations" exist within each social class and, insofar as it is motivated by social aspirations, leisure activity is carried on within these age groups. The lower class teen-age social club, for example, will do all it can to equal the performance of the same kind of club of a higher class.

These various types of participation in leisure, however fitting they are to the situation, are unequally distributed; that is, the benefits of leisure are not shared by all in equal measure. This is partly due to the fact that the benefits of leisure are available on a pay-as-you-go basis, but inequality is present in other forms. Poor families may, for economic reasons, be forced to live in districts where play facilities for children are lacking. In such areas the quality of commercial leisure services may be lower than in others. Inequality may be also due to different types of social exclusion; for example, the discrimination against Negroes in sport and other leisure activity in some parts of the United States. As between large cities and small ones, between cities and towns, towns and small rural places, there is a conspicuous maldistribution of the facilities for leisure—commercial, public, and private.

[26] Clyde R. White, "Social Class Differences in the Uses of Leisure," *American Journal of Sociology*, Vol. 61, No. 1, September 1955, p. 150.

## Women and the Aged

We need not dwell here upon the complaint often heard that modern leisure activities have affected adversely the traditional unity of the family in the home. Most leisure activity takes individual family members by age and sex out of the home for their enjoyment. The wife and mother benefits least in this development. Whether the woman is an employed wife without children or a working wife with children or a housewife with children, her work day is long, despite the modern labor-saving devices which industry offers. Although some of the old drudgery of housekeeping has been taken away, the new ways of life add other duties.

The housewife without children, or the housewife with children who can employ servants, is in a different situation. Of her Hanssen wrote, "The position of the housewife in our culture is so unique and important that one is justified in calling this the epoch of the lonely housewife. She is the center of the family because she is the permanent keeper of the home with its large collection of prestige articles and clothes, all serving as group and ego identifications for the family members."[27] Whether she is an overworked housewife or a lonely one, her principal work usually ends when the children disperse to form homes of their own. Through her work life she had little leisure of her own, but was usually content to enjoy vicariously the leisure activities of her children.

It is true that there are vacations and holidays, but the wife and mother does not always share in these fully. In his study of a German industrial city Kieslich found that of 443 males interviewed, 60 had had no vacation during the previous year; of 426 females interviewed, 269 had had no vacation. That is, 14 percent of the males compared with 63 percent of the females had been without vacations, and 79 percent of the housewives had had no vacation during the previous year.[28] Vacations, Beveridge noted, are very much of an issue involving the wage earners (he uses "holiday" with different meanings):

> To give the wage earner a holiday does not in itself mean giving a holiday to the housewife and mother. . . . Nothing short of a revolution in housing would give the working housewife the equivalent of the two hours' additional leisure a day on five days of each week that has come to the wage earners in the past seventy years, and nothing but a revolution in holiday accommodation can give

[27] Börje Hanssen, "Dimensions of Primary Group Structures in Sweden." See Nels Anderson, Ed., *Recherches sur la Famille*, Tübingen, J. C. B. Mohr (Paul Siebeck), 1956, Vol. I, p. 137.

[28] Gunter Kieslich, *Freizeitgestaltung in einer Industriestadt*, Münster, Institut für Publizistik der Westfälischen Wilhelm Universität, 1956, p. 32.

the housewife with children the essence of a holiday, that is to say change and release from normal duties.[29]

Release from her round of duties usually comes to the wife and mother in her later years, but it is a kind of leisure that mothers often dread and for which some are not emotionally prepared. This applies especially to the one-generation modern family that begins to disintegrate when the children are grown, when what Dumazedier calls the traditional ceremonial functions of the home are no longer needed.[30]

Old age with poverty is no longer the rule in the modern community. The man retires upon reaching a certain age, usually 65, and his wife "retires" when her children disperse. In the modern community, as far as leisure is concerned, retirement segregates the aged into a marginal group. On the subject of retirement, Michelon made this optimistic observation:

> Soul-searching will reveal to the individual that the problem of leisure in retirement is really no problem at all. Rather, it gives the individual the best opportunities of his life, since for the first time he can devote his attention to pleasing himself and those immediately around him. It is an entirely new way of life rather than one of gradual transition. For the first time the individual is truly independent. . . . And if the development of the individual is the end and aim of a free society, retirement is the appropriate time for him to develop in ways formerly denied him.[31]

The phenomenon of old age, however, creates problems which come into prominence when we consider the tendency of most forms of leisure activity to exclude the old. Almost all commercialized leisure activity caters to the young and the strong, which was not true of earlier leisure forms. When we realize that the modern society will include in its numbers more and more old people, the problem assumes a challenging character. Seen in these terms, the recent tendency in the United States for many thousands of retired couples to buy trailers and to begin a life of wandering from place to place acquires new meaning. They have leisure and must find ways of using it.[32]

The aged person who is forced into leisure but who can turn to a cultural interest, as did Charles Lamb a hundred years ago, is in an entirely different

[29] W. H. Beveridge and A. F. Wells, *The Evidence for Voluntary Action*, London, George Allen and Unwin, 1949, p. 275.

[30] Joffre Dumazedier with G. Friedmann and E. Morin, "Les Loisirs dans la Vie Quotidienne," in "Civilisation de la Vie Quotidienne," *Encyclopedie Française*, T. XIV (14.56.10).

[31] L. E. Michelon, "The New Leisure Class," *American Journal of Sociology*, Vol. 59, No. 4, January 1954, p. 373. This article also reports on the efforts of old people to find a new life by going on the road in trailers and living in auto camps.

[32] G. C. Hoyt, "The Life of the Retired in a Trailer Camp," *American Journal of Sociology*, Vol. 59, No. 4, January 1954, pp. 361-370.

position. Lamb had worked for thirty-six years in a clerical position nine to ten hours a day. He was free on Sunday and he had a free day for Easter and another for Christmas. Then without anticipating it, he was suddenly retired with a pension. Although he had other interests to turn to, even he issues a warning:

> For the first day or two I felt stunned—overwhelmed. I could only apprehend my felicity; I was too confused to taste it sincerely. I wandered about thinking I was happy, and knowing that I was not. I was in the condition of a prisoner in the old Bastille, suddenly let loose after forty years' confinement. I could scarcely trust myself with myself. It was like passing out of Time into Eternity—for it was a sort of Eternity for a man to have all his Time to himself. It seemed to me that I had more Time on my hands than I could ever manage. For a poor man, poor in Time, I was suddenly lifted up into a vast revenue; I could see no end of my possession; I wanted some stewart, or judicious Bailiff, to manage my estate in Time for me. And here let me caution persons grown old in active business, not lightly, not without weighing their own resources, to forego their customary employment all at once, for there may be danger in it.[33]

## Commercializing Free Time

Sanders sees the community as a social organization in terms of major systems, of which there are six: family, the economy, government, religion, education, and social class. Under each of these he visualizes various subsystems. For example, under education and public information he lists the school, the press, radio, and television as subsystems. Recreation does not qualify as a major system, but is interspersed with the others. He sees recreation as falling into three subsystems: commercial, public, and private. Concerning the first-mentioned he declared, "There is nothing modest about the American commercialized subsystem of specializing in recreational activities." It strives to give the public what it thinks the public wants, and then proceeds to persuade the public to form a line at the box office.[34]

For a leisure-activity industry to survive and grow it must spread its influence far. Thousands must buy motor boats and camping outfits, tens of thousands every day must buy phonograph records, hundreds of thousands daily must start on vacation journeys, and millions must see the entertainment spectacles. The goal of each leisure industry is to make more people want the

[33] Charles Lamb, "The Superannuated Man," see W. Peacock, *Selected English Essays,* London, Humphrey Milford, 1939 (Oxford University Press, 1903), pp. 215-216.

[34] Sanders, *op. cit.,* pp. 190-191, 326.

same things and still more people do the same things. Thus Pipping sees these industries as powerful standardizing forces, not only in countries but between countries: "Customary entertainments become still more customary."[35] However, it is a form of standardizing that continually changes, and is not tradition making.

The big-business aspects of these industries may be responsible for the major income in some communities (Hollywood for movies; Nürnberg for toys; Monte Carlo for gambling and "high life"; Atlantic City for seashore vacations), but national interests may also be involved. Some European countries, for instance, would be in financial difficulty without the tourist trade. So important is tourism that the Organization for European Economic Cooperation issues a report each year on tourism in Western Europe. These countries in 1956 reported 105,000,000 "tourists days."[36] Counting travel, hotel, and other costs, we may estimate an expenditure of well over 3 billion dollars. The money value of motion pictures, as another example, is evidenced by the efforts of different governments to encourage that industry. In addition to the purely leisure-time industries, there are marginal industries that cater to both free-time and utility uses. Among these are the auto and vehicle makers, makers of clothing, beer, perfume, and so on.

On the economic aspects of leisure activity, Durant observed:

> All forms of leisure have become commercialized, endless devices are offered to the idle person, each to be enjoyed only on condition that he has money to pay. Without money he is condemned, unable to share in the pleasures and pastimes which press on him from all sides. But commercialization does not merely erect a gate through which only those with the necessary fee can pass. It has a profound effect also on the nature of the fare offered. The "machinery of amusement" is run by business men actuated by business motives. Their concern is not primarily with the character of the entertainment or amusement they provide, for it is merely a means to the end of making profits. For them good or bad means profitable or unprofitable. But profits can be secured only by attracting the greatest number.[37]

The commercialization of leisure is nothing new. It has been present in some form or other ever since the rise of cities, whenever a market appeared for such goods and services. The wandering singer, actor, or story teller is

[35] Hugo E. Pipping, *Standard of Living, The Concept and its Place in Economics*, Copenhagen, Ejner Munksgaards Forlag, 1953, p. 159.

[36] *Tourism in Europe*, Paris, Organization for European Economic Cooperation, 1956.

[37] H. W. Durant, *The Problem of Leisure*, London, George Routledge and Sons, 1938, p. 22.

now "in the big time" because technology provides him with a wider audience. The champion in sport can now be a highly paid professional. These are now mass-service markets and effective organization is needed. Talent has a market that hardly existed before. These industries will not be condemned out of existence, but they may need to be regulated if they offend community standards of taste and decency. But the problem of regulation is itself controversial, as we shall note later.

## The Private Area of Leisure

We cannot say what most people do with most of their "unobligated" time when they are not attending spectacles or enjoying some form of commercial entertainment. There are many private organizations to which the individual may belong, but these and the commercial pastimes alike do not take up all of his time. On this subject Hallenbeck ventured the opinion that the home may be more the center of individual leisure activity than is recognized.

> Most young children spend a large part of their time around their homes and in their own back yards, when they have back yards. Some of the modern recreational devices, such as radio, television, phonographs and records, are used largely in homes. Playing games, entertaining friends, parties, and conversation are home- and family-centered. The automobile has encouraged family trips for many purposes. A great many hobbies are carried on by individuals usually at home: reading, creative writing, sketching, painting, sculpturing, gardening, caring for pets, making things of wood and other materials, collecting of all sorts, and many others. Most of women's recreational activities, like knitting, embroidery, and fancy cooking, are home activities.[38]

Private groups organized for leisure or for other purposes including leisure are numerous and varied. They range from the exclusive country club to the "social and athletic clubs" of teen-age groups. They are normally of a membership type. But certain types of private organizations are institutionalized and have their branches in thousands of places. The Boy Scouts and the Girl Scouts are among these, as are the Young Men's Christian Association and the Young Women's Christian Association. In addition there are women's clubs and fraternal organizations for men.

Some types of private organizations concerned with leisure are not as much identified with membership as with service. To some extent, of course, all the national organizations are occupied with rendering leisure-time guidance

[38] Wilbur C. Hallenbeck, *American Urban Communities*, New York, Harper & Brothers, 1951, p. 465.

and service. Perhaps these private organizations make studies of community leisure-time needs, or they may train specialists who serve as playground directors or professional workers in the general recreation field. Not infrequently such specialized workers are in the employ of industries or of organizations that are not directly concerned with leisure-time problems but which, for community service reasons, support some special activity, much as the police in certain cities support boys' clubs. Perhaps the Chamber of Commerce in a city supports a summer vacation program for underprivileged children.

Another type of private organization may be very active in scrutinizing all types of leisure activity in the interest of safeguarding community morals. A women's club may concern itself with censorship of movies; another organization may occupy itself with censoring books. An association for the protection of children may have its watchers in dance halls and night clubs or on the beach to see that the police are on guard against indecent behavior.

These different private activities and organizations in the field of leisure activity are mainly urban, for they serve needs that are more urban than rural. Commercial organization for leisure is found in any sort of community where such business can be conducted with profit, but its operations are bigger and more varied in the urban community because there are more customers.

## Public Services for Leisure

A century ago in the United States public services for leisure were almost nonexistent. Grammercy Park was the first park in New York City, but it was privately owned by the surrounding householders. From 1634 for more than two centuries the Boston Common was perhaps the only city park in the United States. In 1853 Central Park was created in New York City. Until after 1890, the idea of open space reserved for the leisure of the people was regarded merely as another dream of reformers, and the same held for the idea of playgrounds for children. Gradually, however, people other than the reformers, after visiting European cities, came to similar conclusions about parks, playgrounds, and other public facilities for leisure.

The idea that there should be more public service for leisure was moving forward, although still with great resistance, when the Great Depression occurred. Millions were unemployed. The cities were helpless and the federal government took the initiative in using this labor for public purposes. Wecter tells us that through the Works Projects Administration (WPA), the government, cooperating with local communities

> . . . constructed tens of thousands of swimming pools and tennis courts, laid out or improved hundreds of municipal parks, golf courses and playing fields and, in cooperation with schools and extension services, supervised innumerable sports programs, employing more than 40,000 persons as of June 1939, in the role of recreation leaders. About half of this recreation program dealt with physical exercise, including softball, archery and shuffleboard. The rest lent great encouragement to the depression-sired revival of square dances, folk dancing, singing games and amateur drama.[39]

The WPA stimulated the idea of putting murals in schools and other public buildings, once considered a waste of money. The program of mobile libraries to visit country places was also a WPA idea as was, to mention but one other, the idea of low-cost lunches for school children. After the depression experience it was not possible for American communities to return to the old standards.

Public service for leisure holds an intermediate position between leisure activity provided by commercial enterprises and that provided by private groups. Under different conditions the public service is a benefit to the two others. Local government may develop a bathing beach but concessions may be given to commercial enterprises to operate there under regulation. Hotels may profit by locating near the beach—although they are taxed—and thus they share in the public cost of the beach. The city may build a community center at public expense and then assign the responsibility for managing the center to a private group, perhaps a committee whose members represent the various local private groups active in leisure-time services.

Government is the final authority representing the community interest in leisure activity. If boxing or wrestling is to be kept free of corruption, and the job is not being done by private groups, then public authority must be invoked. Whatever the moral standards of the community may be with respect to leisure activity, it is the function of local government to enforce such standards. It is the function of public authority to inspect theaters and other places of leisure activity that they may be safe against fire or other hazards. It is also a public function to protect all types of leisure activity against influence or exploitation by persons of ill repute.

[39] Wecter, *op. cit.*, p. 224. It is now being said that mass unemployment helped to make Americans leisure-conscious. We must add that the mellowing of local public policy toward public service for leisure illustrates how in times of emergency public policy may suddenly change. The forces resisting change tend to weaken in times of emergency and their arguments have less influence than in normal times. Once the change has taken place, it is difficult to return to things as they were.

## Moral Implications of Leisure

Red Smith of the *New York Herald Tribune*, in reporting on a horse race at Hialeah, Florida, on Washington's Birthday 1958, began his story with a quotation from *The Clinton Primer*, published in Philadelphia in 1832:

> Who loves a horse race? Are not too many fond of it? Does it not lead to evils, and too frequently ruin? Never go to a horse race. Mr. Mix had one child whom he called Irene; he had also a good farm and some money. He went to the races with his child dressed in black crepe for the loss of her dear mother. Here Mr. Mix drank freely, and bet largely, and lost all he was worth. At night he went home a beggar; took a dose of brandy, and died before morning, leaving his child a penniless orphan. Never go to a horse race.[40]

The tragedy of Mr. Mix involves two evils: drinking intoxicants and gambling; and, by American puritan standards, horse racing like card playing has been on the taboo list. We must add that dancing was also on this list for generations, and so the reading of novels, except novels with happy endings in which the villain is ignominiously defeated and virtue is rewarded. Theater performances were no less proscribed by these same standards, while actors and actresses were deemed to be a very unworthy class of people.

Crespi reported on a study of card players in Endicott, New York. His conclusions tend to challenge the old taboo:

> The great majority of card players, in opposition to popular misconceptions, play cards because they have discovered it to be an enjoyable and relaxing way of being together with friends and families and one which results in the strengthening of group ties. Another inducement to play cards is that the individual can hope to achieve acceptance into groups which will enhance his social position. We have concluded that card playing in Endicott, and probably in all of the United States, is essentially a group phenomenon and not a manifestation of social disorganization. The prevalence of card playing reflects not moral degeneracy but the struggle of primary groups to maintain their viability in the contemporary scene. Eager for friendliness and easy congeniality, many Americans appear to be incapable of generating such relationships without the artificial stimulation of impersonal, competitive group games.[41]

These attitudes of bias toward different types of leisure forms, being in the main rural attitudes, were reinforced by the fact that the forbidden activi-

[40] Red Smith, *New York Herald Tribune* (Paris Edition), February 26, 1958.
[41] Irving Crespi, "The Social Significance of Card Playing as a Leisure Time Activity," *American Sociological Review*, Vol. 21, No. 6, December 1956, p. 721.

ties were associated with nonrural people. The biases were part of a cluster of anti-urban attitudes. When cigarette smoking began that was also added to the cluster, another trait of the city people, at least of certain types of city people thought to be unwholesome. For rural people to migrate to the city meant to be exposed to all the assumed "slicker" characteristics of urban life. The automobile and good roads, the newspaper and then the radio, all resulting in more rural-urban communication, have tended to weaken the old moral front of ruralism.

It must be remembered that in large measure many of the institutionalized norms of the city, especially as they have become crystallized in the law, are of rural origin. There is a long history to this development, and it extends back to the time when most cities were small towns. Moreover, the urban population is ever being replenished by additions of rural people who bring their norms with them. Most of the reform organizations that stand guard over public morals, however, are not rural, but urban; the money for their support comes from urban people. Religious groups that participate in upholding high moral standards in all leisure activity, and in resisting leisure activity deemed to be detrimental are urban religious groups much more than rural.

Furthermore, these norms or values, which are frequently intolerant of innovation in matters of leisure, had their origin in a way of life which was oriented much more to work. Leisure somehow had to find its justification in work or spiritual up-building. Pieper, who calls leisure a divine gift to be used in spiritual renewal, would not object to a man sitting in his rocking chair, but such leisure is sheer idleness if not used for communion with the Divine. He holds also that one rests "for the sake of work and in order to work, and a man is not only refreshed *from* work, but *for* work." Moreover, such rest becomes true leisure activity only if one is able to capture the inner calm of religious devotion.[42]

In brief, the traditional social values regarding leisure place the conscientious individual in a defensive position when he is occupied with using his free time. Riesman observed:

> Even the adolescent who is engaged in "producing himself" suffers emotional discomfort if he cannot demonstrate that he is at work or training assiduously for narrowly defined work aims. In sum, taking together the young, the unemployed, the postemployment old, the housewife, and the guilty featherbedders,

[42] Josef Pieper, *Leisure, The Basis of Culture*, New York, Pantheon Books, 1952, pp. 51-52, translated from German by Alexander Dru.

not to speak of the "idle rich," we have a great number who more or less unconsciously feel some uneasiness in play—because by cultural definition the right to play belongs to those who work.[43]

## Obligated and Nonobligated Time

In the different definitions of leisure included in this chapter we mentioned the idea that leisure activity is something the individual enjoys outside other obligations. Dumazedier (footnote 3) calls it activity outside the demands of work, family, and society. The same emphasis on individual choice is either expressed or implied in other definitions; the writers of these definitions would agree that one could scarcely enjoy leisure outside his work and other obligations. Dumazedier makes it clear that while leisure may be outside one's obligations, it is still "interior to the social situation."

The satisfaction from leisure activity may be physical, as in sport, or it may be emotional, as in observing a spectacle or feeling some deep satisfaction, but it is rarely nonsocial. In different ways leisure activities serve social needs—recognition, friendship, and so on. The thought is much more helpful, if we are to achieve a philosophy of leisure, when we recognize an interpenetration of work-time activity and leisure-time activity, or we may say obligated and nonobligated activity. Then we will be able to understand that people with very absorbing work may not yearn for a vacation, or that others can turn from factory work to gardening and enjoy it, or that a man and wife may use vacation days to paper the house or do some house painting.

Most people enjoy most of their leisure within the frame of a pattern of obligations, in relation to the many things they must do as social beings. These include, as enumerated by Dumazedier, duties around the house, adorning one's person, performing the rituals and ceremonial functions of family life, performing as one should with reference to friends and neighbors, going to meetings, to church, and so on.[44] One may attend a wedding or a baptism as a family duty, but that does not prevent the individual from enjoying himself eating, drinking, and making merry. The individual may use his free time for study, but in connection with his study he may have a hobby. These efforts may be very much a leisure activity, even if the individual has the ulterior

[43] David Riesman with Nathan Glazier and Reuel Denney, *The Lonely Crowd*, New Haven, Yale University Press, 1950, p. 329. For those who may not know, featherbedding applies to a trade union practice of requiring that if a nonunion worker is engaged for a special task a union member must also be hired who is paid but does no work.

[44] Joffre Dumazedier, Friedmann, and Morin, *op. cit.* (14.54.4).

motive of one day advancing himself to a better job. Thus people may find within their obligations, to quote Marshall, "the qualities and virtues which might well find a place among the ends of the good life itself."[45] And this does not prevent the individual from making a wide range of choices about his use of free time.

## Leisure and Social Control

In the changing community the problem of regulating leisure activity is never solved. There are clusters of problems and groups differ in their views about each of them. Those who concern themselves with moral implications of the movies, for example, may complain about features which are of little concern to that group having an interest in the art side of the cinema. Both groups are interested in the subject, but from different viewpoints. For another example, old people and young people will differ about questions of behavior, and each group is likely to have conflicting ideas about what is right and wrong.

There is a growing group of people, many of whom are active in a professional way in leisure activity, which does not agree—at least most of them do not—that public morals are in jeopardy, but which feels that the physical man is on the verge of decline. He sits at ball games, at the movies, before television, or he sits drinking, and eating. He takes no part in leisure activities, being merely a looker-on. These critics see a condition growing which they call "spectatoritis," and they feel that a leisure revolution is needed to force more people to participate in activity. This is another call for guidance and regulation. Another group of experts answer that the "spectatoritis" fears are psychologically not warranted.[46]

These different groups with different viewpoints—and many others could be named—ultimately come forward with proposals that involve some responsibility for public authority. Perhaps more facilities or services are called for, or facilities and services of a different kind. In most democratic communities public authority functions with reasonable tolerance, patience, and fairness; perhaps slowly, but this is to be expected when sections of the public are hardly able to agree on what is to be done, how much is to be done, how it is to be done, and so on.

Actually, the major burden of regulation of leisure in the urban com-

[45] T. H. Marshall, *Citizenship and Social Class,* New York, Cambridge University Press, 1950, p. 129.

[46] For such a questioning of the "spectatoritis" argument, see the views expressed by Eric Larrabie in Staley, Ed. *op. cit.,* p. 67.

munity is not carried by public authority; it results in the give and take between the many kinds of private organizations, the press participating in various ways. In this relationship, one that involves mainly the elite groups, both competition and cooperation may be found, but the basic social values are recognized and sustained. Out of such a relationship of diverse groups, which is usual in the urban social organization, the atmosphere is likely to be freer and more tolerant than one finds in communities, particularly small places where social relationships are dominated by a single private organization. This is illustrated in the case of a Dutch village studied by Gadourek. The dominant group is the Protestant church. The situation is similar in other Dutch villages in which the Catholic church is dominant, but these observations apply with less force to cities of Holland.

> We have mentioned already that most sport associations and clubs are organized by the church members as such. The people of other denominations are not only unwelcome, but often openly refused membership or entrance. These social barriers are especially strong whenever the sport association or the club activity offers an opportunity for social contacts between young persons of different sex. Tennis and dancing parties are strictly denominational affairs. This division of all recreational and social life is not only confined to social strata. We notice two football clubs in Sassenheim as well as two lawn tennis clubs. For community leaders, the local elite, there exists a "neutral" club house (*Societeit*), but this is visited predominately by Roman Catholics and the more liberal-minded among the inhabitants; the more orthodox members of the Dutch Reformed Church and the Calvinists do not visit it, as a rule.[47]

Sunday, Gadourek adds, is for rest. While in Catholic villages Sunday football is permitted, it is forbidden by the Protestants. People will not ride buses on Sunday because that encourages the idea of the driver working, so they walk to church. If one drives his car to church he might be suspected of going somewhere for sport or pleasure.[48]

Probably the most effective force for regulation, whether in the village or the city, is the pattern of obligations by which each individual is surrounded. These obligations are the guiding forces with which he lives and to which he is responding through all his waking hours. He may resist some obligations, but he is not likely to resist all of them, for these influences symbolize group relationships. It is in these relationships that he enjoys prestige and gains his status, and only as he behaves in relation to them does he feel the importance of being an individual making his own choices.

[47] Ivan Gadourek, *A Dutch Village*, Leyden, H. E. Stenfert Kroese, n.v., 1956, p. 95.
[48] *Ibid.*

The same thought is better expressed by Foote in an article on the play implications of sex. In many of its expressions sex is furtively associated with play and that gives rise to fears, according to Foote, that granting play status to sex would lead to dangerous excesses. Those who entertain such fears forget that play develops strict rules of its own.[49] Stone, commenting on the observations of Foote, adds the thought that few leisure activities could be enjoyed without rules, that rule making and rule keeping are the natural conditions of participation in all leisure pastimes. It is the making and keeping of these rules that confer dignity on leisure activity.[50] We must add the observation that, were this not true, it would be difficult for sport to enjoy such universal interest.

Such interest in leisure activity permeates all society. There is no East-West cold war, for example, in sport or in the leisure arts, nor is there cross-border bickering about the rules. This does not mean that there is no need of control, because leisure activity is the greatest of influences for social change. Regulation and control may be needed, especially since so much service for leisure is commercial. How much control there should be and what the nature of that control should be would not only vary from one part of the world to another, but these control needs would vary from time to time. It is control that adapts to the needs and conditions of leisure.

## Summary

We seem to be living in a time when the wider implications of leisure are being recognized and increasing thought is being given to the relationships between leisure time and work time, or between work as one kind of activity and leisure as another. While the two concepts in earlier times were indistinguishably interwoven, there is a tendency in the modern community for leisure and the values associated with it to be separated from the values associated with work.

Leisure has been defined as time during which the individual is free from occupational, family, and social obligations and is privileged to occupy himself in ways suitable to his own wishes for his relaxation, diversion, and personal

49 Nelson N. Foote, "Sex as Play," *Social Problems*, Vol. I, No. 4, April 1954, pp. 159-163.

50 Gregory P. Stone, Comments on "Careers and Consumer Behavior," by David Riesman and Howard Roseborough. See Lincoln H. Clark, Ed., *Consumer Behavior*, New York, New York University Press, 1955, Vol. II, p. 26.

development. While the term is sometimes used synonymously with "recreation," we have used "leisure activity" because of its wider applicability, "leisure" being used to indicate free time. Although it is indicated that choices of leisure activity belong to the individual, it needs to be recognized that in different ways the choices made by the individual are socially conditioned.

Leisure under ruralism is much more oriented to the cycles of nature and the demands of work than is possible or necessary under urbanism. Rural leisure is more integrated with all other phases of living. It becomes natural for the rural man to be suspicious of leisure activity that does not have such an orientation. These values tend to be woven into the religious norms for work and living. These attitudes are very American, but not peculiarly American. They were very much present in the development of American urbanism. These attitudes express a fear of leisure activity that does not have a moral purpose.

Under urbanism, work for most people is separate from the home and tends increasingly to be separate from the personality of the individual. Leisure interests and activities, on the other hand, are very much identified with the personality of the individual. Like work, leisure emphasizes and induces individual participation. The individual finds much of his leisure activity outside his home. Under urbanism, leisure is generally enjoyed in money terms. The level and the type of leisure activity enjoyed depends largely on one's ability to pay. Thus types of leisure activity and levels of taste tend to be associated with social class levels. For these reasons leisure activity may be motivated by the social aspirations of individuals or groups.

In efforts to determine how much of the individual or family income is spent for leisure purposes we are confronted with many ambiguities. Spending for leisure may be intertwined with other types of spending: social spending, spending for education, and types of utility spending. These ambiguities evidence the extent to which the phenomena of leisure are involved in all phases of living, even with work.

It was recognized that the benefits of leisure are not enjoyed equally by all sections of the community. For example, the benefits of the shorter work day and work week as well as the benefits of paid vacations are not fully shared by housewives. But there are other marginal groups, such as the very poor and the aged, and in many communities certain minority groups are held in positions of disadvantage.

It is nothing new for leisure activity to be commercialized, but in the modern community the inducements to commercialization are greater than in earlier decades. The trend toward commercialization has been enhanced by technical advances which have made several large leisure industries possible, which are urbanizing in their widening influence. While government is the final authority for the regulation of leisure activity, private organizations in the urban community are considered equally, if not more effectively, as control and guidance agencies, even though views differ regarding the nature of guidance and regulation. Perhaps the most important control influence for the individual is the network of obligations, social and other, with which he is surrounded. It is in this network that his individual choices of leisure activity take on meaning and meet with social approval or disapproval.

CHAPTER 15

# Some Economic Aspects of Urban Living

IN THIS CHAPTER we are to consider the social implications of earning and spending as they concern the people who live in the urban industrial society, in particular those who are city dwellers. We found in the previous chapter that leisure interests and activities tend to take people out of isolation. In many respects, leisure is a consumption preoccupation, but it relates to earning as well as to spending. Work is a production preoccupation, but it relates to spending as well as to earning. Both production and consumption are activities which link people together* in ever-widening circles. Whether the modern man is earning his money or spending it, he is engaged in activity which involves him in many ways with others, both with people he may know and with strangers who pass him on the street.

To earn or to spend is to be joined in a vast impersonal network of economic activity, but through earning and spending one is also joined in various social networks which may also be far-reaching. One acquires social status in relation to his earning and spending. At different periods of his life the individual becomes a different kind of earner and spender. He is, moreover, confronted with various social expectations with respect to his earning and spending and these social expectations differ with social classes as they also differ as man moves from one stage of his life cycle to another.

## Capability and Self-sufficiency

Urban people live in a world of money values. While an urban man may be very rich, his visible possessions may be few indeed; his wealth is written

* For a theoretical statement of the relationship between sociology and economics, see Talcott Parsons and Neil J. Smelser, *Economy and Society*, Glencoe, Illinois, The Free Press, 1956.

on pieces of paper. This is very different from the rural man who owns 80 acres, 160 acres, or some other amount of land, a resource that can be cultivated to yield crops year after year. So long as he keeps his land he knows what his occupation will be and the source of his income. Also, within reasonable limits, he knows what his income will be through the years. In the lives of most urban people we find no such constant economic elements, no fund of natural resources to which they may retreat.

The urban man must use money in exchanging his skills and services for all the material things which come from the outside, from which he makes all the visible properties of the urban scene. He uses a part of these materials from the outside in making finished goods which are exchanged for more materials from the outside. Or he sells services, as when he entertains the rural man with urban-made songs and stories, or when he exports the results of his scientific research or his achievements in technology.

Although isolated from the natural sources of supply, the urban man is not without resources. Such resources as he may have are in himself—his strength, his skill, his general knowledge, and particularly his special capability. More than is necessary for the rural man, he must spend time and money in preparing himself in order to earn money. Allowing for exceptions, the rural man can function in his earning and spending roles with relatively less education. If the farmer's son or daughter are educated beyond a certain utility point, that is likely to be the stimulus for going to the city in search of another occupation.

Modern cities cannot function at the rural level of training. One can be a laborer in the urban labor market with eight years or less of schooling, but even a laborer finds it useful to have more. The skilled worker needs not less than eight years of education, and three or four years of special training besides. The clerical worker needs not less than twelve years of schooling to which he may add another two years of special training. One can hardly become a technician or a superintendent or occupy some other specialist position with less than sixteen years of education. For some highly specialized position additional years of special education may be required. Thus as the urban man invests in his preparation he is able to enter his life's work at one or another occupational level.

We may speak of the "worth of a man" at different points in his life, and the term is sometimes heard, "the economic value of a man." Assuming that an urbanite at the age of 25 has been 20 years getting educated, we can very well ask the question; what is he worth? During much of this time of preparation, years when he might have gone to work, he has depended on others for support, which means that the community has an investment in him also.

Once the urbanite becomes an income earner, to the extent that he proves himself useful in his job or position, his worth increases. He acquires added experience and skill, and perhaps he is given responsibility. In this process his worth to himself and to the community increases, for he acquires degrees of indispensability. Worth in this sense rises with the years, as we shall consider later, and then begins to decline. His period of learning may be from less than eight years to more than twenty, while his period of earning may extend normally from thirty years to more than fifty, after which he moves into a period of nonwork, nonearning, and decline.

## Capability, Income, and Place Size

On the basis of their study of Bureau of the Census figures for 1950, Duncan and Reiss found that education, occupation, and income tend to rise with the size of the community. Level of education, level of occupation, and level of income are also higher for all urban places than for rural places.

For all persons 25 years of age and older the median number of years of school completed was 8.9 for incorporated places under 1,000 population and 8.5 for farm population. It rose by size of place to 10.8 years of schooling for places of 1,000,000 to 3,000,000 and was 10.3 for places of 3,000,000 or more. For males between 25 and 44 years it was 8.7 years for farm people, 11 for persons in places of 1,000 and less, and 12 or 12.1 for all urban places from 10,000 to more than 3,000,000. For females it was 9.7 on the farms, while for women in all urban places from less than 1,000 upward from 12 to 12.2 years of schooling were reported. For all persons of 45 years and older there was little difference between large and small places; the education level ranged between 8 and 9 years. For the age group of 25 to 44 years the level of education was higher by nearly four years than for the older age groups. For this same group it was higher in the larger cities and it is higher for females than for males.[1]

Again citing Duncan and Reiss, we note that professional, technical and kindred workers, sales workers, and private household workers are present in about the same percentages regardless of size of city. But managers, of-

[1] Otis D. Duncan and Albert J. Reiss, Jr., *Social Characteristics of Urban and Rural Communities*, New York, John Wiley & Sons, 1956, p. 89.

ficials, and proprietors are relatively more numerous in the larger places. The same holds for craftsmen and foremen. For service workers the percentages are higher in the larger places, but for laborers the trend is reversed: 8.5 percent for places of 1,000 or less, down to 4.6 percent for very large cities. All occupations, except farm labor, 70.9 percent, are scarce in rural areas.[2]

Such comparative figures on education and occupation levels by size of place are informative, but they might be more so if some of the hidden detail could be brought out. The demands of the labor market on education tend to be more exacting for larger than smaller places, and the hierarchy of occupations is more complex and diversified in larger than in smaller places. Moreover, these figures are not for cities as such, but for "urbanized areas." The urbanized area includes a large variety of towns and smaller cities clustered around a central city, and these tend to affect the generalizations made.

The same generalizations apply to the following figures on comparative income, also from the study by Duncan and Reiss. These figures show the 1949 earnings for all males and females 14 years of age and older by place size. The first group shows median income for urbanized areas:

| | *males* | *females* |
|---|---|---|
| 3,000,000 and over | $3,078 | $1,603 |
| 1,000,000–3,000,000 | 3,026 | 1,471 |
| 250,000–1,000,000 | 2,779 | 1,215 |
| Under 250,000 | 2,692 | 1,121 |

The next group of towns and cities includes those that are outside of urbanized areas. Again we see the trend of income rising with place size:

| | *males* | *females* |
|---|---|---|
| 25,000 and over | $2,554 | $1,003 |
| 10,000–25,000 | 2,484 | 926 |
| 2,500–10,000 | 2,354 | 839 |
| 1,000–2,500 | 2,268 | 749 |
| Under 1,000 | 1,935 | 626 |

Employed persons in rural areas are divided into two groups. For the rural nonfarm group the median 1949 income was $2,029 for males and $718 for females. For the rural farm group the median income was $1,379 for males and $459 for females. It should be noted in each of the groups listed above that incomes for female workers were considerably lower than those for male workers. It should also be noted that in urbanized places exceeding 3,000,000

[2] *Ibid.*, p. 96.

the median income for females was 52 percent of the median income for males. This comparative percentage drops for each size of place until for places of 1,000 and under the median income for females was 32 percent of that for males. This would seem a strong incentive for small-town women to migrate cityward.[3]

## A Measure of Community Well-being

We have seen that Americans in larger places are better fitted with education and that a higher percentage are in the upper occupation levels than is true of small places. They also have more money to spend. A thousand male workers in urbanized areas of 3,000,000 and over in 1949 earned $3,078,000, while a thousand female workers in these same places had a combined income of $1,603,000, making a total of $4,681,000. But for cities of 10,000 to 25,000 the total for a thousand male workers and a thousand female workers for 1949 was $3,410,000. A thousand each of male and female workers in farm communities in 1949 had a combined income of $1,838,000. In other words, the per capita amount of money in circulation increases with the size of the place.

In terms of such figures we come to understand why the big cities are able to erect more costly buildings, have better paved streets, own more parks and playgrounds, and operate better equipped schools than is possible for some small cities.

Not only is the big city wealthier per capita in money terms; it is richer also in terms of human resources. It is mainly in the larger cities where the high-level technicians find the best market for their services. Specialists in different fields find it almost necessary to settle in the large city. Intellectuals, artists, musicians, writers, and other creative persons have little choice but to go to the larger places. These different people of high competence not only add worth to the community where they live, but they are also important national resources.[4] The more advanced a community becomes in the application of technology and science and in the efficiency of its work the larger becomes the ratio of skilled, professional, and technical workers to the unskilled workers.

[3] *Ibid.*, p. 104.

[4] How much of a resource technical and professional persons are deemed to be is evidenced by the fact that all such persons in the areas occupied after World War II by the Russian military were gathered together and taken to Russia where they were put to work under the most favorable conditions. Most of them are still being retained there.

As we already noted, unskilled labor in American cities ranges from nearly 9 percent in very small towns to under 5 percent of the labor force in the larger cities. Between 1910 and 1950 the unskilled part of the American industrial force decreased from 36 percent to 19.8 percent of all industrial workers, and this decrease continues. At the same time the skilled, technical, and otherwise specialized part of the labor force has increased from 64 percent to 80.2 percent of the total.[5] Thus as industry moves in the direction of the "automatic factory," an increasing percentage of the work force moves from lower income to higher income groups. Paralleling this development has been the increase in productivity. Using the years 1947-1949 as 100, the index for industrial production has increased from 34 in 1913 to 125 in 1954, nearly four times. The productivity per worker has also increased in proportion.[6]

High levels of education, a more skilled labor force, and higher levels of income coupled with higher productivity in goods and services are the essential elements in community well-being, and these elements tend to be present in greater abundance in the larger urban places. But the urban communities can enjoy these advantages only as they are able to increase a demand for them elsewhere. Markets must be found outside for these goods and services which are symbols of a kind of life that is urban, and as these goods and services are spread abroad, the urban way of life follows in their wake.

Another evidence of this growing well-being is the shift in the use of the labor supply. Of the 1870 urban labor force, 75.4 percent was needed for the production of physical goods, but in 1950 only 47.1 percent of the urban labor force was so occupied. In the meanwhile the part of the labor force producing services increased from 24.6 percent in 1870 to 52.9 percent in 1950.[7] It is this increase in the service industries that has perhaps contributed most to the evolution of community well-being in the United States.

Versluys reminds us of resistant elements in the cities of the Far East where the supply of unskilled labor is so great that industrialization, with all the change that goes with it, has difficulty getting established. Hand-work industry lingers on, and modernization that might increase the productivity of labor is strongly resisted. The result is that very high percentages of the labor force represent unskilled workers. This means that the median urban income is low. Instead of there being, as in the West, three or four lower-level incomes

[5] *Economic Forces in the United States of America,* Washington, Department of Labor in cooperation with the International Cooperation Administration (4th ed.), June 1956, p. 29.

[6] *Ibid.,* pp. 45, 55.

[7] *Ibid.,* p. 25.

to one higher-level income, the ratio is from 15 to 20 lower-level incomes to one higher one.[8]

## The Individual and Income

We would find in the cities of the Far East, as in the West, various levels of income, but the distribution from lowest to highest would be very different. In the cities of the West higher percentages would be found in the middle groups, both for income and occupation. Katona and Mueller[9] found the following distribution for American incomes as of October 1955:

| | *percent* | | *percent* |
|---|---|---|---|
| Under $1,000 | 7.5 | $ 5,000–$7,499 | 20.5 |
| $1,000–$1,999 | 11.2 | $ 7,500–$9,999 | 8.2 |
| $2,000–$2,999 | 12.4 | $10,000 and over | 6.0 |
| $3,000–$3,999 | 16.9 | Not ascertainable | 2.7 |
| $4,000–$4,999 | 14.6 | Total | 100.0 |

Needless to say, this distribution changes from year to year, even from month to month. It is easy to visualize how the distribution of incomes might change in a period of economic recession and what changes to expect as the market for labor moves into full employment. Here we are confronted with the phenomenon of a competitive labor market in which the individual, as he succeeds or fails, moves upward or downward on the income scale. This state of affairs is largely urban, but it includes in its embrace all people with incomes, even the farmer, who are indirectly affected.

At this point we return to the question raised at the opening of this chapter concerning the economic worth of a man. We noted that the resources of the urban man are found largely within himself. How he fares in the struggle for income depends in part on the state of the labor market, which may be more favorable for him under one set of circumstances than another. But his success or failure also depends in part on what he has to offer at a given time and place and under the particular circumstances.

The recognized marketable wares of the income earner are his education, his special training and skill, and his experience and his ability (physical and mental) to perform the work. These are the minimum requirements of the

[8] J. D. N. Versluys, "Introduction," *The Social Implications of Industrialization and Urbanization* (Five Studies in Asia), Calcutta, Unesco Research Center, 1956, p. iv.

[9] George Katona and Eva Mueller, *Consumer Expectations. 1953-1956,* Ann Arbor, University of Michigan (Survey Research Center), 1957, p 120.

work itself. However, various large employers—imitated by small ones—may by management mandate establish other requirements. These other requirements often are different for different occupation levels. Work requirements may set minimum levels for education, experience, skill, and the like. A worker may be excluded for physical reasons—he is too short, too fat, too weak—and the management mandates may also concern personality and appearance qualifications. The record of the applicant is examined: How did he perform on other jobs and why did he leave other jobs?

For some of the requirements established by management mandate there may be real or imagined economic reasons, as when it is decided that certain positions should be reserved for women. It is reasoned that women can perform the work as well as men, and female labor is cheaper. But the management mandate may extend beyond, as when Negroes are preferred or excluded, or there is an exclusion of foreign-born or other minority-group applicants.

It is in such a network of requirements, which may change with labor market conditions, that the individual approaches the task of earning his income. Not only is the situation a changing one, but the individual changes—for example, as he gains more skill and experience. This would be a positive change, but in some cases the change may be negative, as when he begins to age, or as when he accumulates "a bad record" on previous jobs. Thus his prospects for gaining a higher income or having to accept less are ever present, whether he is holding a position or seeking one.

## The Earning Cycle

Moore calls attention to some of the problems that workers in the less developed countries discover when they enter industrial employment. They find that they must think in terms of security. They find that industrial work is largely a matter of going repeatedly through cycles of feeling more secure or less secure. They also learn about promotion or advancement.[10] This does not mean that insecurity was absent in the tribal or somewhat primitive agricultural backgrounds of these workers. It existed, but there it was a group matter, a worry of the family, and the family was large. Here in the industrial situation the individual stands alone. Moreover in the village he did not know advancement, except that which comes with the cycle of life. But under in-

[10] Wilbert E. Moore, *Industrialization and Labor,* Ithaca, Cornell University Press, 1951, pp. 113-119.

dustrialism, if one would earn more money he must advance to a better position, again largely a matter of individual effort.

Not only does the urban man live in a continuously competitive situation, but it is accepted as part of "the game of life"; one must work and one must get ahead if he can. In his youth the idea takes root in him that he must prepare for the future, and the future for youth usually means the time when he will go to work and earn money. That is his entry into an active life. He has already developed the idea of looking ahead, but now he looks ahead to a better job, to a higher income, to gaining prestige. If he is a serious person, he looks even beyond that, to old age and retirement. He sees the evidence all about every day that the labor market will one day not need him.

Rowntree in his first study of poverty in York, England, concluded that the industrial workers early in this century lived their lives along a poverty line. The individual began life at the line and advanced above it as he was able to earn money. After marriage and the arrival of children he gradually sank to some point below the line. As his children went to work and married he was relieved of burdens and he rose slowly. But by then he was getting old and his earning ability was vanishing. He died at the poverty line.[11]

Although the expression, "poverty cycle," is not used in the industrial Western countries today, the concept of a cycle is used in other connections. One may very properly speak of the urban worker as having a life cycle with respect to his status in the labor market. Riesman and Roseborough suggest that:

> To a degree, for the factory and office workers, the "poverty cycle" that B. Seebohm Rowntree found in York, England, a half-century ago still holds, though at a far higher level: an early peak is reached, followed by a plateau and a slow decline—modified, to be sure, by the secular rise in the real income, especially among factory workers.[12]

They continue with mentioning that the cycle is different for men and women, especially in the younger years when the man may spend his money freely while the woman may save hers for her hope chest. Something akin to such a cycle is to be discerned at different economic levels, even though no poverty line can be drawn. There are points at which the family, whatever its economic level, feels itself at a poverty line, and feels humbled or disgraced if it must

[11] B. Seebohm Rowntree, *Poverty, a Study of Town Life* (2d ed.), New York, Longmans, Green, 1922.

[12] David Riesman and Howard Roseborough, "Careers and Consumer Behavior." See Lincoln H. Clark, Ed., *Consumer Behavior,* New York, New York University Press, 1955, Vol. II, p. 6.

sink below such points. Whatever that point is for the factory worker, the store clerk, the accountant, the supervisor, the manager, or the man with the independent occupation, he must exert every means within his power to remain above it. If he should rise to another economic level, that only means that another "point of no return" will be established somewhat higher than the earlier one.

Since the cycle concept relates to rising and falling movement or rising and falling status, we should expect it to appear in different forms in any society, each form oriented to the life of the individual or the family. Yang speaks of the cycle of the village family in China in which family status is linked with land ownership. The cycle of slowly acquiring land and finally losing it may continue through three or four generations.[13] In that struggle the family is the continuing unit and does not disintegrate when the primary source of its security is lost. It begins again to get land, and in the meantime the family members work for hire and pool their gains in the family fund.

The family life cycle under urbanism is oriented more to the earnings of the chief breadwinner. All the factors involved, the dominant ones being economic, tend to limit the life cycle of this modern family to the life span of the married pair. The principal exceptions are the families that have considerable amounts of property. The propertied families retain longer the traditional institutional character, but even within these families the individual members have their cycles of rising and then falling economic utility. This observation applies particularly to the males of the wealthy families.

## What the Urban Man Accumulates

In the preurban society individual accumulation hardly exists. If there is any accumulation, it is done by the family, and if the individual from such a community goes to the city the family continues to claim a share of his income. Schapera observed, however, that Africans who settle in the towns very soon begin to resist the demands of their families in the village, at least the demands of secondary kinfolks.[14] Yang, describing the economics of Chinese village life, reports that the idea of an individual accumulating wealth

[13] Martin C. Yang, *A Chinese Village*, New York, Columbia University Press, 1945, pp. 132 ff.

[14] I. Schapera, "Migrant Labor and Tribal Life in Bechuanaland," *Social Implications of Industrialization and Urbanization in Africa South of the Sahara*, Paris, UNESCO, 1956, p. 219.

or keeping his earnings hardly exists. "Everyone works or produces for the family as a whole."[15]

For the urbanite, accumulation takes various forms, and physical wealth is only one of them. As already mentioned, he may accumulate knowledge and skill, an activity which he can continue until late in life, so long as his mental self does not succumb to the aging process. He can accumulate by saving a margin of his income and become a holder of liquid assets. A third form of accumulation is prestige, acquiring a good name in some particular field. For the corporation the equivalent is known as "good will." These are among the values that the unemployed man, according to Bakke, tries to hold, even though he has lost his job. These values and status relationships "cannot be designated as primarily economic."[16] Even the seemingly unimportant individual tends to accumulate an identification with certain phases of the life about him, and he gains recognition to some degree or other.

The assets which the urban man accumulates are in part personal, in part social, and in part economic. If his economic accumulations are large, his efforts to gain social and other forms of prestige are made easier. He is often supported in these efforts by his wife who may be a considerable social asset to him if he belongs to the middle or upper economic and occupation groups. For the individual whose work is at the supervisory or management level of industry, who is a professor, an entertainer, or a member of an independent profession, the matter of getting ahead is in part a social effort. In these areas the man who earns the money may be very much aided by his wife who keeps the home and spends the money.

## Participation in Community Life

In the preurban community the individual belonged to a group, his family or lineage group, and this group was organized to protect and advance the interests of its members. In the urban situation the family, by comparison, is a relatively weak support for the individual. Conversely, it leans on the individual member, usually the husband and father, who is the income earner. He must find support by joining groups of another kind and in relation to these he functions in a community citizenship role. Whether he joins the medical society or a trade union, he enjoys through such an affiliation, as

[15] Yang, *op. cit.*, p. 76.

[16] E. Wight Bakke, *Citizens Without Work*, New Haven, Yale University Press, 1940, p. 3.

Marshall points out, a kind of citizenship, for citizenship is manifest in the behavior of individuals, and especially of groups, in relation to the total community. Such economic-oriented groups have "created a secondary system of industrial citizenship parallel with and supplementary to the system of political citizenship."[17]

In and through such groups—the trade union, the chamber of commerce, the professional organization, even neighborhood, social, or sport groups—the individual is reinforced. These many organizations speak for their members and, as we noted in previous chapters and will repeat later, they form a network which bears much of the responsibility for the management of community life. A point of interest here is that reports on the behavior of natives in the new industrial cities of Africa show that even these very illiterate people try to render themselves more secure by forming associations. Clément describes the predicament:

> The loosening of the urban immigrant's ties with the primary groups, while to a certain extent it emancipates him, temporarily restricts his participation in social life. It means that he is enrolled in and attached to an artificial "patchwork" community where, in view of his precarious social position, his role, during the period of adaptation bears no relation to that which he played in his earlier, coherent and stable group. His resultant feeling of isolation is accompanied by a sense of insecurity, due to the relaxation of the traditional social supports and controls, and further accentuated by the daily need to grapple with the diverse problems in an unfamiliar social setting whose administrative structure, because it is not easily comprehensible, he tends to regard as threatening or arbitrary.[18]

Thus the native joins associations, groups of natives from his own tribe or village. They help one another find jobs or to keep jobs, or they help a member in trouble. They are not groups through which the individual might participate in the wider community life as he might through a trade union. The Belgians have tried without much success to get these same natives to join a trade union. But a trade union is a secondary organization, and these people need time and experience, also more education, before they are ready for that level of organization.

When we look at Sweden we find a situation almost the extreme opposite

[17] T. H. Marshall, *Citizenship and Social Class*, New York, Cambridge University Press, 1950, p. 44.

[18] Pierre Clément, "Social Patterns of Urban Life," phase of study with V. G. Pons and N. Xydias, "Social Effects of Urbanization in Stanleyville," *Social Implications of Industrialization and Urbanization in Africa South of the Sahara*, Paris, UNESCO, 1956, pp. 485-486.

to the African community in the Belgian Congo. Sweden is highly industrial and urban, and enjoys a high level of living. Eighty-five percent of all workers and 90 percent of all workers in industry belong to trade unions. In 1956 the consumer cooperative associations claimed as members 15 percent of the population and 44 percent of the working population. Through these two types of secondary organizations, especially strong in the cities, the earners of income participate not only in the social, economic, and political life in their local communities, but they are also collectively a political force in the nation.

The individual income earner in Sweden is perhaps less of an individualist than the individual income earner in the American city. He may be called a "group man" in the secondary group meaning as the newly urbanized African in the Belgian Congo city is a "group man" in the primary group meaning, but the Swedish group man is oriented to the industrial urban way of life. Some of the results for the Swedish income earner, and this includes workers up to the management level and all but a small part of the upper clerical and professional workers, are that between 1930 and 1954 the income earners' share of the national income rose from 70 to 77 percent of the total. During this time the number of smaller bank accounts also increased. Between 1930 and 1953 (1953 values) bank accounts under $100,000 increased by more than 60 percent, while the number of accounts of $250,000 and above decreased by nearly 60 percent. While the individual income earner in Sweden through his exercise of his civil rights has taken much of the insecurity out of his "cycle of life," there is still plenty of room for individual competition. If he wants to advance to a higher income and to higher status he must exert himself much as the American income earner.[19]

## Comparability of Real Income

Urbanism is inherently international, and whoever approaches the subject in terms of one country alone will have difficulty if he does not follow some phases of his subject into wider areas. This applies very emphatically to the phenomenon of leisure. It also applies to other types of spending and to income earning. The labor market in one modern country is in a sensitive stimulus-response relationship with the labor markets in other modern countries. As these countries come increasingly into contact, their respective labor

[19] Hugo Hegeland, "The Structure and Functioning of Sweden's Political Economy." See Rudolf Frei, Ed., *Wirtschaftssysteme des Westens,* Basil, Kyklos Verlog, 1957, pp. 218-227.

markets become increasingly similar, and the fact that the trend is in the direction of similarity is the best evidence we have of the world-wide network of urbanism.

International comparisons, therefore, become increasingly important. European engineers since World War II have visited the United States to learn how American cities are supplied with water and what is the per capita cost of water supply. They have been interested also in learning the per capita cost of sewage disposal. American traffic engineers have studied the German system of automobile highways in terms of cost and efficiency. Americans have studied the British system of social medicine to learn not only about unit costs but about per capita services rendered. American and European economists and statisticians have frequent meetings to develop uniform terminology as well as methods for comparing wages and prices or for reporting employment and other types of practical information.

International comparisons, as Blaisdell has emphasized, are sometimes difficult, particularly comparisons that relate to the per capita incomes in the different countries:

> The underlying reason for this is that the only adequate measure that we have had devised for the overall activity of a community is the common unit of exchange of that community. I refer, of course, to the unit of currency—the dollar, the pound sterling or the franc. They are national units. However, when one strives to equate these units of measurement and then compare the income standards, the comparisons take on less and less validity. Even the most vigorous attempts to get rid of the technical difficulties inherent in this process have not proved entirely satisfactory.[20]

The values involved are measured in terms of money, the life blood of urban life, whether the city is in the United States, Great Britain, France, Germany, or elsewhere. For the purpose of comparing internationally both income and living costs, some students of the problem have resorted to the device of using time worked rather than money as the measure. For example, knowing the money wage per hour, and the money cost of consumer goods, it is possible to indicate comparatively how many hours of labor a worker, say a steelworker, must give to buy a pair of shoes in Pittsburgh, Sheffield, Essen, and so on.

Kravis made such a study for 1950. He found that in order to buy a

[20] Thomas C. Blaisdell, Jr., "Problems of Evaluating the Effectiveness of Development Measures," *Economic Development and Cultural Change,* Vol. II, No. 4, January 1954, p. 288.

pound of wheat flour the American worker had to give 4 minutes of labor, the German worker 11 minutes, the French worker 17 minutes, and the Swiss worker 19 minutes. In each case the worker was in the same occupational group. In 1950 a pound of butter cost the American worker 31 minutes of labor, but for the Swiss worker the cost was 117 minutes of labor, for the German worker 129 minutes, and for the French worker 169 minutes. While the American gave 33 minutes for a pound of coffee, it cost the British worker 66 minutes, the Swiss worker 97 minutes, and the French worker 159 minutes of labor; but the German worker had to toil 631 minutes for his pound of coffee in 1950. This high cost in Germany, which encouraged a black market for coffee, was largely due to a high tax which made coffee a luxury item. Kravis found that the time cost of potatoes was the most uniform item: 2 minutes in the United States, 3 minutes in Great Britain, 4 minutes in Germany, 5 in Switzerland, and 9 in France.[21] This method of measurement is used in some countries to determine the real income of different occupation groups.

## The "High Authority" Studies

The United States Department of Labor makes periodic studies of commodity costs in different cities and regions. This is done with reference to different-size family groups and different economic levels. The result is a cost of living index, which is a recognized measure of real wages. European countries have done likewise, but it is now becoming necessary to devise some international method for comparing real income. Such studies are now being conducted by the European Coal and Steel Community, the "High Authority," for its industries in Belgium, France, Germany, Holland, Italy, Luxembourg, and the Saarland, now part of Germany. When we ask, for example, about the wage of a coal miner we find that a German coal miner received in 1955 only 60 percent of his listed wage, a Belgian coal miner received 69 percent, a French coal miner only 52 percent and an Italian only 40 percent.

Here we are confronted, however, with factors that make comparison difficult. The Italian worker, for example, who received 40 percent of his wages in cash also received other benefits: coal, electricity, housing, and cash allowances according to the number of his children (not called wages). The employer retained the worker's taxes and social insurance cost, about 25 per-

[21] Irving B. Kravis, "Work Time Required to Buy Food," *Monthly Labor Review*, Vol. 72, No. 2, February 1951, p. 154.

cent of the overall wage in Italy.[22] The European worker does not feel that he is less an individual because the employer does so many things for him,[23] it is his way of meeting the problems of industrial urban living.

A central objective of the High Authority is to develop within the "Community" a common market for goods and services, and for labor. Workers would be free to move from country to country, as Americans do from state to state. Food and other commodities would move easily across borders. Wages and prices would become more uniform. Some standardization has already begun; in 1955 the cost to the employer per hour for a steelworker, figured in Belgian francs (Bf. 50 to $1) was 41.34 in Germany, 44.34 in Belgium, 42.55 in France, 42.99 in the Saar, 34.84 in Italy, 50.87 in Luxembourg, and 37.19 in Holland.[24] These rates include family allowances and money handled for the worker by the employer.

The problem of the High Authority study is to learn also how the coal miner or steelworker spends his earnings and what he pays for commodities. Much as has been done on budgetary research in the United States, the High Authority made a list of some 220 articles commonly purchased by families. Minimum specifications were written for these: a pound of meat of certain kind and quality, a described type of work shirt, wine of a specified grade, etc. By this means an international "basket" was visualized, each "basket" containing all of the 220 items, all recognized as essentials in each of the "Community" countries. This uniform "basket" is then tested against market prices in the different countries, making it possible to compare real income from country to country.[25] Moreover, it provides a yardstick for the employer

[22] "Les Salaires et les Charges Sociales dans les Industries," *Informations Statistiques*, Luxembourg, Communauté Européenne du Charbon et de l'Acier, Vol. III, No. 6, November-December 1956, pp. 14-15.

[23] Some European industrial plants have gift-giving traditions. Costs are charged to wages but the gifts are expected and accepted by the workers. An example known to this writer is a German brewery in which since guild days the employer has made gifts of money when a worker marries, when his first child is born, when he has been with the firm 10, 20, 30, and so on, years. A loan is made if a worker builds a house. At Christmas there is a party for all family members. When a worker retires he can come to the plant each day for his liter of beer. These are regarded as rights and are in the contract between the plant and the works council.

[24] "Comparison des Charges Salariales dans les Industries de la Communauté," *Informations Statistiques* Luxembourg, Communauté Européenne du Charbon et de l'Acier, Vol. III, No. 4, July-August 1956, p. 10.

[25] *Comparison des Revenus Réels des Travailleurs des Industries de la Communauté*, Luxembourg, Communauté Européenne du Charbon et de l'Acier, Etudes et Documents, 1956, p. 60.

groups and trade unions in their bargaining for wage rates. These efforts at comparisons, with different currencies, wage systems, price systems, and tax systems, remain complex. The search for parities is the more complex because more than prices and currencies are involved.

> Even when we possess the results of this further work and the statistics on family budgets, *we shall still not have the complete answer* to the problem of country by country comparison of real incomes. Depending on the social, economic and political structure of the different countries, free State or other benefits in kind are, in some cases playing an important part (e.g. National Health Service benefits in Britain): these also require separate study and statistical classification in respect to the Community countries. We have only to recall the different positions which working-class families may be in as regards their means, their reserves and the durable consumer goods they may own, the different systems in respect of old-age pensions, the differences in the classifications of miners' and steelworkers' incomes in the income structures of the various countries, to realize the magnitude of the task involved in establishing comparability even for incomes and means, let alone standards of living generally.[26]

## The Life Cycle and Earning Power

It is being said of the "Welfare State" that under it the worker can approach old age without fear. When his earnings stop he turns to his pension

Table 8. Percentage income distribution by head of unit, United States, 1952*

| *Age group* | *Under $2,000* | *$2,000-$3,999* | *$4,000 plus* |
|---|---|---|---|
| 18-24 | 32 | 50 | 18 |
| 25-34 | 12 | 40 | 48 |
| 35-44 | 13 | 35 | 52 |
| 45-54 | 19 | 30 | 51 |
| 55-64 | 28 | 33 | 39 |
| 65 and over | 63 | 21 | 16 |
| All ages | 25 | 34 | 41 |

* Figures from *Federal Reserve Bulletin*, September 1953, Supplementary Table 14, p. 16.

[26] "Consumers' Purchasing Power Parities," *Statistical Information*, Luxembourg, European Coal and Steel Community, Vol. II, No. 5, August-September 1955, p. 3. Since English is not an official language of the High Authority, only an occasional document is translated.

and such accumulations as he may have, although he ceases to be of interest to the advertising man, unless he is one of the few old people who in his work years was able to accumulate real property and other assets. Fisher has brought together some comparable figures for the United States and Great Britain. Table 8 shows the low, middle, and high income earners in the United States for 1952 in a percentage distribution according to major age groups. We note that the high point for low incomes (63 percent) is for the age group of 65 years and older, while for the middle incomes there is a gradual decline from 50 percent for the 18 to 25 group to 21 percent for the oldest group. For the incomes of $4,000 and over the figures rise from 18 percent for the young group to 52 percent for the 35 to 44 group and drop to 16 percent for the old.

Table 9 shows the 1951-1952 income distribution for Great Britain. A

Table 9. Percentage income distribution by head of unit, Great Britain, 1951-1952*

| *Age group* | *Under £200* | *£200-£399* | *£400 plus* |
|---|---|---|---|
| 18-24 | 38 | 50 | 12 |
| 25-34 | 6 | 39 | 55 |
| 35-44 | 6 | 32 | 62 |
| 45-54 | 9 | 35 | 56 |
| 55-64 | 24 | 36 | 40 |
| 65 and over | 67 | 22 | 11 |
| All ages | 24 | 35 | 41 |

* Figures from H. F. Lydall, "National Survey of Personal Incomes and Savings," Part II, *Bulletin of the Oxford University Institute of Statistics*, XV, Nos. 2 and 3, Table 11, p. 46.

striking point is that the percentages for all ages are much like those for all ages for the United States. The massing of income in the low level (67 percent), however, is higher than for the United States, and in the high incomes the distribution goes higher (62 percent) and drops more rapidly to 11 percent. The behavior of the middle column corresponds to that in Table 8.[27]

Fisher also notes that the "Distributions of ownership and size of liquid asset holdings are very similar to age" in the United States and Great Britain. In other words, the earnings and spendings of a man tend to rise and fall with his cycle of life changes.

[27] Based on Janet A. Fisher, "Family Life Cycle Analysis in Research on Consumer Behavior." See Lincoln H. Clark, Ed., *Consumer Behavior*, New York, New York University Press, 1955, Vol. II, p. 34.

## The Idea of Saving

Saving is a phenomenon of modern society, and so inherent in the capitalist system that it has come to be regarded as a basic virtue of Western cultures. Doubtless, as Samuelson reminds us, it was an essential virtue on the frontier, as it was when industry was getting started and had to rely on private savings. But we have reached a time and an economic situation in which it may or may not be a virtue to save. It depends on the individual and the circumstances. In time of war, for instance, when consumer goods production must be curtailed in favor of defense production, saving becomes a national service.[28]

But people do save, and the urge to save seems to be somewhat greater among Americans than among Britons. This is seen in comparing Tables 10 and 11. Table 10 shows that, with the exception of persons 65 and over (46 percent), about two out of three Americans are savers. Even though the amounts are small in many cases, this voluntary practice seems deep-rooted. On the other hand, nearly a third of the Americans are dissavers, which means they spend more than they earn, and this holds for the younger age groups as well as for the age bracket 65 and over.

**Table 10. Positive, zero, and negative saving by age of head of unit (in percent), United States, 1950***

| *Age group* | *Savers* | *Zero savers* | *Dissavers* |
|---|---|---|---|
| 18-24 | 60 | 6 | 34 |
| 25-34 | 61 | 3 | 36 |
| 35-44 | 65 | 4 | 31 |
| 45-54 | 63 | 6 | 31 |
| 55-64 | 66 | 6 | 28 |
| 65 and over | 46 | 19 | 35 |
| All ages | 61 | 7 | 32 |

* Figures from *Federal Reserve Bulletin*, September 1951, Table 8, p. 1063.

Table 11 shows a somewhat different percentage distribution for Britons. The percentages for savers are lower than for Americans, especially for the two age groups 55 and older, and the percentages for dissavers are also somewhat lower. When we count together the zero savers and the dissavers, we

[28] Paul A. Samuelson, *Economics*, New York, McGraw-Hill, 1951, pp. 215, 285.

Table 11. Positive, zero, and negative saving by age of head of unit (in percent), Great Britain, 1951-1952*

| *Age group* | *Savers* | *Zero savers* | *Dissavers* | *Not ascertained* |
|---|---|---|---|---|
| 18-24 | 46 | 30 | 23 | 1 |
| 25-34 | 56 | 9 | 34 | 1 |
| 35-44 | 61 | 6 | 30 | 3 |
| 45-54 | 63 | 6 | 28 | 3 |
| 55-64 | 57 | 6 | 30 | 7 |
| 65 and over | 41 | 27 | 28 | 4 |
| All ages | 55 | 13 | 29 | 3 |

* Figures from H. F. Lydall, "National Survey of Personal Incomes and Savings," Part IV, *Bulletin of Oxford University Institute of Statistics,* XI, Nos. 10 and 11, Table 63, p. 365.

note that for both Table 10 and Table 11 the high percentages are found within the age group 65 and over.[29]

The idea of saving is not only modern, but it is also quite urban. It grows largely out of the belief that the individual must look after his own economic welfare, that while he is young he should put something away for old age. Such an idea in any primitive or early rural society would be impractical and would also be regarded as antisocial. During summer one might store food for the winter, but not for several winters in advance. As the Thorners observed in Pakistan and India, millions upon millions do well to live from day to day.[30] They need to move far economically before reaching the point where savings can begin, and this situation holds for many other peoples.

In the Herskovits report on a Trinidad village our attention is called to the fairly widespread practice among different peoples of engaging in group saving through a form of gambling. This is a feature of modern lotteries. In Trinidad groups of natives make contributions to pools. These collections may continue for a week or several weeks. The rules are made by the members. Finally lots are drawn and the winner takes all.[31]

Very few people in the industrial urban society are really long-term savers, and much of the saving reported in the various consumer surveys is

[29] Based on Fisher, *op. cit.*, p. 35 for Tables 10 and 11.

[30] Daniel and Alice Thorner, "India and Pakistan." See Ralph Linton, Ed., *Most of the World,* New York, Columbia University Press, 1949, p. 2.

[31] Melville J. and Frances S. Herskovits, *Trinidad Village,* New York, Alfred A. Knopf, 1947, p. 76.

of the short-term type. People save for a vacation or to buy something. Actually the savings of most people are not sufficient to support them for more than a few days or weeks. Saving to buy a house or an automobile is really a form of deferred spending; depending on what is purchased, it may be wise spending.[32]

A recent unpublished report on consumer income, "Spending and Saving in Germany," by Otto Blume of the Institut für Silbsthilfe at Cologne, contains some interesting figures for that country. The data were secured by an "opinion poll" type of survey. Regarding the "feeling of security," 52 percent of the respondents expected no change of income in 1958, 79 percent of the employed interviewees felt safe in their jobs for the next three years, while 53 percent felt permanently secure. As for possessions and purchases, 12 percent owned automobiles and 3 percent planned to buy, 19 percent owned washing machines and 4 percent planned to buy, 14 percent owned refrigerators and 4 percent planned to buy. During the previous year 28 percent spent money for furniture and during the next year 14 per cent planned to buy. About 47 percent planned no such expenditures for 1958.

Under 60 percent of the interviewees in towns of less than 5,000 owned houses, land, stores, or other real property, compared with about 28 percent for towns of 5,000 to 50,000 and about 17 percent for persons in places over 50,000 inhabitants. Many living in the small places are workers in larger places.

Concerning savings, 71 percent of all interviewees were savers, but 39 percent were saving for some particular form of spending—house, education of a son, dowry for a daughter, etc.—while 21 percent were saving for an unforeseen emergency. Thirty-one percent of all the savings accounts were less than DM 300 ($71), 42 percent were less than DM 500 ($120), while 49 percent were less than DM 800 ($190). Some, about 19 percent, would not reveal the amount of their savings. Well over half of these savers did not have sufficient money in their accounts to meet family expenses for much more than a month. As for the future, 97 percent of the white collar workers felt that their insurance would be sufficient for their old age, an opinion that was shared by 62 percent of the hand workers, 53 percent of those in the free occupations, 41 percent of the people in trade, and 41 percent of the rural people.

[32] George Katona, *Psychological Analysis of Economic Behavior,* New York, McGraw-Hill, 1951, p. 151.

## Saving and Economic Worth

While some Americans who work for wages or salaries are able to save money, others are less successful. Lansing and Morgan found that of all spending units (families and single individuals) in the United States in 1954, 11 percent reported that they had only sufficient liquid assets to equal one year's salary, and the median liquid assets for all spending units was only $350. For young single people, childless young couples, and older couples with children, only 5 percent, 3 percent, and 6 percent, respectively, reported assets equal to one year of income. Median face value of insurance was highest ($4,700) for young married couples without children, next highest ($4,600) for married couples with children, while for older married couples with no children under 18 years the median face value for insurance in 1954 was $1,700. But 18 percent of these older couples reported liquid assets in excess of one year's income.[33]

People who lived on earned incomes and who have been able to save may have accumulated through the years other property, all of which helps to indicate their net worth. The Federal Reserve Board in 1950 made a survey to determine what holdings were owned by the different income groups. The results are presented in Table 12. We note that all the income groups owned

Table 12. Spending units holding various types of assets as percentage of all spending units within each specified income group (in percent), 1950*

| *Previous year's money income before taxes* | *Liquid assets* | *Auto-mobile* | *Home or farm* | *Other real estate* | *Business interest* | *Corpo-rate stock* |
|---|---|---|---|---|---|---|
| Under $1,000 | 44 | 24 | 50 | 9 | 3 | 3 |
| $1,000- 1,999 | 54 | 37 | 32 | 11 | 5 | 3 |
| 2,000- 2,999 | 68 | 54 | 40 | 12 | 5 | 5 |
| 3,000- 3,999 | 74 | 63 | 46 | 18 | 6 | 6 |
| 4,000- 4,999 | 86 | 74 | 55 | 18 | 10 | 9 |
| 5,000- 7,499 | 94 | 82 | 62 | 26 | 15 | 15 |
| 7,500 and over | 99 | 89 | 65 | 44 | 18 | 36 |
| All groups | 69 | 55 | 46 | 16 | 8 | 7 |

* "1950 Survey of Consumer Finances," *Federal Reserve Bulletin*, Washington, Federal Reserve Board, 1950, p. 1592.

[33] Based on John B. Lansing and James N. Morgan, "Consumer Finances over the Life Cycle." See Lincoln H. Clark, Ed., *Consumer Behavior*, New York, New York University Press, 1955, p. 46.

some portion of the six different kinds of holdings. Liquid assets were held by fairly large percentages of all groups, from 44 percent of the lowest income bracket to 99 percent of the highest. For each type of asset there is a rising ownership trend from the $1,000 and under group to the $7,500 and over group, the only exception being ownership of a farm or home. A point to be kept in mind is that large numbers of urban people who may have considerable other assets prefer renting to owning a home.

Table 12, like any other table dealing with incomes, prices, savings, or other data that change from day to day and from place to place, has the value of showing relationships between types of data. The data may change in details but the relationship tends to be fairly continuous. This particular table shows how net worth tends to rise with the amount of income. Net worth is:

> The measure of the financial resources of the consumer . . . —excess of his assets over his liabilities. His worth reflects the accumulation of savings over a period of years—both the low incomes and financial reverses of some years and the high incomes and savings of other years. Information on the net worth of consumers, therefore, provides additional insight into the operation of our economy over a period of time. The pattern obtained with this more comprehensive measure varies considerably from that obtained when the saving of a single year is considered.[34]

## Spending and Consuming

"The stages of development in the life cycle," states Clark, "are many and varied, and each stage is related to the individual's tastes and consumption activities."[35] Much of all the behavior of the individual that is not directly associated with his work is closely associated with spending for consumption. Experts in the field of consumer behavior make it clear that there are many kinds of consumer behavior. For any individual, consumption behavior varies with his income, with his age, with his family status, and with other changing elements in his life.

The economist knows that "as consumers' incomes rise, they will spend more on consumers' goods. And they will also save more."[36] The experts on consumer behavior, many working in connection with the advertising business,

[34] "1950 Survey of Consumer Finances," *Federal Reserve Bulletin*, Washington, Federal Reserve Board, 1950, p. 1585.

[35] Lincoln H. Clark, Ed., *Consumer Behavior*, New York, New York University Press, 1955, p. vii.

[36] Alvin H. Hansen, *Business Cycles and National Income*, New York, W. W. Norton, 1951, p. 145.

are finding that spending and consumption in the lives of most people are ever changing. Thus Riesman and Roseborough speak of family buying in relation to a "standard package." Whether it is food, clothing, vacations, furniture, books, or a car, there are levels of spending for each social class. Whether a man is a skilled worker, professor, technician, or manager (low, middle, or high), the norms of the group or class tend to determine what possessions he will have in his home. This "standard package" may vary within limits; it "allows for both expansiveness and expressiveness, even while it represents one's integration into the society and allows, once it is bought and paid for, further goal-directed moves in preparation for an open-ended future."[37]

Whyte studied an American upper middle class suburb, "average income, about $5,700; average husband's age, 31; children, one and one on the way; politics, 68 percent Republican," most husbands and wives college graduates, some with advanced degrees. Most of these people had come from a lower level and were concentrated on moving higher. They were as residentially mobile as were their job advancements, 35 percent moving from the community annually. Whyte reported that the possessions of these suburbanites were precisely suited to the mobility and occupational prospects of this group. Consumption practices were geared to their objectives and expectations.[38]

Stone, commenting on the "standard package" idea, noted how the family living room changes with changing circumstances of the family. Such changes are due to basic changes in social position.

> The same is true, I suspect, of most of the items of the standard package—the car, the home, the table setting, foods and furniture. I know it is true of clothing, for every major change in social position—birth, entrance into school, graduation, marriage, parenthood, and even death—involves a change of wardrobe. I mention these things to indicate how personal mobility—changes in one's social circumstances—and the items of the package are inextricably interlinked.[39]

There is an ever-present element of pursuit in the spending and consum-

[37] Riesman and Roseborough, *op. cit.*, p. 17. They note that as the family gains new status it acquires new articles and sheds old ones, especially ones that might reflect negatively on the origins of the individual or family. P. 8.

[38] William H. Whyte, Jr., "The Consumer in the New Suburbia." See Lincoln H. Clark, Ed., *Consumer Behavior*, New York, New York University Press, 1955, Vol. I, pp. 1-14.

[39] Gregory P. Stone, comments on "Careers and Consumer Behavior," David Riesman and Howard Roseborough. See Lincoln H. Clark, Ed., *Consumer Behavior*, New York, New York University Press, 1955, Vol. II, p. 25.

ing activities of the modern urban man. The pursuit differs with circumstances. For most people it is no longer the quest for food, shelter, and clothing. For most people in most modern countries those needs have been met. Above that level the pursuit assumes another form. Spending and consuming becomes involved with various forms of striving for prestige. One is under a great compulsion to gain status or to avoid losing status. In these efforts the income-getting and the income-spending activities of the individual or family come to be very much interrelated.

## The Social Spending of Money

As mentioned above, in the preurban society the individual is not preoccupied with either spending or earning, and he is not likely to be much of a holder of property. He has no need of saving. All of the influences and institutions which subordinated the individual to the group, actually the family, had a proper place in that kind of a society. It must be recognized as an arrangement that served well and survived long, a sort of primary group welfare state; the individual was one with the group and the group served the individual. During the course of the emerging urbanism, under which people are learning various forms of collective living, the old patterns have been replaced by various compromise relationships between groups and individuals. Some of these must be mentioned because they are pertinent to the economic aspects of urban living.

It is sometimes charged that industry and government encroach upon the home and take away various of its functions, with the result that the family is being forced through a disorganization and reorganization process. This is true, but it is one of those truths which needs further examination. Certain other truths must be passed in review. One is that if man wants to live in great agglomerations he must find mass ways of earning his living. To do this he has exploited industry, being himself somewhat exploited in the process, and industry has served him well. But industry cannot function in the framework of the traditional family system, and the traditional family system was not organized to meet the difficulties inherent in the industrial urban system.

If man wants to live in large agglomerations he must also find new ways of keeping clean and safe and healthy. Instead of every man going to the river for his day's supply of water, he makes use of government to bring pure water into his house and he pays for it with taxes. This is social spending. Instead of each family carrying its smelly slops to the river and dumping them at night (as only the wealthy families could in the early days of New York City), he has the government dig sewers. This is another collective service that he buys through his taxes. By the same means he gets libraries and museums, parks and playgrounds, public concerts, and so on.

By collective spending the modern man uses government to maintain an extensive educational system. It is an expensive item, but this is not a service taken from the family; it is something very different, a service beyond the performance capacity of the family. Education is one of many mass services performed through government by the collective buying method.

Perhaps the most important, and to some the most revolutionary, are the programs of government (different levels of government cooperating) for social security. Here is collective saving for meeting the risks of life, accident, sickness, unemployment, and old age. While the subject of security will be considered more fully in the next chapter, it is mentioned here to illustrate how the modern urban man is learning to use collective methods of meeting some of his economic problems, and at the same time is learning much about participation in community life.[40]

In other words, one phase of economic urban life is collective in civic terms while other phases are collective in social terms, as people join groups to protect and to advance their interests. Very little of it is purely individual. This means that living in the modern community calls for many kinds of participation in the life of the community which, as noted earlier in this chapter, is the essence of the emerging concept of citizenship. This is a cosmopolitan citizenship more than a merely local one.

## Summary

The urban man lives in a money economy, the implications of which affect his social behavior as well as his earning and spending. Much of his time and effort, as of the satisfactions and discouragements of living, relate to his money-getting efforts and the spending of money for his well-being and advancement. He is occupied not only with advancement, but with achieving

[40] Private organizations for collective saving and collective spending are varied and many. There are the many forms of insurance for collectively sharing risk. In some countries the private cooperative for buying or selling has a growing importance. Even the profit-making corporation is an instrument through which goods are produced cheaper and in greater abundance. Depending on how it is managed and controlled, it serves in a collective way the economic needs of the community.

security against the risks of urban living. He is also occupied with the challenges involved in moving upward (or downward) from one status to another and with his final exit from income earning.

It was noted that, in addition to the accumulations he makes from his earnings, the urbanite functions in the urban environment as an individual; not as the preurban man whose individual identity is absorbed in the lineage or other group. His resources are largely contained within himself. For some individuals in the economically secure classes there may be the assets of family prestige and fortune. But the assets of the rank-and-file individual are his education (much or little, general or specialized), his work experience, his appearance and personality, his native physical and mental capabilities, and his network of contacts. He accumulates within himself a personal worth, his equipment in his efforts to earn a livelihood. This "equipment package" changes from year to year, from experience to experience, becoming more enriched or less so.

To the many differences among communities may be added the differences in the clusters of persons found at the various capability levels. Thus we note that median income tends to rise with the sizes of places. Not only do the most qualified find the most favorable market in the larger places, but here is the most varied market for income earners of special capability. The reverse is found in the diminishing percentages of unskilled by size of place. Higher income-earning groups tend to be higher living-level groups.

There is a growing tendency to think of the urban man as having a life cycle, both with respect to income earning and income spending. During the years between his entrance to and his exit from the labor market, whatever the level of his income or occupation, his fortunes rise and fall. Even though the point of beginning is not at the poverty line, it is the beginning of a cycle which rises and falls, except for the individual who accumulates a competence. Moreover, it is being recognized that the modern family also moves through a cycle of several stages, each stage of which is marked by a different pattern of earning-spending-consuming-saving. This is akin to recognizing the modern family as a one-generation unity. In these terms, the life of the individual and the family is continuously occupied with adjusting to status mobility and to other types of mobility.

Earning and spending for the urban man are not only highly related money-handling activities; they figure in all the value making of urban living, a way of living that must be continuously oriented to change. Urban living becomes a kind of "pursuit" out of which different people derive different degrees of satisfaction as they move from stage to stage. But in this process a new type of social order takes form, one in which the individual tends increasingly to identify himself with secondary organizations which, in one way or other, are security insuring.

Instead of finding his economic security in primary lineage groups, so characteristic of preurban society, the urban man is gaining a different kind of security through secondary organization. Especially in his economic interests, his behavior becomes increasingly civic. In fact, in all of his interests he finds himself pressed into the role of a citizen who functions in the community by making use of its facilities and by using the different resources of private groups.

CHAPTER 16

# Welfare in the Modern Community

HOW DO THE PEOPLE in the modern community take care of themselves, and what happens when they are unable to take care of themselves? We are moving away from that kind of industrial urbanism which assumes that "Every man takes care of himself, and the Devil takes the hindermost." And the Devil did take the hindermost unless some form of human benevolence intervened. However praiseworthy human benevolence may be, though, it is sentimentally based and often morally motivated. Nor does it extend any substantial promise for people of minimal means when they think of the future. With the spread of urbanism, increasing numbers in the world population did find themselves in a condition of economic uncertainty with respect to the present and the prospect of insecurity when they look ahead.

It was not so in most preindustrial societies, where the economic position of the individual was integrated with that of the group. While his status in the group from infancy to old age changed with his ability to contribute to group survival, he was not cast aside when unable to contribute as a producer. Such an idea of welfare breaks down before the advance of industrialism, but we seem to be entering upon a new type, one not oriented to the primary group organization but to the organization of the secondary society. That is the main subject of this chapter.

## Welfare and Social Security

Whoever uses the word "welfare" needs to make his meaning clear. For different reasons, many economists, sociologists, and social workers have purged the term from their vocabularies. However, in spite of the ban, the essential meaning of the concept is in their thinking. Foote, for example, writing about community services, speaks not of welfare but of balancing means and needs:

> At the risk of oversimplification, we may take as an axiom that need and means are rarely if ever in perfect proportion. The young and the old are less able to pay for what they need than the middle-aged, the sick less than the well, the black less than the white, and the poor less than the rich. And need varies as much as means, both quantitatively and qualitatively. Many devices are employed to adjust one to the other, but no device is universally or permanently satisfactory.[1]

Much as means differ, so need changes for each individual and family from time to time and from one situation to another. If temporarily or for a long time people cannot meet their need requirements, they may have no choice but to suffer want, unless help comes from individuals or from some organized private or public community service. Insofar as need is a matter of community concern we may say that community welfare is involved. It would be difficult to think of any other reason than welfare for community services.

Ducoff and Hagood define "rural welfare" in terms that seem to have a more general application. Their definition covers "rural levels of living and income; protection against economic insecurity; educational, health and recreational services available to rural people; and even their opportunities for aesthetic, intellectual, and religious activities and experiences."[2] The idea of welfare, of course, was inherent in the early and unsentimental reform efforts of Jeremy Bentham. Dicey summed the Benthamite objective as:

> . . . the transference of political power into the hands of a class which it was supposed was large and intelligent enough to identify its own interest with the interest of the greatest number—the promotion of humanitarianism—the extension of individual liberty—the creation of adequate legal machinery for the protection of the equal rights of all citizens.[3]

Welfare in broad terms means the promotion of the common decencies, which if neglected may coarsen and brutalize human feelings, as, for example, doubtless was true of public hangings and whippings in Bentham's time. It

[1] Nelson N. Foote, "Community Services," *Annals of the American Academy of Political and Social Science*, Vol. 314, November 1957, p. 47.

[2] Louis J. Ducoff and Margaret J. Hagood, "Rural Welfare. See Carl C. Taylor, Ed., *Rural Life in the United States*, New York, Alfred A. Knopf, 1950, p. 178.

[3] A. V. Dicey, *Law and Opinion in England*, London, Macmillan, 1948, p. 185. (New York, St. Martin's Press)

means equality for individuals. In many ways it means safeguarding the interest of the community in the individual and the family, and this safeguarding may assume various forms. In the interest of community welfare the local government may operate a hospital or children's playgrounds. The local community may extend aid to the aged, and it may also sponsor open-air concerts. It may provide free or low-cost lunches for school children, and it may also maintain parks and beaches.

Whereas welfare, as the term is used here, is a very general concept, social security is specific. Social security is concerned with specific types of need for which some particular service is available, a public or a private service which operates at the community level. As there are different categories of need, so the services are different. Need may concern the family or the individual; it may be temporary or long term, and it may concern the subsistence of people, their health, their conduct, or other matters. People may have their ideologies regarding welfare policy, but when they turn to social security they are confronted with needs which often call for specialized programs.

The idea of welfare, as seen by Hofstee in connection with rural life in Holland, is concerned with psychological as well as material values. Welfare is realized when the community enjoys a sense of well-being, and one feels secure in his social position and at ease with his total environment. This to Hofstee is mental and emotional security which would be reflective of stability in the material things. Such a condition prevails in rural Dutch communities, but Holland is a very urban country, and Hofstee recognizes that the old integration of rural values tends to be replaced by a new type of integration in which even the security system of the farmer is oriented to various national programs.[4]

## Social Security and the Risks of Life

Durand, a Frenchman, begins a book on social security with a discussion about risk. He sees the main function of social security as one of eliminating the risks in modern living. If it does not eliminate risk, it may ease and indemnify some of the losses. Security is mainly concerned with those risks which relate to the continuity of employment and the level of living. Other

[4] E. W. Hofstee, *Rural Farm Life and Rural Welfare in the Netherlands,* The Hague, Ministry of Agriculture, 1957, pp. v, 349-355.

social risks have to do with the employability of the worker; he may be injured and, of course, he grows old.[5]

It has been largely with respect to risk that views about different phases of social security have differed. Many people have argued and still argue that giving help to an unemployed man is to deprive him of the stimulus to effort which comes from the possibility of and the need for finding work. They hold that fear will force him to venture in one direction or another until he finds work: "Give him relief and he sits there until you bring him a new job."

Risks as considered by Durand is concerned with the hazards of life. It may concern one's falling down or falling back, but the risk that is glorified by the critics of social security concerns venturing into new fields. One gambles with insecurity. As Lynd observed, the individual sometimes cherishes the idea of accepting risks. He noted that the human personality

> . . . craves novelty (the learning and doing of new things), provided this can be taken on the personality's own terms, i.e., "in its stride." It craves risk as exhilarating—when it *is* exhilarating. But risk is exhilarating only at the points of peak energy storage in the individual's rhythms of personal living; and when risk is continuous or forced upon one the personality is put under unwelcome strain which invites discomfort, demoralization, and regression. The human personality dislikes to "go it blind" into important risks, but prefers to have its options implemented by the fullest possible information as to the precise nature of the risk and as to the best chances of minimizing that risk.[6]

Risk of the venturing type is much more idealized by Americans than by Europeans, but even in the United States the venturing type of risk is becoming less frequent. Americans also have become security minded. Whether venture risking is evident in large amount or not depends on the level of the economy in a country. Venturing is more likely in a country blessed with abundant resources. In an overcrowded country the meager resources there may be much venturing but much of it may be on a very minor scale. Abe tells us that in 1954 Japan had 172,613 manufacturing enterprises, 92.6 percent of which employed less than 50 workers each. Most of the workers in these smaller plants, so insecure that they cannot be unionized, are frequently or potentially unemployed. An oversupply of people have accepted the risk of opening stores and restaurants:

[5] Paul Durand, *La Politique Contemporaine de Sécurité Sociale,* Paris, Librairie Dalloz, 1953, pp. 16-17.

[6] Robert S. Lynd, *Knowledge for What,* Princeton, Princeton University Press, 1948, p. 195.

> These traders mostly take no holiday even on Sundays, and carry on their business from morning to night, even delivering dishes of food to the customers' houses. Thus Japan is one of the countries in which customers receive the most accommodating service. Not many customers, however, are aware that they actually pay for these services; in other words, consumers maintain potentially unemployed people with their own private money.[7]

Most people are not risk takers in the venturing sense. This verdict has been expressed by Gagliardo, but others have found the same to be true. He noted that poorly paid workers who are sure of their income and who need not fear unemployment are less concerned about security and less prone to unrest than are better paid workers whose jobs and incomes are irregular and uncertain.[8]

## The Ideal of the Self-sufficient Man

According to Lord Bryce, long-time British ambassador to the United States, Americans, as he knew them in the 1890's, were staunch believers in the principles of *laissez faire*. He called them the "most self-reliant of peoples." Yet he found they were not averse to asking help from government, although they were firm against government controls. Americans then were as individualistic as they were optimistic, for the land frontier had not yet vanished and some of the great industries were yet to be born. There was a sudden panic in the 1890's and "General" Jacob S. Coxey led a march of the unemployed on Washington. The so-called "industrials" came riding on freight trains from all parts of the nation. They were described by the press in the most uncomplimentary terms as people wanting something for nothing. A goodly number did reach Washington where their demand for public work went unheeded. Whether he said it or not, the word spread that President Cleveland answered the plea with, "It is not the responsibility of the government to support the people; it's the responsibility of the people to support the government," and the answer was applauded as a very proper reply.

It was not considered improper at that time for a corporation to ask support from the government. But, although the government was a work provider, under the American ideal an unemployed man could not ask for a public works program. Most people believed, at least those not unemployed,

[7] Gen-ichi Abe, "Competition and Monopoly in the Japanese Economy." See Rudolf Frei, Ed., *Wirtschaftssysteme des Westens*, Basil, Kyklos Verlag, 1957, p. 124.

[8] Domenico Gagliardo, *American Social Insurance*, New York, Harper & Brothers, 1949, p. 8.

that any jobless person could find work if he looked wide enough and far enough. While they conceded that help should be given to widows and orphans, to the blind, the crippled, and the aged, they also believed that every worker should prepare for old age so that he would not need to be dependent. All respect went to the self-sufficient man, but the attitude toward those who could not stand alone was one of pity, if not one of contempt. Life was regarded as a continuous struggle in which some will rise higher, but all must keep trying. The idea of relief was to put a cushion at the bottom upon which people might fall without getting hurt.[9]

This ideally self-sufficient man was assumed to be a family member, and if the family was not self-sufficient financially, it was at least assumed to be a somewhat united group in which the members came to the aid of one another in time of need. The next step in this line of thinking was that the individual and his family belonged to a community, and the families in a neighborhood were assumed to be willing to help one another. That the facts were contrary to these assumptions, as was demonstrated again and again during the Great Depression, did not make much difference to those who held these views. The argument of individual responsibility and of family and community responsibility continued to be repeated.

It has generally been true that most communities, in one way or another, are able in normal times to meet the occasional problems involving local welfare. But in the Great Depression the demand was overwhelming, and European countries have had similar experiences.

## Socialism and the Welfare State

Over the past several decades in the United States most proposals for social legislation have been met with charges of socialism. Even parcel post service was called the first step of the government in competing with private enterprise—a socialistic move. The charge has been made again and again in opposition to public aid for low-rental housing, and it was proclaimed with vigor by the opposition to the social security program adopted in 1935 but

[9] These views are still staunchly supported by some. At the Corning Conference in 1951, when trade unionist Harry H. Cook defended public social security, Don C. Mitchell, an industrialist, said there should be more emphasis on "consciousness of self." He said that people who "sought security ahead of opportunity" were not the kind that built the frontier. Another industrialist, George M. Humphrey, held that security through giving removed the "fear of consequences" in venturing and might dampen incentive for effort. See Eugene Staley, Ed., *Creating an Industrial Civilization,* New York, Harper & Brothers, 1952, pp. 157-160.

now accepted generally as a necessary service. Today when social legislation is proposed, less is said of socialism; instead it is argued that such legislation will lead to a "Welfare State."

It was the fear of socialism in Germany that led to three pioneering social security laws in the 1880's. Germany under Chancellor Otto von Bismarck established the first social insurance programs. Bismarck's reasoning in part was a common-sense conclusion that industry would be helped by the stability which such programs might afford. But, by instituting them one after another, he was able to hurt the political chances of the Socialists, and the work of organizing was made more difficult for the emerging trade unions.[10]

It would be a mistake to assume that the social security program of Great Britain, considered by many to be a welfare state, was due to the effort of the Socialists, although they had a part in it. But socialism in England, Cole reminds us, is something very different from German socialism:

> It thought, indeed, in terms of social evolution, as Marx did; but its evolutionism was from Darwin, Spencer and Huxley, and not from Hegel and Marx, and was thought of as a gradual and continuous process, and not as involving catastrophic change. In effect, it rooted itself firmly in the established British ways of thought, and visualized socialism as arising rather by a natural and gradual development of British institutions and tendencies than by any process of revolutionary upheaval. It set itself not to smash the capitalist State, but to turn it, in accordance with the needs of the times, into a "Welfare State." Its policy was rather to use and adapt than to destroy things as they were.[11]

These British "institutions and tendencies" had also become part of the American social heritage, but socialism did not develop in the United States. It had nothing to offer the frontier-fostered American individualism. Having no well-stocked frontier on which to feed, British workers turned to other means for their security. Even before the beginning of this century England was already more than 80 percent urban and industrial and the great mass of workers there were in deep poverty and almost without security law.

The Scandinavian countries may also be called welfare states. Over a period of more than three decades Denmark, Norway, and Sweden have either been under Socialist governments or Socialists have had a strong role in the formation of public policy. Per Albin Hansson, long-time Prime Minister of Sweden, has said that the objective there is to make Sweden a "home for the

[10] Durand, *op. cit.*, pp. 58-59. (The sequence was sickness insurance, 1883; accident insurance, 1884; old-age and invalidity insurance, 1889.)

[11] G. D. H. Cole, *A Short History of the British Working Class Movement.* London, George Allen and Unwin, 1948, p. 289.

people" and "to provide security for the necessities of life for all in a human way, without infringing on the human integrity of anyone." For 1952 the welfare program was equal to 10.7 percent of the national income of Denmark, 9 percent of the national income of Norway, and 10.2 percent for Sweden. With the new universal health program in Sweden, the welfare cost in that country will increase to about 12 percent of the national income.

Since 1906 Finland has been the scene of a continuing reform drive under socialism. This has also been a period of rapid urbanization and industrialization. New symbols and conceptions about social class have developed and welfare has come to be regarded as practical economics. A study of a village rural community and an urban community was made by Sariola. While the purpose of the study was to determine social class views of Communists, Socialists, the Agrarian party, and the Conservatives, we take from it the following description of services found in both the rural and the urban community to illustrate some of the elements of the welfare state:

> Both communities engage in a wide range of social welfare activities and employ, besides administrative personnel, such functionaries as physicians, nurses, midwives, inspectors in various fields, home economics instructors, home nurses, youth workers, architects, constructors, etc. Schools give free lunches, health inspection, dental care, vaccinations, etc. The extensive social security measures include maternal, child, unemployment, disability, old age assistance, etc., that conform to the national social welfare policies. Strong and specialized cooperative agencies are found in both communities in such fields as merchandising, dairying, cattle breeding, milling, farm machinery, credit, transport, etc.[12]

## Situational Factors and the Quest for Security

In a sharp criticism of Socialist welfare-promoting efforts, Dicey observed that "it is a highly probable opinion that the poorer citizens of all civilized countries have arrived at a stage of education which makes it easy for them to perceive the possible benefits for wage earners to be derived from the interference of the State, and at the same time to be victims to the easily propagated delusions that all wealth possessed by the rich is so much stolen from the poor."[13] It is his view that much of social legislation in modern countries is

[12] Sakari Sariola, "Defining Social Class in two Finnish Communities," *Transactions of the Westermarck Society*, Abo, Abo Tidnings och Tryckeri Aktiebolag, Vol. 11, 1953, pp. 134-157.

[13] Dicey, *op. cit.*, p. lxix.

the result of Socialist pushing and prodding. This view does not explain social legislation in the United States, although it may in Australia. To explain social welfare legislation in such terms leaves too much that is unexplained.

One further consideration is the possible influence of industrial urbanism itself in the emergence of problems and in stimulating organized efforts to deal with them. As these problems affect sections of the population differently, people may be expected to react differently; but they also react differently to the same set of problems. As their interests are differently affected, they find themselves aligned in different organizations; employers, for example, in the manufacturers' association and workers in a trade union, while the independent occupations may form their own associations—medical association, bar association, associations of merchants, and others.

The Socialist movement as an ideology and its various political parties evolved out of the industrial urban situation, but so did the various Christian Democrat parties in Europe, and these in some countries have had as much to do with promoting social welfare as the Socialists. In the United States efforts for social legislation are associated mainly with the Democrats, but in some states and urban areas this leadership may be carried by Republicans. The "conservatives," "reactionaries," "liberals," "progressives," or "radicals" are names given to or assumed by groups that take form in the same industrial urban situation. Whatever the leadership for or the leadership against social welfare developments or reforms, we must acknowledge that there is a clearly evident trend in the direction of all that these terms imply.

Mostly during the first half of the twentieth century, and mainly during the latter half of this period, measures for social security (dealing with old age, accidents, disability, maternity, widows, children, health, unemployment) have been established fully or in part by all the industrial urban countries. In all of these countries public responsibility for housing, slum clearance, and urban redevelopment has been accepted, although not to the same degree. Programs and facilities for leisure and services for recreation are being provided. Measures to equalize income, whether through progressive taxation or other means, are found today in all of these countries. And when we consider these developments further we find that they all stem from various efforts being made to meet problems that emerge in each country out of the industrial urban situation, problems that stem from technological and economic change. The central challenge was stated in the President's Committee on Recent Social Trends in the United States in its 1934 report:

> All of these problems may be summed up in the question: How can society improve its economic organization so as to make full use of the possibilities held out by the march of science, invention and engineering skill, without victimizing many of its workers, and without incurring such general disasters as the depression of 1930-1932? [14]

A point of interest about these social welfare trends is that they have ceased to be sentimentally motivated. True, the pioneers for many of these causes were called "sentimentalists" or "reformers," but that generation did its unpopular task and passed on to be succeeded by a generation of professional workers. The idea of "charity" has also passed, and about this Landis says:

> It was only as urban industrial society reached the point where the secondary group replaced the neighborhood and where secular thinking replaced religious motivation that public welfare on a larger scale came into its own. When the idea of public responsibility for the security of the citizen came to be a recognized philosophy, there was but a step to the removal of public benefits entirely from the realm of charity and the classification of them as rights, and duties of the state to its citizens. As a consequence, with the development of a national social security program, the term "poverty" has tended to pass from the vocabulary and the term "security" to replace it, with an appropriate terminology to cover benefits to the citizen. In place of charity have come pensions and annuities, unemployment compensation, aid to dependent and crippled children, and other categories which are not considered charity in the old sense, but normal and justifiable benefits to those who qualify.[15]

## Public Services and Private Services

The role of private agencies in all welfare developments is highly important, even though their share in the total cost of all services is indeed minor. This was not always true, particularly in the United States where in the 1920's and earlier much of the social services in the larger communities was the work of private organizations. These were found in different special fields as well as in the general services. Special fields included family and health agencies; agencies for the aged, the crippled, the blind; agencies for immigrants and for travelers in need; agencies in the field of recreation, and many others. Much of the burden they tried to carry has since been assumed by public agencies.

[14] *Recent Social Trends in the United States,* New York, McGraw-Hill, 1934, Vol. I, p. xxviii. Committee members did not suspect that the depression was yet to deepen and to continue another five years.

[15] Paul H. Landis, *Social Policies in the Making* (2d ed.), Boston, D. C. Heath, 1952, pp. 428-429.

For example, public housing and federal aid for the aged have obviated the need for private agencies to campaign for or provide these services.

In the United States and Great Britain, where voluntary service agencies are more varied and numerous than in other countries, these groups are finding new and no less important roles. They concern themselves with higher service standards. But what of countries that have "welfare state" public services but no voluntary agency tradition? Eugen Pusic of the School of Public Administration in the University of Zagreb, Yugoslavia, sees the social welfare services as a professional approach to the problems of people in need. These problems become more complex in the urban society. Social welfare service has a conscience function.

> On the one side there were the general risks, inherent in the industrial and urban society, and it was clearly recognized as the responsibility of the community—and that meant under existing conditions, the responsibility of the state—to provide a measure of basic security against these common risks. On the other side were the problems where extraordinary personal and environmental elements coincided to make a person or a group of persons dependent on the help of society. In both kinds of problems society had to step in, but by different methods. The first kind, general social security, required general measures chiefly of an economic nature. Here the legal method was of paramount importance, and generally economic planning was the main tool with which to bring about full employment, increase living standards and put a floor under everybody's economic existence. That was really the prerequisite for the practice of social work, for the organization of social welfare as it came to be understood.[16]

In this process, continuing Pusic's thought, social work became less occupied with whole problems and was able to focus attention on the functioning of public programs, to give more thought to individual needs. "The concept of social welfare in a sense became narrower, but gained in depth." Social welfare workers, previously in a marginal role, were now in a position to "define the social problem."[17] The new role may be called one of keeping

[16] Eugen Pusic, "Social Welfare in a Welfare State," *International Social Work,* Vol. I, No. 2, April 1958, p. 8.

[17] *Ibid.* Asa Briggs, noting the entry of social work into new fields, observed, "To the charitable urge to help the needy, the social urge to guarantee minimum standards and the democratic urge to reduce inequalities has been added the administrative urge to secure coordination and comprehensiveness." Although Briggs was thinking of Great Britain, the thought also applies to the American scene. "The Social Services," see G. D. N. Worswick and P. H. Ady, *The British Economy 1945-1950,* Oxford, Clarendon Press, 1952, p. 365.

the public conscience. In the United States the task of the private social service agency has lost none of its importance as a factor in local community life.

In the complex urban community many problems arise which do not normally fall within the specified categories entitled to public aid. It may be a stranger in need, or a case calling for emergency aid. There are always such marginal problems which the private agency can handle until the proper public service can be arranged for.

In the Western countries it has been the private agencies and the professional associations that have pressed for higher qualified personnel in all public welfare posts. As a result, the standards have been raised. The idea is no longer accepted that any well-meaning person can hold any welfare service position. Some observers, however, feel that this pressure for professionalization may react negatively for all concerned, that many positions can be held by less well-trained personnel working under professional supervision.[18]

## Specialization and Complexity in Welfare

Not too long ago the aged widow, the orphan child, the chronically sick person, the mentally ill, and other social problem cases were all sent to the same poorhouse, an institution designed to quiet public conscience and to get such people out of sight. But social reformers like Dorothea Dix and others did not allow the public conscience to rest. One reform and then another resulted in the establishment of different types of institutions, some public and some private. The trend toward specialization resulted in the formation of a variety of private services, a development which faithfully reflected the complexity of urban life.[19]

Before the United States government entered the field of welfare service the governments of some states had gone far, but even greater advances toward the specialization of such service had been made by the larger urban communities. Specialization as it applies to welfare must be seen in different contexts. If seen from the point of view of the individual or the family, it may concern a division of labor between services. The same family at the same time may present a variety of problems. The teen-age boy may be having

[18] Eveline M. Burns expresses the view that the professional groups in the welfare field are sometimes excessive in demanding that all public welfare positions be given to highly trained persons, although persons of lower qualifications might do as well. *Social Security and Public Policy,* New York, McGraw-Hill, 1956, p. 261.

[19] See, for example, John W. Riley, Jr., "The Changing Pattern of Destitution in an Urban Area," *Social Forces,* Vol. 20, 1941.

trouble in school (truancy problem), his teen-age sister may be getting out of control in the family (a juvenile protective agency problem), the father is unemployed and sick (problem for the family relief agency, also for a health agency). Finally, another agency may be called on to rehabilitate the father for employment. Thus the treatment of a single family problem called for the services of several different agencies, public and private. Of the public agencies, some might be local and others financed and managed by the state.

This trend toward specialization and complexity was not diminished by the entrance of the national government into the welfare field. Now that an increasing share of the burden was to be carried by public agencies, the work of dispensing services was rendered more complex, for the whole task was divided for policy, finance, and administration between three levels of government, federal, state and local.

What the federal government did, in the United States as in Great Britain and other industrial countries, was to assume responsibility for certain large sections of the total welfare problem. The central purpose in Great Britain, as set forth in the Beveridge report, was "to win freedom from want by maintaining incomes." This meant carrying on a variety of programs at the same time and establishing a national policy that was wider in scope than the mere list of services.[20]

Quite aside from the philosophy behind the different national welfare programs for social security, the problems of rendering the special services and those of dividing the responsibility for administration between agencies and levels of government had to be worked out.[21] In the following sections some of these services well be considered from the point of view of both policy and administration.

## Accident Insurance and Safety*

As already mentioned, a national accident insurance program was established in Germany in 1884. In Great Britain a limited workmen's compensation law was passed in 1897 which was amended in 1900 and in 1906 replaced

[20] William Beveridge, *Social Security and Allied Services*, London, Stationary Office, 1942, p. 153. And for an excellent survey of this in the United States, see Wayne Vasey, *Government and Social Welfare*, New York, Henry Holt, 1958.

[21] Burns, *op. cit.*, pp. 155-156. She calls attention to some of the problems of welfare administration as between levels of government.

* For this and the following five sections the writer is grateful for help from Dr. Gabriele Bremme, Unesco Institute for Social Sciences at Cologne.

by a law covering all employed persons. In France the first law for workmen's compensation was passed in 1898 to be amended at various later dates including 1938 and 1946. In the United States workmen's compensation is not a national program. This kind of insurance had to be struggled for state by state. The first adoption came in 1911, while the last state to institute such legislation was Mississippi in 1948.

Industrial accidents and industrial diseases, as much as slum housing and certain forms of unemployment, represent a type of problem specifically associated with industrial urbanism. That it received earlier attention in European countries suggests that there were feelings of urgency there about it. Various difficulties stood in the way of writing legislation. There was the notion that a worker normally was assumed to accept some risk, and the idea that insurance to indemnify injury might in some way encourage carelessness. Furthermore, with many kinds of industry, many kinds of working conditions, many ways of organizing work, and many degrees of capability among workers, it was found difficult to plan and administer an equitable insurance scheme. Moreover, employers differed in their ability to assume responsibility.

Thus, some of the early attempts at legislation did no more than to indicate employer responsibility, and the task was left with the worker to secure compensation through the courts.[22] This had been true in both France and England, but the worker risked losing his job if he brought suit against the employer. The same was true for most of the earlier workmen's compensation laws in the United States; they imposed on the injured worker the burden—and cost—of getting compensation. That condition has been remedied in most states.

Countries have different methods of levying the cost of accident insurance. In France, Germany, and Great Britain the employer pays the cost of the insurance, and this holds true generally in the United States. In the United States a good share of the employers carry their insurance with private companies, although in some states they are insured in the "state fund."

Whereas the basic problem regarding unemployment insurance is that government must find ways and means of insuring full employment, the basic problem in these insurance progams against industrial accidents and diseases is to reduce the incidence of risk. In most countries the idea of industrial safety is promoted by inspectors who go from plant to plant. This means that the

[22] Florence Peterson reports how employers usually defended themselves by proving the accident was due to the worker's own carelessness or the carelessness of a fellow worker. *Survey of Labor Economics,* New York, Harper & Brothers, 1951, p. 804.

incentive for safety comes from outside the plant. In the United States the incentive for safety is largely within the plant. In part this is because, under most of the insurance arrangements, if the accident rate in a plant rises the employer must pay a higher insurance premium. For this reason employers are stimulated to be "safety conscious." This may help to explain the comparatively lower accident rates in American industries.

## Old-age and Survivors Insurance

Every society has been confronted with the problem of age, but none of the methods used effectively in primary societies seem to meet the needs in the industrial urban society. The idea that the problem should be left to the family and the local community turns out to be quite unrealistic. And so it is with the idea that, since every man knows that old age will overtake him, he should prepare so that he will not arrive at the end of his working years in poverty. This notion assumes that every urbanite can render his own old age secure simply if he wishes to.

The fact is that few people who live and work in the industrial urban community, however steady their work life and however sane their way of living, are able to provide for their own old age. This applies especially to persons who have diligently carried the responsibility of family rearing while being employed at lower income levels. Always there have been those who have been indolent and prodigal, but, by and large, they are the exceptions. A recognized consideration is that the problem of need in old age is real.

A second consideration is that the proportion of old people in the United States, as in other Western societies, tends to increase. The number of persons 65 years of age and older in the United States increased from 3.8 percent of the population in 1890 to 8.2 percent in 1950, and it is estimated that the increase will reach 12 percent by 1980.[23] Burns, quoting estimates from the *Analysis of the Social Security System,* prepared by the Census Bureau, notes that in 1950 there were 18.4 million persons 60 years of age and older in the population, and the estimated number in 1980 will be 30.1 million.[24] Even though the number of productive persons, absolutely and percentage wise, will increase between now and 1980, the number of dependent aged will also increase.

[23] *The States and their Older Citizens,* Chicago, Council of State Governments, 1955, pp. 9, 119. Also Wilma Donahue and Clark Tibbitts, *Growing in the Older Years,* Ann Arbor, University of Michigan Press, 1951.

[24] Burns, *op. cit.,* p. 106.

In France as of 1836 there were 296 children under 10 years and 90 persons 65 years and older for each 1,000 persons between the ages of 10 and 64. Over the years for each 1,000 persons between 10 and 64 years of age the proportion of children has decreased while the proportion of persons 65 and older has increased. In 1951 for each 1,000 persons between 10 and 64 there were 221 under 10 years and 158 over 64 years.[25]

In Great Britain, where all are insured against old age, the estimated population in 1955 was 49,968,000 (112 females to 100 males) and 5,765,-000 people were in the age 65 and over group (150 females to 100 males). This is 11 percent of the total population. Considering that women may be pensioned at the age of 60, this means that the number of pensionable persons in Great Britain in 1956 was about equal to half the 11,700,000 children under 15 years.

Seldon gives us an idea of the prospect for this most industrial and urban of countries. He estimated that, whereas the number of persons in Great Britain of pensionable age was 5.3 million (about 5.7 in 1957), it will be up to 7.4 million in 1960 and to 9.6 million in 1980. He does not estimate in terms of percentages because of the hazards in predicting future birth rates and death rates.[26] Seldon's estimate was given in connection with an examination of the future cost of pensions and other public services. He writes in his summary:

> Worst of all, the notion that the social services are here to stay for all time is not ennobling but degrading. For they are meant to help us when we are in need. But what is there ennobling about free or subsidized milk, or education, or medicines, or housing when you can afford to pay for them yourself and when you know they are being paid for by someone else? What self-respecting man wants his neighbour to pay his milkman? . . .
>
> This is how we should see State pensions. They have a part to play in a developing society, but they should be abandoned as soon as we can dispense with them. We should use them as a crutch, to throw away as soon as we can walk.[27]

The problem of old age is not merely economic. As Tibbitts reminds us, old age in our modern society is for many a form of social isolation. The old become problems to themselves and others. Because they are lonesome, they

[25] Durand, *op. cit.*, p. 171. Figures cited are from J. Bourgeois-Pichant. See also A. Sauvy ("Le Viellissement des Population et l'Allongement de la Vie," *Population* [Paris], Vol. 9, No. 4, October-December 1954, pp. 675-682) for a discussion of effect of birth and death rates on the proportion of aged in the population.

[26] Arthur Seldon, *Pensions in a Free Society*, London, Institute of Economic Affairs, 1957, p. 18.

[27] *Ibid.*, p. 36.

complain, and even when their married children are cooperative, the children can help little because they have their own lives to lead.[28]

Old-age insurance in those countries where it prevails creates unsurmountable administrative problems. Such systems once adopted, however, are not likely to be abandoned. Economically, we may say that the problems of old age are being met, but the social and psychological problems of old age have yet to be solved. Living under urbanism tends to isolate the aged at the social fringes of community life, a very different condition from the status of the aged in primitive life.

## Social Security and Health Service

The American Social Security Act of 1935 made no provision for meeting the health needs of the people. All attempts to secure such legislation since 1935 have met the strong resistance of the medical profession. A similar resistance was offered in Great Britain, but it could not prevail against the demands of organized labor and other pro-social-medicine groups. The British Health Service Act became law in 1946 and went into effect on July 5, 1948. A beginning had been made in Great Britain but, as Lewis observed, it was not meeting the needs.[29] The new service included everyone and, until the revised health service of Sweden went into force in 1955, it was the most complete in any country. Regarding the Swedish service we quote the Minister of Social Affairs, Labor and Housing:

> The new health insurance covers all the inhabitants of the country, irrespective of age, health and sex. It does not matter what one's financial position is nor what he does for a living. Even those who do not work are included. Both Swedes and foreigners living in Sweden, a total of 5,000,000 persons, are members of the health insurance associations. And the 1,700,000 children of these members are covered through their parents' insurance.
>
> The inclusive character of the Swedish insurance is what distinguishes it from compulsory insurance plans in other countries where it is usually limited to wage earners or other defined groups. The basic principle behind the Swedish scheme is simply this: since everyone runs the risk of becoming ill, everyone should be assured some protection against the economic consequences.[30]

[28] Clark Tibbitts, "Retirement Problems in American Society," *American Journal of Sociology*, Vol. 59, No. 4, January 1954, p. 307.

[29] Ben W. Lewis, *British Planning and Nationalization*, London, George Allen and Unwin, 1952, p. 189.

[30] Ernst Michanek, "Compulsory Health Insurance in Sweden," release issued by the Swedish Institute, Stockholm, 1955.

As with the British health service, the Swedish patient selects his doctor. Medical fees are fixed but urban doctors are paid a higher rate. Contributions depend on the size of income. As in Great Britain, certain costs, like private hospital rooms, are paid for by the patients. In both countries it has been necessary to set certain minimum limits, for example, in Britain, dental work beyond a certain cost point is paid for by the patient.

For the general practice service in Great Britain 21,000 of the 23,000 general practitioners are included. All hospitals and all types of specialist service are included in the program. Local functions under the health service include "ante-natal, post-natal and infant welfare clinics (also providing for dental care); midwifery; health visiting; home nursing; vaccination and immunization; ambulances; prevention of illness; care and after-care of the sick; local mental health services; and also domestic help when needed on health grounds."[31] The costs for such services are shared by the national and local governments.

Objections often heard in the United States against social medicine are that a public system would be operated by politicians and that the relation of confidence between doctor and patient would be destroyed. We need not consider the charge that medicine would fall under the control of politicians; that does not happen in local health departments. However, a word needs to be said about the doctor-patient relationships. As Durand has observed the development of medicine in France, the personal relationship is already changing because of changes within the profession. Like other professions in the modern community, medicine becomes more technical and clinical. It is no longer a matter of the doctor carrying all his medicines and equipment in a little black bag. Much expensive equipment is needed and doctors must become specialists, making treatment more expensive, or they must join in groups. The patient sees the doctor very little, but talks to his secretary or the nurse. Medicine tends to become more impersonal.[32] However, in the 1955 report of the British Ministry of Health an opinion is expressed by the Committee on General Practice of the National Health Service, submitted by a medical man, Sir Henry Cohen. The Committee found that two-thirds of the patients identified a single physician as the "family doctor," although this is not always possible because of diverse conditions common to the medical practice; pa-

[31] H. W. Stockman, "History and Development of Social Security in Great Britain," *Bulletin of the International Social Security Association*, Vol. X, Nos. 1-2, January-February 1957, p. 58.

[32] Durand, *op. cit.*, pp. 451-452.

tients must sometimes visit specialists. The verdict was that the National Health Service had not interfered with doctor-patient relations, but there were situations that needed correction. One important item in the report concerns the medical fee, a problem that once marred doctor-patient relations, and was often a worry to both doctor and patient. That difficulty no longer exists.[33]

## Aid and Allowances for Children

The Social Security Act of 1935 committed the federal government to a program of sharing with states and localities the cost of aid to dependent children "who had been deprived of parental care by reason of the death, continued absence or physical or mental incapacity of a parent" but who lived with a relative in his or their own home. Local communities might extend aid to other dependent children, but the federal government contributed only to children in broken homes. This was done under the same policy in which federal aid was extended to the destitute aged, the blind, or the crippled. The difference is that the per-child monthly grants are less than half the per-person grants for the destitute aged. White observed that Congress is likely to be more influenced by the latter group because old people can vote and organize themselves into pressure groups,[34] as they did during the Great Depression.

This American program for aid to children is concerned with aid to broken families and not with aid to all children of the poor. It should be mentioned that a form of aid is given for all children in the income-tax deductions allowed to parents. There has been little demand in the United States for children's allowances, so customary in other countries. It is not yet recognized that, unlike in rural life, children in the urban community are an economic burden. Thus parents, however poor, are responsible for their own children, however many. Canada since 1954 has paid family allowances for all children under 16 years. Such allowances are also made in most European countries, although the methods of payment and the amounts vary from country to country. In Belgium, France, and Italy the allowances are paid by the employer along with wages, although he is reimbursed by the state. The idea is that these allowances may encourage parents to have more children.[35]

In Great Britain family allowance payments are paid to the family· gov-

[33] *Report of the Ministry of Health,* London, Stationary Office, 1955, p. 54.

[34] R. Clyde White, *Administration of Public Welfare,* New York, American Book, 1950, p. 179. The number of children receiving such aid rose from about 404,000 in 1936 to 1,691,733 in June 1955.

[35] Durand, *op. cit.,* p. 92.

ernment is the direct giver and sharer with parents in the cost of children. The grant is not mixed with the earned wage. Said the Beveridge report "a national minimum [of subsistence] for families of every size cannot in practice be secured by a wage system, which must be based on the product of a man's labour and not on the size of his family."[36] Where such allowances are paid by employers as part of the wage, even when employers are reimbursed, as most of them are, the illusion is created that workers with more children are receiving higher wages for their work.

In general, public authorities assume most of the cost for the education of children, and the same authorities is some countries are assuming some of the cost for rearing children. Children are deemed to be an important national resource. Churches, anxious to increase membership, have helped to promote this policy,[37] but in some of these countries there is general concern about the low birth rate and it is hoped that these allowances will help meet that problem.

## Some Implications of Unemployment Insurance

When a country adopts an unemployment insurance program it must also assume other responsibilities related to it. When a worker is unemployed, it is possible that another job is available in the community, but he may not know where to find it. Thus a supplementary service to unemployment insurance in most advanced countries is an employment exchange. In Germany the labor exchange keeps a record on every worker. If a worker changes from one job to another, the *Arbeitsamt* must be notified. Any employer hiring a worker must inform the *Arbeitsamt*. Workers drawing unemployment benefits in any country are expected to hold themselves in readiness for a call from the employment exchange or labor office.

Moreover, a country with an unemployment insurance program also needs to have a public policy for promoting full employment. Government can bolster a declining labor market in different ways: (1) It can lower the interest rate on money needed for industrial development. (2) It can loan public money to private enterprises. (3) It can increase government purchases. (4) It can open up public works projects or even work relief projects paid for

[36] Beveridge, *op. cit.*, p. 154.

[37] Francis J. Corley, SJ., "Are Family Allowances Doles?" *Social Order*, Vol. VI, No. 2, February 1956, pp. 68-70. The article deals with the family and community responsibility for children.

with public funds. The payment of unemployment compensation is itself a device for putting money into circulation, for this money is immediately spent on consumer goods. It is now well recognized that an unemployed worker is not the only loser, for the productive value of his work is lost to the community. The cost of unemployment is shared by all.

Thus the establishment of unemployment insurance brings with it various controls over the labor market, or at least efforts to sustain the labor market. Many problems arise and much controversy is involved in determining what that control should be. While employers also benefit from an unemployment insurance program, they are usually critical and often claim that such a program encourages malingering. Within limits that is true, but, as Bakke has demonstrated something must be done for the unemployed man else he loses his self-respect, and in such a situation the expedient of work relief is better than sheer idleness.[38]

How the family fares in a period of mass unemployment will vary with families, but in a period of enduring depression families at all economic levels are inconvenienced. This is emphasized by Burgess and Locke in a summary of studies that were made during the Great Depression,[39] in particular one made by Caven and Ranck in Chicago which concerned families at different economic levels.[40]

Most of the families in the Chicago study had used their savings and some had had to forfeit their life insurance. Other families went into debt. Most of these families had experienced unemployment, and this called for such sacrifices as giving up movies, absenting themselves from church, and excluding themselves from various community activities. In some cases families had to move to cheaper apartments which involved living in more crowded quarters. Needed medical and dental services were postponed, as were plans for the education of the children. In some cases related families "doubled-up" in the same apartments, and this was frequently the cause of friction, particularly if one of the family heads was unemployed. It was also disclosed by the Chicago study that most of these families experienced tensions. There were cases of separation and of divorce as well as cases of nervous breakdown.

These and other types of family disturbance represent the social costs of

[38] E. Wight Bakke, *Citizens without Work*, New Haven, Yale University Press, 1940, p. 280.

[39] Ernest W. Burgess and Harvey J. Locke, *The Family, from Institution to Companionship*, New York, American Book, 1953, pp. 564-565.

[40] Ruth S. Caven and Katherine H. Ranck, *The Family and the Depression*, Chicago, University of Chicago Press, 1938, pp. 539-543.

unemployment. It must be remembered that these studies were made at a time when there was no form of unemployment or other social insurance. The American social security program had not yet come into force. How families fare in respect to security under a welfare program, or without such a program, is illustrated by Rountree's studies of 1936 and 1950 in York, England.

Rountree divided York families into eight categories from *A* at the bottom to *H* at the top. On the basis of tested criteria, he determined that families having less than a specified income would be regarded as being below the poverty line. This included the lower two categories *A* and *B*. In the 1936 study Rountree found that 31 percent of the "working class population" were in the two "below the poverty line" categories, while in 1950 only 3 percent of the 18,099 families in York were in the *A* and *B* categories. Due allowances had to be made in comparing conditions during the depression year of 1936 with the full employment year of 1950,[41] but even when making these comparisons, Rountree concluded that there was considerable evidence of improvement. For example, 29 percent of the *A* and *B* families in 1936 were below the poverty line because of unemployment, while 33 percent were in poverty because of low wages and irregular employment. These two reasons were nonexistent in 1950 because of unemployment insurance, while most of the aged were out of the povetry categories because of their pensions.[42]

The American unemployment insurance program is administered by the states although the funds are administered for the states by the federal government. The details of administration must necessarily be handled by offices in the locality. Administrative policies tend to vary from state to state, being most liberal in the more industrial urban states. Standards may be low in the less industrial urban states and, furthermore, because there tends to be less tolerance in the smaller and more rural places toward the unemployed, the federal administration has found it necessary to set various minimum standards for insuring fair treatment of the unemployed workers.

## Assistance and General Welfare

No matter how effective the social security program of a country may be, the different phases of the program—health service, accident insurance, old-age insurance, unemployment insurance, and so on—deal only with legally

[41] B. Seebohm Rountree and G. R. Lavers, *Poverty and the Welfare State,* London, Longmans, Green, 1951, p. 34.
[42] *Ibid.,* p. 95.

defined categories of need. Various minor, and often temporary, categories of need are not met by these programs, and persons protected by one type of insurance may at times need additional help. For this reason there must still be the "catch-all" assistance agencies, private as well as public, at the local level. In the United States these are the financial responsibility of the local community, although they sometimes receive financial help from the state. Federal aid is given for certain categories—the aged without insurance, dependent children, the blind and the partially or totally disabled; these include the major part of the general relief burden for which local communities assume responsibility in part for financing and almost entirely for administration.

White observed that persons who apply for general relief are indeed insecure since they have "no contractual right to services." The law may impose a duty on the state or local community, but local officials have wide discretion in deciding if relief will be given, to whom it will be given, and under what conditions. Local American communities differ widely in their policies and practices.[43] In most European countries the policies and practices for general relief are defined by national authority.

## Full Employment and Welfare Costs

Economists differ about the idea of welfare in economics. Scitovski associates it with public policy. It has a collective utility function "which expresses everybody's perferences relating not only to his personal satisfaction, but also to the state of the entire community."[44] Talking about welfare, he implies, is to talk about what *should be,* which puts the concept outside the realm of what economists believe *economics should be.* The idea of welfare had little place in the evolution of the capitalist economy. It begins when governments decide that the accumulation of private wealth and power must be curbed to some degree.

Limiting the accumulation of private wealth and power usually takes the form of measures through which the community shares in the accumulation. Says Pigou, "As everybody knows, transfers of money income from the better-to-do to the worse-to-do sections of the community must in practice be ac-

[43] White, *op. cit.,* p. 320.

[44] Tibor Scitovski, "The State of Welfare Economics," *American Economic Review,* Vol. 41, No. 3, June 1951, p. 311.

complished, if they are at all large, with the help of steeply-graduated taxes."[45] The assumption behind such measures is that the community has shared in the creation of the wealth and should share in its blessings. But Pigou warns:

> Nonetheless, it remains true that transfers *may* indirectly damage potential income so much that in the end they make against rather than in favour of economic welfare. Badly constructed schemes for giving poor people a "fairer share" of the national cake *may* even make the cake so much smaller that the absolute amount which they receive is actually reduced. These considerations do not, of course, warrant our standing still and doing nothing at all. But they do suggest that in getting forward we should move with reasonable care and probe for hidden minefields.[46]

It is not for us here to examine the merits of the trend, but we do need to recognize that in the modern societies the trend is in the direction of the equalization of income. By this means, says Friedmann, the modern states are able to finance their "costly social service schemes."

> Taxation is one of the most important weapons by which the State can mitigate the two objectionable aspects of unrestricted private property: first, the inequalities of wealth, and secondly, the power to use property for private profit and without regard to community purposes.[47]

After calling attention to some of the more urgent social problems in the United States, Hansen cited the need for the expansion and improvement of social security, health services, national school lunch programs, minimum wage legislation, federal aid to education, low-cost public housing, slum clearance, and urban redevelopment. Such measures in connection with a proper system of taxation would help to redistribute income and would promote a greater equality of opportunity.

> Greater equality in the distribution of "earned" income could thus be achieved. Moreover, greater equality in the distribution of property income could be attained by a steeper progressive rate and lower exemptions on inheritance taxes, and by a wider distribution of ownership of government bonds, savings accounts, life insurance and home ownership in the lower and middle income groups.[48]

[45] A. C. Pigou, "Some Aspects of Welfare Economics," *American Economic Review*, Vol. 41, No. 3, June 1951, p. 301.

[46] *Ibid.*, p. 302.

[47] W. Friedmann, *Law and Social Change in Contemporary Britain*, London, Stevens and Sons, 1951, p. 26.

[48] Alvin H. Hansen, *Business Cycles and National Income*, New York, W. W. Norton, 1951, p. 572.

It is also recognized that redistribution of income measures are not more basic to general economic welfare than is full employment. In a country where everyone has work and a high level of living, the burden of social welfare is light. The opposite tends to be true when employment diminishes: When economic welfare goes down, the cost of social welfare rises.

## Summary

This chapter is concerned with the place of welfare in the modern urban community and with its wider implications, services for the total community as well as for individuals in need. While many welfare services are provided by private organizations, the emphasis here is on public services provided by the different levels of government, although for most of these the point of contact with the people is in the local community. Particularly in the larger urban communities, or as influenced by these communities, welfare services tend to increase in number and variety, as they also tend to become more professional and specialized.

That the subject of welfare has always been and continues to be controversial may be some indication of its importance. Economic, social, and political viewpoints come into frequent clash over issues of welfare, with the result that most opinions one hears, whatever the welfare problem, tend to be colored by one type of bias or another. For example, the traditional American faith in the self-sufficient individual tends to continue, however much it has been contradicted by the realities of industrial urban life. In the preindustrial and preurban societies welfare was not an issue, but the old primary group ways of achieving security and welfare fail to operate with the same effectiveness in the secondary life of the modern community.

It is now being recognized that the individual working and living in the modern community is confronted not only with many new risks, but with risks of greater magnitude than ever before existed. Moreover, it is being recognized that certain of these risks must be met by new types of collective effort. Only in this way can secondary society achieve what primitive peoples were able to do collectively in primary society.

It is often alleged that the different programs for collective security being established in the more advanced countries are so many steps in the direction of a "Welfare State" in which men will be deprived of initiative. The drive for such collective security has often been called socialistic. In actual fact, the drive is present in countries that are not socialistic, the United States being such a country. It is true that, although collective security measures were adopted later in the United States than in other countries, American security measures have been developing faster than in other lands. The political and other "isms" may have influenced the development of collective security in some countries, but the real force is the problem-making industrial urban way of life by which the demand for collective security is set.

Public welfare in the United States is finding expression in a variety of programs, all of which are becoming firmly established and none of which has been slipping backward. These programs include old-age insurance, unemployment insurance, workmen's compensation, and the various assistance programs; aid to dependent children, to the blind, to the handicapped and to the noninsured aged. Other programs include public housing and slum clearance and the various aids to education. The main weak point in the American program for welfare is the lack of a health insurance system. The role of the local communities, both for administering and financing, in part or entirely, is probably greater in the United States than in other industrial urban countries. Communities, however, differ widely in the amount and quality of these services, but in general these services are most effective as well as most humane and professional in the large cities. It is mainly in the large cities, for example, where highly developed services for leisure may be found.

Welfare services which concern all the people, such as those relating to leisure, are needed at all times. Others, such as unemployment insurance, serve a rescue function in economic emergencies. Still others serve individuals in need, but these needs may multiply in hard times. The real, or the most demanding of, welfare challenges to a country are those of affording full employment to the largest number of employable people and to enable them to enjoy a high level of living. Under full employment most types of welfare aid are not needed, but it is then that reserve funds are set aside.

While local communities can help, they cannot, under the modern global network of interdependence in production, insure full employment. Whatever can be done by civil authority in cooperation with private enterprise, must be initiated at the national level. The local labor market is linked with other labor markets. But even the best organized national plans for insuring full employment may be influenced by work and welfare in other countries. This is merely another reminder that industrial urbanism is global.

CHAPTER 17

# Social Change and Conformity under Urbanism

IN HIS STUDY of St. Denis, a French-Canadian village, Miner reports that a man away from home is known by the parish from which he comes and in the parish by the family to which he belongs. And the groups that make up the community are mainly the families. On the work side, life in St. Denis is oriented to the seasons and the crops, and most activity is regulated by the rigid seasonal schedule of work. Although individuals from St. Denis go to other places, even to American cities, and later return, the daily round of living is both severely patterned and isolated, and more or less contentedly so.

St. Denis appears to be the kind of a place where a near total social control is maintained and where there is a near total conformity on the part of individuals to an established set of values and a fixed way of living. And there was, at least up to the time the study was made (1937), a near absence of social change. On all three counts—control, conformity, and change—St. Denis was then quite the opposite from the urban community of today. What the urban society is like with respect to these three concepts is the subject of this chapter.[1]

## Use of the Terms

"Social control is a phrase generally used to designate those social behaviors which influence individuals or groups toward conformity to established or desired norms."[2] Social control is exercised by groups both formal and

[1] Horace Miner, *St. Denis, A French-Canadian Parish*, Chicago, University of Chicago Press, 1939. Citation here is from excerpts in Ronald Freedman and Associates, *Principles of Sociology*, New York, Henry Holt, 1956, pp. 293-299.

[2] George A. Lundberg, Clarence C. Schrag, and Otto N. Larson, *Sociology*, New York, Harper & Brothers, 1954, p. 698.

informal, by institutions or other agencies—inventions, for example. As we shall see, social control is exercised in different ways by a great variety of agencies, but its purpose is to influence conformity.

Conformity is the favorable or desired response to control, as when children conform to the guidance of their parents, behaving as the parents feel they should. The Boy Scout, for another example, conforms when he behaves according to the code of the organization. It is usually implied, and it generally follows, that conformity concerns more than merely behaving as expected; one must also think and believe in this conformity.

There are many types and methods of control, and there are many reasons why control is imposed. Much of this can be summed up in the words "institution" or "institutionalized." To the extent that the ways of behaving of a group or society are repeated for generations and become formalized into ritual or crystallized into the language and social habit generally, they may be described as institutionalized. When a cluster of interests and ways of behavior crystallizes in relation to some basic phase of the culture, the result is an institution. This term can be applied to marriage, to education, to public health, to the police and court system, and so on. In the modern society we find that most institutions are rooted partly in folk tradition and partly in the law. Insofar as they become involved in social control, this control is exercised partly by primary group pressures and partly by law.

As there are different kinds and degrees of social control and different degrees of conformity, so institutions differ with reference to their firmness and rootedness. We may think of various areas of behavior and thinking as in process of being institutionalized. Trade unionism may very well be such an example. In our society, whether the institution is well established or only taking form, it is also in process of change and perhaps resisting pressures for change.

We have been using the term "social change" frequently in the preceding pages. In simplest terms, it involves substituting one way of behaving or—and the two are interrelated—one way of thinking for another. Change may result from developments within a culture, such as the appearance of inventions, or it may be due to influences from outside the particular group or society, for instance, the arrival of strangers or the use of products from elsewhere. Social change is so named if it affects social relationships. If the Guatemalan Indians bring in shoes from the outside, that may influence behavior on the part of others as well as the wearer. As Tax said, shoes are a symbol of the outside. Besides, if one wears shoes he must also wear socks and that belongs to a way of life not in the Indian tradition.[3] It is social change if the wearing of shoes and socks (and behaving as one should who wears shoes and socks) affects relationships between the individual and his family or other groups.

Social change involves relationship adjustments between the group and the individual, and it may also involve such adjustments between groups within the community. Dube noted that in the villages of India the organization of the group life (generally family groups) is such that the government meets various types of subtle resistance in its efforts to encourage improvements, even such an improvement as digging a well in order to get good water. Public officials are suspected. But if the well is dug, or other improvements made, it means that various established social relationships are changed.[4]

Organized groups, especially if long organized, are likely to respond negatively at first to innovating influences, but later step by step they yield to these influences. So long as they resist, they tend to put pressure on the individual to conform.

## Change and Social Order

All cultures are in process of change, although the rate of change may be slow in some cases and fast in others. It is slower among the Guatemalan Indians than among the French Canadians, and it is slower among the rural French Canadians than in the cities of Canada. Similarly, the rate of change is slower in some cities of Canada than in others. Looking at these French-Canadian villages, Siegfried saw much to admire. Here was social stability and order, isolated from industrial urbanism and the secular way of life. The attitude of the villages was one of resistance against the encroachments of any influence that might disturb the social order. Siegfried found the social order very much under church control, a nearly perfect social as well as physical isolation that has hardly been disturbed for generations. He concluded that "without their priests the French Canadians could not have remained intact."[5]

The kind of social order that prevails in the isolated French-Canadian village or in the Guatemalan Indian village is something very different from

[3] Sol Tax, "Changing Consumption in Indian Guatemala," *Economic Development and Cultural Change*, Vol. V, No. 2, January 1957, pp. 147-158.

[4] S. C. Dube, "Some Problems of Communication in Rural Economic Development," *Economic Development and Cultural Change*, Vol. V, No. 2, January 1957, pp. 129-146.

[5] André Siegfried, *Canada, an International Power*, London, Jonathan Cape, 1947, p. 63. See also Horace Miner, *St. Denis, A French-Canadian Village, op. cit.* Miner confirms the static and isolated character of these villages, but since his study was made some of these French-Canadian villages have been disturbed by industrial development.

that which prevails in the modern urban community. In the extremely rural community or the primitive community, social order is easily maintained if it is not disturbed by outside influences. Otherwise it is indeed frail, and this is demonstrated when the money economy is introduced. If a single major element in the primary order is disturbed the entire balance is immediately vulnerable. If certain values in the value system are changed the entire system is shocked. The pyramid of primary group authority is such that no part can weaken without affecting the whole. The opposite is true of urban society.

> This development in the value system of the modern world, of which the industrial city is one of the most characteristic symbols, has sometimes been expressed by stressing the presence of universalist norms, as against the predominance of particularistic, ascription-oriented norms in the less developed societies. This transition in value structures is accompanied by stresses and strains which tend to produce features of social disorganization in various degree.[6]

The more universal character of the social order in the city makes it possible there for groups with their particularistic norms to be identified with the wider community life without disturbing the equilibrium of that wider life. Thus a great city like Paris or New York can include within its population people of different cultural groups which, even while resisting change, are in process of change. Urbanism affords a type of order not easily disturbed by change and this affords a protective covering for social change in the constituent elements. A penetrating report on this type of change in the cities and villages of Java has been made by Willner. The cities are a mixture of foreign and native peoples. The latter include those who had higher prestige before the revolution and those that are rising because of the revolution.

> Members of both these groups voice alternatively and even simultaneously the desirability of "modernization" and "return to our traditional values" while increasingly adopting consumption, recreation, and association patterns barely distinguishable from those of the European community with which they increasingly fraternize. They inhabit new houses in new suburbs, drive large cars and purchase, however prohibitive in price and regardless of quality, imported in preference to domestic articles. Mixed sports, receptions, and even cocktail parties are replacing the traditional family of clan gathering with its segregation of the sexes.[7]

[6] Bert F. Hoselitz, "The City, the Factory and Economic Growth," *American Economic Review*, Vol. 45, No. 2, May 1955, p. 172.

[7] Ann Ruth Willner, "Social Change in Javanese Town-Village Life," *Economic Development and Cultural Change*, Vol. VI, No. 3, April 1958, p. 229.

Willner's study concerns the village more than the city. She finds that revolutionary change is also under way in the village, but there it is more disturbing to social order.

## Social Control and Tolerance

Smith and Parenton observed that in the urban community, because of social differentiation, special solidarities arise. Here is specialization of labor and of interests and there develops a control through indirection.[8] The new control that emerges tends to be more tolerant than strict, but for that very reason it tends to be effective, at least in long-range terms. The urban situation, using the term mentioned above by Hoselitz, is one in which universalist norms prevail. This means that the controls are also somewhat universal, more impersonal, and less intimate and exacting than those of the small primary community. This is implied by Hill in the report on a study of Eddyville, a southern town in process of being industrialized. Here we see how control tends to be dispersed as it relates to the young:

> Perhaps it is a function of the age of the respondents, but more likely it is a characteristic way of life that these young Eddyvillers report a great deal of trouble and conflict, and in particular struggle with people in authority. Life (for them) is a great battle between superior and inferior forces. This is understandable of people who live in a town so characterized by recognized differentials of power. In general, the few leaders dominate the masses, the men dominate the women, the whites dominate the Negroes, and the adults dominate the children. It is no wonder, then, that much fantasy is occupied with situations involving subordinate and superordinate individuals. The boys tend to see this kind of world as a hostile one—the powerful figure being out to injure or thwart the lesser. The girls, however, tend to see the world as more beneficent. Powerful figures are generally helpful and succoring, regardless of the final outcome of the situation.[9]

[8] T. Lynn Smith and Vernon J. Parenton, "Social Cohesion and Social Control." See Joseph S. Roucek, Ed., *Social Control*, New York, Van Nostrand, 1947, p. 72.

[9] Reuben Hill and Associates, *Eddyville's Families*, Chapel Hill, North Carolina, Institute for Social Sciences, 1953, p. 111. Mimeographed. On this subject of children being under strain because of parental controls, see Matilda White Riley and John W. Riley, Jr., "A Sociological Approach to Communications Research," *Public Opinion Quarterly*, Vol. 15, No. 3, 1951, pp. 445-460. Regarding hostility toward parents, they observe: "When the social structure imposes undue strain upon the individual in such ways as this, we should expect him to be highly productive of fantasies, and therefore to select a kind of media material, such as little animals or violent action (movies, picture books, etc.) which would foster such fantasies." P. 455.

P

The system of control in this once isolated rural town now tends to be less exacting than formerly, allowing for more "free play," which we may call tolerance between the hierarchy of authorities. The idea is well illustrated in the administration of the antisaloon law in certain cities of Kansas at the turn of the century. While it was unlawful to operate saloons, in some cities saloons did operate because businessmen and others argued that "grass would grow in the streets" if there were no saloons. The problem was met by permitting the saloons to operate, but the operators were arrested every month or so and fines were imposed on them. The same device was used in Kansas City, Kansas, according to Garwood, for operating bawdy houses:

> Around 1900 there were 147 houses, paying fines averaging $3,250 a month, each fine being in fact a recognized license fee. In the houses were 554 inmates, whose earnings averaged $48.35 a week, which they divided equally with the madams.[10]

Tolerance is evidenced in the strictness with which the codes are enforced. In the pattern of social control there are different codes, of which MacIver mentions four types. These include the legal, which are enforced by civil authority; the moral, which vary with persons and groups and which are culturally inherited and enforced by group pressure; the communal, or general local rules for behavior; and the associational, which are the rules of formal and informal voluntary groups.[11] Social control is an interlocking, although not necessarily a systematic, division of labor between the various codes, and it involves the imposition of sanctions—rewards or punishments—to some degree or other.

In the urban community, tolerance is evidenced not only with respect to the control mechanisms but also with respect to thinking and attitudes. The young woman from the rural village who "made a mistake," and who for that reason is shunned, may be able in the city to "begin life anew." She may even marry more favorably than she could have in the village. Her case would exemplify another aspect of the tolerance prevailing in urban communities. It must be added that even in that rural village, aside from the moral codes, may be found other types of tolerance in social control.

In times of crises in the community or in community group relations, social control may become more strict, but there is a tendency to relax when

[10] Darrell Garwood, *Crossroads of America, The Story of Kansas City*, New York, W. W. Norton, 1948, p. 179.

[11] Robert M. MacIver and Charles H. Page, *Society, An Introductory Analysis*, London, Macmillan, 1952, pp. 140-141.

the emergency has passed. This greater strictness in a time of emergency is generally expected and approved. Moreover, in such times of emergency conformity to controls is likely to be prompt and full. That controls can tighten or relax evidences the virility of the social order.

## Fluctuation in Conformity

Just as control may be lax at times and strict at other times, so it is with the conformity of people to these controls. In addition to the tendency for control to be more strict in times of crisis, it may also become more strict if sections of the population seem unduly lax in their conformity. For example, if there is a wave of juvenile delinquency in the community and if the peace is disturbed by the antics of teen-agers, there will be demands by the press, churchmen, and community leaders that parents become more strict, that something be done by the educators, or even that the police show a little more vigilance. But strictness, supervision, and vigilance call for an extra expenditure of time and energy as well as continued attention. Once a peak has been reached there is a tendency again to lapse into laxness.

We can well pause here to ask about the utility of a measure of tolerance in the enforcement of social controls, and also of a degree of laxness in conforming to these controls. A full and continuous enforcement of all the codes of conduct with full and faithful compliance, in theory at least, would constitute a firm barrier against social change. Such a condition in the urban community, a "still-life" state of things, would be unthinkable. Social change calls for a degree of nonconformity as it also calls for a measure of tolerance in control. We cannot say how much nonconformity or how much of one type of control or another is necessary. The thought is injected here merely to call attention to the need for some "free play" in both control and conformity to clear the way for social change. In a changing society many rules out of the past that no longer apply tend to be evaded. Thus Homans speaks of the "existing level of obedience to a norm."[12]

Hayek recognizes that although some codes of conduct are not enforceable and may not be strictly obeyed, there tends to be a sufficient amount of conformity to "make the behavior of other people predictable to a high degree."[13] If in a changing society the behavior of an individual is in general

[12] George C. Homans, *The Human Group*, London, Routledge and Kegan Paul, 1951, p. 299.

[13] Friedrich A. Hayek, *Individualism and Economic Order*, London, Routledge and Kegan Paul, 1949, pp. 23-24.

of a conforming type, he can work effectively with others even though his behavior may be nonconforming in certain instances.

How much conformity is needed to meet the demands of social order in the community and to insure a balance between stability and social change, is a question for which there would be a different answer in each situation. Tumin noticed that when Guatemalan village youth returned after working in the town, wearing town clothes, swaggering, talking new talk, and showing disrespect for local tradition, they were a disturbing influence. Exposed to more freedom than they could socially digest, they found themselves outside the bounds of tolerance as far as the village control system was concerned, whereas in the town they would attract little special attention. As for the girls:

> It is significant to note by contrast that the departure of the Indian girls from traditional ways proceeds more slowly. The primary reason is perhaps that it is more possible for a girl without a dowry to get married than it is for the comparable boy. Moreover, even if she remains unmarried beyond the expected time, she can continue to be of measurable service in the household beyond the point at which the boy can be of measurable service on the increasingly small familial plots of land.[14]

In the urban community where change is more the order of the day, the greater attention given to change usually results in more moderate expectations regarding conformity. Of this interest in change, Moore says "It is fostered and rewarded, and failing to change is often condemned and penalized."[15] The fully urbanized man is one who, whether by intuition or judgment or a combination of the two, becomes his own censor of personal conduct, although he is influenced and guided more than he may recognize through his contact with the variety of primary and secondary urban groups with which he shares one interest or another. Conformity for him is determined more by present situational factors and less by traditional ones.

The opposite is often true for nonurban people, especially those who transfer to urban life but remain under the influence of the old codes. Wilson observed that years may pass before the African who migrates to the town is able to emancipate himself from the control of his family and tribe in the village. He must send them money and he must make frequent visits to the village. He must receive and support family members from the village who migrate to the town. He must marry according to the rules of his tribe and

[14] Melvin M. Tumin, "The Dynamics of Cultural Discontinuity in a Peasant Society," *Social Forces*, Vol. 29, No. 2, December 1950, p. 139.

[15] Wilbert E. Moore, "Creation of a Common Culture," *Confluence*, Vol. IV, Part I, No. 2, 1955, p. 231.

rear his children in the traditions of the tribe. What happens, as Wilson notes, is that this man becomes "detribalized." As years pass his visits to the village become less frequent. He finds ways of evading the tribal codes which have no application to his urban life. He conforms in a perfunctory way to the traditional requirements and if severe pressure is brought against him he may rebel against all of the codes.[16] His own children may grow up having little or no contact with the tribal culture. Their problem is one of conforming to the evolving urban culture in which they live.

## Disorder in the Urban Social Order

There are a number of ambiguities that confront us when we associate such terms as "social order," "social control," or "conformity" with urban life. We may be asked, for instance, where is the social order or can we point to social control, and we may be reminded of many examples of nonconformity. Lynd, for example, sees such incongruities as these: "A dangerous undemocratic vacuum exists in our culture between the individual citizen and the authority at the top, between the worker and the corporation that hires him, between the person and the city in which he lives."[17] These indictments, and they are many, can all be admitted, and still we can say there is social order in the modern urban community, there is social control, and there is conformity. There is, of course, social change, very much of it, but if our attention is on social change only, it may be difficult for us to envision order, control, and conformity.

We are not likely to see order, control, and conformity if we approach the modern urban community with a rural orientation; that is, if our expectations are rooted in concepts that flowered in preurban societies. We need to recognize that modern urbanism is, for the most part, new. We can here find no better statement of the conditions basic to modern urban life than that prepared by Freedman and associates. They list nine elements:

1. Large masses of men are functionally interdependent. . . . Their lives are bound up together regardless of their individual wishes.
2. They tend to live in, or organize their lives with reference to, communities with large and densely settled populations.

[16] G. Wilson, "An Essay in the Economics of Detribalization in Northern Rhodesia," summarized in *Social Implications of Industrializaton and Urbanizaton in Africa South of the Sahara,* Paris, UNESCO, 1956, p. 153.

[17] Robert S. Lynd, *Knowledge for What?* Princeton, Princeton University Press, 1948, p. 217.

3. The population is heterogeneous. . . .
4. Large numbers of functions of the family in folk or feudal societies are dispersed among specialized associations. . . . The place of work tends to be separated from the home. The family group is no longer the basic unit in the division of labor.
5. The overwhelming majority of the people who are interdependent do not know each other or see each other. . . .
6. Much of the common information shared by the population is not derived from face-to-face communication between persons but from *impersonal* mass communication media. . . . Individuals are linked to common objects of attention through these impersonal agents rather than through a direct sharing of experience and information.
7. Even most of those who know each other or are in face-to-face contact do not have common membership in primary groups. Such membership has been the basis for the forms of social organization under which most men have lived in the past. Necessarily, in the urban community interdependence far transcends the limits of the small groups which meet problems in terms of common experience.
8. The fact of dense settlement and interdependence of a large population necessitates a high degree of specialization in the population.
9. The culture of urban societies is subject to rapid change.[18]

We must then expect in the urban community an urban type of social order, which is different from and more tolerant than a rural social order. We must see social control in terms of urban realities, as outlined above; a control that is more impersonal and indirect, but not ineffective. Conformity is also present but it tends to be less impelled by outer pressures and more the product of an inner orientation to social controls. In the following sections we will look more closely at control and conformity in the urban society.

## Control in the Secondary Society

To a high degree social control in the secondary urban society is rational. Ford, listening to the life story of a Kwakiutl chief on Vancouver, learned of the various charms that had been used in his infancy to render him safe against evil. He learned that these Indians generally conformed to the wishes of their dead ancestors.[19] Miner reports on the role of magic in the daily life of the Timbuctoo people. Magic even today figures variously in the interpersonal

[18] Ronald Freedman and associates, *Principles of Sociology*, New York, Henry Holt, 1956, p. 449.

[19] Clellan S. Ford, *Smoke from their Fires*, New Haven, Yale University Press, 1941, p. 46.

relations of people. One worker fears to encroach upon the occupation of another. If a man builds a house, no other may repair that structure, or some evil will befall him.[20] Beals found an extensive use of magic among the Sierra Tarascan villagers. It has a place in the codes. One uses magic to win a lady or to frighten a rival suitor. He may use it in business to outsmart a competitor, but the competitor may have a stronger charm. Folk rules about the use of charms are elastic, but charms are generally respected.[21] Hellman reports that the charm maker is an important man among South African natives. He not only sells charms but he gives advice to people in trouble.[22]

While primary group controls use such devices, they are largely also of the face-to-face kind, all of which may be found in the secondary society, although they tend to be replaced in part by controls of a secondary type. These include the rules of secondary groups, various uses of the civil law, and different types of habitual routine, such as the impersonal behavior of large crowds. Two other forces which must not be underestimated are the domination of mechanisms and the pervading influence of written records.

The industrial urban community, as well as any community coming under the influence of urbanism, integrates the individual into a social agglomerate, often through chance contacts rather than firm, enduring ones. To Brownell this is often a "fictitious solidarity of more or less massive groups." As he sees it, in this new pattern of mass living the individual must join with people he does not know and may not want to know, but he must share in mass efforts to protect his rights and advance his interests.[23] One must become oriented in a quasi-anonymous way to much of the life about him and, whether it is good for man or not, the world's population is increasingly entering this way of secondary living.

We have seen how the family as a primary group is being transformed under the influence of urbanism. Although it seems now to be assuming another form, the family is still the basic primary group in the urban community. All other types of primary groups found in the urban community tend to be transient. There are neighborhoods in the urban community, but

[20] Horace Miner, *The Primitive City of Timbuctoo*, Princeton, Princeton University Press, 1953, p. 55.

[21] Ralph L. Beals, *Cheran, A Sierra Tarascan Village*, Washington, Institute of Social Anthropology, 1946, p. 156.

[22] Ellen Hellman, "Rooiyard, A Sociological Survey of an Urban Native Slum" (in Johannesburg), summarized by Meran McCullach, *Social Implications of Industrialization and Urbanization in Africa South of the Sahara*, Paris, UNESCO, 1956, pp. 170-190.

[23] Baker Brownell, *The Human Community*, New York, Harper & Brothers, 1950, p. 20.

they are rarely the vigorous kind that town planners wish to "restore." Nevertheless, although the old-fashioned type of neighborhood is rarely found in the urban community, because families are constantly moving in and out, neighborhoodlike relationships do exist in great measure, and they are also a sort of primary group.

In the urban environment friendships develop between people in school, at their work places, or in their leisure activities. In this way small groups may take form. Often, too, primary groups may form between members within the secondary groups. This may happen at the trade-union meetings, at the meetings of the businessmen's association, or at the meeting place of a political party. Especially in a secondary organization, the few leaders who comprise the elite may become a very tight primary group.

Most of these primary groups, the family excepted, experience a great amount of change, particularly in the urban community. There are different types of change. A person may "graduate" from one set of primary group contacts and try to form others if he changes from one income level to another. Changes in one's family status may cause him to leave certain primary groups for others. The most frequent reason for change is physical mobility; one moves from one location to another and it becomes inconvenient to retain old contacts. However, in spite of such moving about by individuals, the primary groups persist, not in competition to secondary groups, but supplementary to them.

## *Control under Law*

Social control in any society finds its ultimate sanctions in the law, although law may differ very much from one society to another. As far as our subject is concerned, the important difference is that between law in the modern secondary Western type of urbanized society and law that is or tends to be primary-group-oriented. The latter may be tribal law, or it may be ecclesiastical law that had its origin in a primary group type of society. Such law survives with difficulty when confronted with the secondary way of life in modern communities. Kelsen observes:

> In a primitive and comparatively homogeneous society, law and morality coincide. In a civilized society a differentiation of law and morality takes place as a result of the differentiation of society. The law of civilized society is more or less in conformity with the ethical conviction of one or the other group within society. But every social order is virtually a moral order—"moral" from

the point of view of one or the other group, i.e., relatively, not absolutely, moral.[24]

Systems of law, whatever their origin, have the common function of maintaining order, although they may vary greatly about the nature of that order and to whom the law applies. The tradition-rooted law, especially if religiously oriented, traces back to some great law-giver whose utterances were final and absolute, and may not be changed, however much situations may change. Such law is generally not only sacred and immutable, but it usually applies to one people or religious group only and is designed to promote the interest of that group against all outsiders, who are regarded as unclean. However, even a law that has acquired a religious character had its beginnings much earlier. MacIver wrote:

> Every society at every stage of civilization, rests on a firmament of law that is vastly greater and much more intricate than any ever devised by any government, one that is too great and too intricate to be completely overturned even by the most revolutionary of governments. We must recognize this elementary fact if we are to understand the nature of government and the authority of government. This firmament of law is composed of various interfused elements, the composition varying with the kind of society.[25]

This, MacIver notes, is true not only of primitive law but of law in the most complex societies. At the one extreme the law is intent on what Kelsen called "the exaggeratedly scrupulous observation of customs and usages inherited from, and watched over by the ancestors,"[26] while at the other extreme the law is detached, impersonal, and relatively free of obligation to special groups or creeds. In the industrial urban society the law needs to be impersonal and detached, no more the servant of one class or group of interests than another, but it also needs to be responsive to social, economic, and technological change. It is with reference to these demands for change that the preurban and preindustrial legal systems meet with difficulty, for they must not change, at least with respect to social matters.

We think of government as the maker and enforcer of law. If this is not wholly true, at least government is the ultimate authority in the community. Actually every member of the community shares the enforcement function, partly by obedience to the law and partly in the example he sets. But not

[24] Hans Kelsen, *Society and Nature,* London, Kegan Paul, Trench, Trubner, 1946, p. 55.

[25] Robert M. MacIver, *The Web of Government,* New York, Macmillan, 1947, p. 65.

[26] Kelsen, *op. cit.,* p. 20.

every member of the community may be fully obedient to the law, sometimes one person may have difficulty with another, or groups may come into disagreement; in these cases the presence of a constituted authority serves to keep things in balance. On the whole, as Pound observed, the community is self-regulating; law has a very small part of "the whole task of social control; and the criminal law does a small part of that portion which belongs to the law."[27] He called attention to the large part of the work that is performed by the clubs and associations, the organized groups.

In the complex modern community the policeman is only one of the symbols of the law, but even the policeman has come to be much more than a man catcher and frightener of children. Very little of his daily work is concerned with making arrests. As for the law in general, Litwak explains that it is more occupied with other things than punishment. He mentions, for example, social therapy, as when the police support boys' clubs; still another function of the law is education, which may take different forms.[28] Practically every function performed by government is to some measure a legal activity, for which reason most of these functions are said to lie within the "police power" of the community or the state. This can be said of the health service, the water department, the control of private construction, the regulation of parks and playgrounds, and so on.

No less important than these activities that can be named, all of which have something to do with social control in the community, is what might be called the "standing guard" function of the law. The individual citizen can take comfort in the feeling that, although he may never invoke the law, it is always there at his service. He knows, too, that it can be invoked against him. The corporation, itself a creature of the law (through its directors), knows the same. Corporations and organizations may even in various ways share the work of the law as they exercise control over their employees or influence the conduct of their members.

## Associations in Jonesville

Warner and associates made a study of organized life in "Jonesville," a small Middle West city, which is characteristically American as far as club and association life is concerned. Many of the associations are branches of

[27] Roscoe Pound, *Criminal Justice in America*, New York, Henry Holt, 1930, p. 5.

[28] Eugene Litwak, "Three Ways in which Law acts as a Means of Social Control, Punishment, Therapy and Education," *Social Forces*, Vol. 34, No. 3, March 1956, pp. 217-223.

the many national organizations that have branches in the thousands. They found 133 associations for adults, one for each 35 of the population; actually one for every 16, since a little more than half of the adults in Jonesville are not "joiners." The average membership of these formal groups was between 60 and 70. Of the 133 associations, 27 were for men only, 69 for women only, and 37 were mixed. Of the associations for women, 28 were church connected and 27 were social; and of the entire 133 associations, 52 were church connected, 39 were social, and 12 were economic or professional. The remainder were civic, patriotic, political, fraternal, or social service. In addition to the adult associations, there were 43 clubs and associations for youth.[29]

Jonesville was found to be one of those American places that thinks it has shed its small-town character, but is still not quite urban. Striving for growth and recognition, it eagerly adopts all the trappings of industrial urbanism that seem to proclaim progressiveness, but the real life of the place is lived in the hundreds of primary groups. The formal groups are the instrumentalities for getting things done. They, or the elite leading them, are in fact the government of Jonesville. They are the secondary groups that symbolize the emerging urban way of life in a small American city.

## Secondary Groups as Private Government

In the urban community, to quote Merriam, the formal private groups "sit at the table with government." Public and private groups have overlapping functions, and control in the one segment tends to be control of the other.[30] Hunter made a study of a southern city near an air base. One of the questions dealt with was the part played by the leadership of this city in getting the air base located there, when it might have located near some other city? But the results of his study had wider importance. It was clear that the affairs of the city were fairly well under the direction and leadership of a combination of organized groups, but it was equally clear that these groups were largely under the direction of a hierarchy of individuals, a sort of interlocked directorate of leaders in both public and private life. These key persons, and not only in their own organizations, tend "to set policy for other institutions and associations."[31]

[29] W. Lloyd Warner and associates, *Democracy in Jonesville*, New York, Harper & Brothers, 1949, pp. 116-120.

[30] Charles E. Merriam, *Systematic Politics*, Chicago, University of Chicago Press, 1945, p. 124.

[31] Floyd Hunter, *Host Community and Air Force Base*, Human Resources Research Institute, Maxwell Air Force Base, Alabama, 1952, pp. 3-5.

The method of this type of research is to have a number of informed persons name the ten or more top influential individuals in the town or city. By progressive interviewing it is possible to arrive at a list of "key influentials," persons who can be relied on to get things done. Who are these people? Miller, following the method of Hunter, found that in a Pacific Coast city the order by status of the 12 key influentials was: Manufacturing executive, Wholesale owner and investor, Merchantile executive, Real estate owner-executive, Business executive (woman), College president, Investment executive, Bank executive, Episcopalian bishop, Mayor (lawyer), Lawyer. With two exceptions, all were leaders in business and industry.[32]

In this same study Miller examined an English city of comparable size (about 400,000 population) and here he found a different array of "key influentials." The rank order was: Labor Party leader, University president, Manufacturing executive, Bishop (Church of England), Manufacturing executive, Citizen Party leader, University official, Factory owner, Labor leader, Civic leader (woman), Lawyer, Society leader. The city council of the English city includes 110 members among whom are five of the 12 "key influentials."

> The Council is the major arena of community decision. Issues reach it directly, are investigated by Council committees, and are decided upon by a vote taken in the full Council. Community organizations play important roles in debating the issues, but these are definitely secondary or supplementary activities. The community value system condemns any pressure tactics on the Council as "bad taste."[33]

Who are the 110 members of the city council of the English city? Obviously most of them are members of various secondary groups in the community, and there is probably not a single secondary group that does not have at least one of its members on the city council. The "key influentials" of the city have memberships in different secondary groups, and five of them are members of the city council. Thus the control mechanism of the community is a network of key personalities operating through a network of private organizations, and public officials are members of these groups also.[34]

[32] Delbert C. Miller, "Industry and Community Power Structure; a Comparative Study of an American and an English City," *American Sociological Review*, Vol. 23, No. 1, February 1958.

[33] *Ibid.*, p. 15.

[34] The leading role of the lawyer in American community life is well known. The nature of his profession tends to force him into leader roles. In the Eddyville study it is reported that there are seven lawyers in the place. "One is a major figure in the local bank. Of the 12 persons in Eddyville most prominent in real estate promotion or finance, five are lawyers. Each of the two law firms has one or more of the partners active in

The secondary organizations take on the character of private government because they are, like government, rule-making and rule-enforcing groups.

## Control Functions of Secondary Groups

Whatever else the secondary association stands for, it stands for the conventional virtues: obedience to law, order in the local community, the wholesome family life, clean sport, civic pride, and so on. No matter what the association—labor union, a veterans' organization, the local medical society, the Rotary Club, the local board of trade, or something else—in addition to its private objectives, it has its rules of good conduct. While, on the one hand, it may protect and promote the interests of its members, on the other hand, it sets high standards of citizenship for its members. One organization or another may "play politics" or try to influence public officials in order to get some benefit for itself, but in broad terms these organizations are the staunch defenders of the common decencies in public and community life.

Organized groups, depending on their nature, usually set certain standards of conformity for their own members. One must meet these standards before attaining membership, and violation of these standards after joining may mean expulsion. Reasons for expulsion vary from one organization to another, as do the effects of expulsion. Being expelled by a medical association may mean the loss of one's occupation, while being expelled from an exclusive social club may mean a loss of social status. A worker may be expelled from a trade union for reasons which may have nothing to do with the quality of his work or his behavior as a family man and citizen. For example, he may refuse to join in and support a strike.

Most secondary organizations, with the warm support of most members, engage in activities relative to community life in general. A trade union may try to "swing the labor vote" in favor of a particular political candidate or a particular public policy, while other organizations may try to influence the vote in the opposite direction. A church may become so preoccupied with matters of public policy that it establishes a political party of its own. Such are to be found in most European countries. Moreover, a church may organize

---

town, county or state politics. Of the seven lawyers, three are office holders. Two of these are delegates to the State legislature from Eddy County and the third is the town mayor." Lawyers or members of their families are the leaders in business and in the local social life. They lead the dominant conservative faction in the community, and are responsible for the town's poor public services because they resist tax increases. Hill and associates, *op. cit.*, pp. 33-39.

"Christian" trade unions or form pressure groups within trade unions. These organizations at the national as well as at the community level may use their influence to have persons of their viewpoint appointed to different public bodies or to key positions in government.

That different organized groups try in different ways to influence community life is quite consistent with the principles of democracy. They may try to influence community behavior or public policy in some way or other, but their avowed aims, at least, are not destructive. While they may try to secure advantages from civil authority, as MacIver emphasized, they all "come to look upon government as their common protector." And although groups may try to bend the law to their purposes, the law remains "their code, consecrated by tradition, sustaining and promoting the common well-being in which they share."[35] Large groups may even employ specialists in order better to influence public thinking and behavior or the conduct of public administrations.

These special units organized to influence legislation of public administration are called lobbies, and generally the word has negative connotations. Finer, among many others, thinks of the lobby as a "whole vast corpus of associations of all kinds, *in so far as* they are trying to influence public policy in their own direction; while themselves (unlike political parties) unwilling to assume direct political responsibility for governing the country." He recognizes many kinds of lobbies.

> For all this, there is great value in the system. It has positive advantages. It provides continuous consultation all the time and at all levels between the "interested publics" and the public authorities. It brings knowledge and also emotions—favourable and unfavourable—to bear upon projected policies. In short, it embraces two democratic processes: the right to participate in the framing of public policy, and the right to petition for the redress of grievances. Without it, our constitution could not effectively operate.[36]

Such secondary organizations do enjoy sway power, and one need not belong to them in order to be influenced by their viewpoints and activities.

## Control through the Mass Media

The effectiveness of secondary organizations in the modern society is dependent largely on the access they have to the press, the radio, and other

[35] MacIver, *op. cit.*, p. 80.

[36] S. E. Finer, "The Lobbies," *The Twentieth Century* (special number on "Who Governs Britain?"), Vol. 162, No. 968, October 1957, p. 377.

mass methods of communication. They have helped to broaden the use of these media by which millions are brought within the same reading, hearing, or viewing circle. Symbols and slogans have always been used by organized groups to concretize ideals or objectives, and these may have a powerful influence for crystallizing sentiment. Consider, for example, the flag used by some colonists during the American Revolution, the symbol of a coiled snake ready to strike. Across the flag underneath the serpent was the slogan, "Don't tread on me." This effective propaganda symbol and slogan could not be spread abroad at that time as readily as today, but it was spread by messengers on horseback. They served much the way as symbols and slogans since that time and before; namely, to influence people in their thinking and behavior. Symbols serve the same function as stereotypes; they serve to put an idea into concrete form and thus they facilitate communication. A symbol, according to Albig, is a social utility.

> To arrange the variety and complexity of human experience in intelligible terms, capable of classification and remembrance, the mind must create symbols. These symbols are a simplification and a concretion of complex and sometimes abstract realities. . . . Language, figures, images and other concretions provide classificatory systems of referential symbols. Thinking in symbols is an inevitable basis for thought in common.[37]

The mass media and the increasing effectiveness with which they are used are developments of this century, as Domenach observes. They emerged mainly from the urban life, out of its problems and insecurities as well as its technology.[38] To Young, this urban way of life is especially susceptible to various types of mass influence, for there is a lack of organization:

> Today, under urban conditions especially, we have a mass of isolated individuals, interdependent in all sorts of specialized and external ways, yet lacking any close sense of personal intimacy and emotional security, and often having no central unifying value or purpose. In a sense, modern urban culture has produced a crowd-minded society, with transitory contacts, emotional reactions, irrationality, and values quite different from those of more stable cultures.[39]

This describes the secondary society in some respects and its responsiveness to mass influences. However, some observers are unwilling to believe that urban society lacks stability, that it is without unifying values or purpose, or

[37] William Albig, *Modern Public Opinion*, New York, McGraw-Hill, 1956, p. 76.

[38] J. M. Domenach, *La Propagande Politique*, Paris, Presses Universitaires de France, 1950, p. 11.

[39] Kimball Young, *Handbook of Social Psychology*, London, Routledge and Kegan Paul, 1946, p. 408.

that it is lacking in personal intimacy. It is true that the urban way of life is characterized by a high degree of anonymity, which is considered by some as being pathological. Actually, some degree of anonymity may be normal and necessary for such a way of mass living.[40] If there is a difference between rural and urban societies in this respect, it is found in the greater amount and variety of stimulus to which the urban man is exposed. It is especially of urban society that Brembeck and Howell were thinking when they wrote:

> Today we find ourselves in a society where the competition for votes, customers, joiners, sympathizers, readers, contributors and so on is most intense. We live in a country with perhaps the greatest persuasion density in the world, and where the manufacture and manipulation of public opinion through persuasion is recognized as a legitimate and intricate science.[41]

The press, radio, movies, television—each is a medium of influencing the individual in his thinking and conduct. As Brembeck and Howell have concluded, the media vary in effectiveness with different people.[42] Even if we do not know how or to what degree, it is certain that these media do influence public opinion. Much of this influence is propaganda offered in various guises by interest groups. Doob observes:

> Public opinion and propaganda represent a significant segment of behavior. As terms, they pervade the vocabularies of many peoples and, when a word is in common usage, it usually refers to a phenomenon that affects men and women, or at least the fact of its being employed indicates that it is serving some kind of function. No matter how public opinion is defined, moreover, it is clear that many people and their beliefs are involved, it is equally clear that here is a man-made and man-directed force that is affecting large groups of people.[43]

In our examination of the subject we can do little more than to call attention to the function of the mass media. The subject has a wide scope insofar as the mass media are means of control, but these media also present the problem of being themselves objects of control. Thus wrote Bogardus:

> It is not too soon to develop techniques of social control of the means of mass communication which will allow them to remain in the hands of the people

[40] For example, see Joel Smith, William H. Form, and Gregory P. Stone, "Local Intimacy in a Middle-Size City," *American Journal of Sociology*, Vol. 60, No. 3, November 1954, pp. 279-282.

[41] Winston L. Brembeck and William S. Howell, *Persuasion, a Means of Social Control*, Englewood Cliffs, N. J., Prentice-Hall, 1952, p. 3.

[42] *Ibid.*, p. 431.

[43] Leonard W. Doob, *Public Opinion and Propaganda*, New York, Henry Holt, 1949, p. 3.

> and yet safeguard the people against malign manipulations of opinion-making. Only thus can instruments of communication develop scientific validity and at the same time contribute to the welfare of the human race. A modern Jeremiah may safely declare: Woe to the world if television comes into general use before methods for its adequate social control are created and made practical. As the instruments for manufacturing public opinion in any desired direction multiply in effectiveness, the greater will be the need for increased and improved measures of social control.[44]

The task of controlling the media offers much more of a challenge in regulation than is consistent with the freedoms of a democratic country. The problem of control raises questions about definitions and who will write the definitions. Perhaps the safest control lies in the mutual censorships and limitations that rival interests impose upon one another.

## The Control Influence of Records

In the autumn of each year the Burgomaster of Darmstadt invites a hundred or so leading citizens to join him in the *Grenzgang*, which means to "walk the border" of the city. Such an annual walking of the border was customary in early times when land maps did not exist except in people's heads. All land and other borders were marked by stones and it was necessary for the interested parties periodically to see that the stones were in place. Today this historic ceremony affords local leaders of business, industry, labor, government, and church an opportunity to spend an informal day walking and talking and settling many small matters "off the record."

Modern man is not only able to map the lay of the land on paper, but he also uses paper in connection with most of his activities that need to be recorded exactly. Formerly such information was entrusted to memory; in fact, among rural peoples or tribal peoples where records are only coming into use, memory is still relied on.

Ours is a contractual society and the dealings of a man may often be with strangers, or with people elsewhere whom he has never seen. Written records are used throughout a man's lifetime and, furthermore, a contract may be binding after a man's death. Much that is called property is no more in fact than claims to values that are written on paper, but in the contractual society they are valid and binding.

If a man joins a formal organization, he expects the organization to

[44] Emory S. Bogardus, *The Making of Public Opinion*, New York, Association Press, 1951, p. 94.

have a written constitution, else he does not feel secure in it. This organization must also keep records of meetings, finances, and activities. The employer must keep a payroll which shows every hour and day of service rendered by every worker. When a man pays his taxes he wants a record of it. When he receives a letter from a public office or a business firm he knows that the senders have retained carbon copies. In one way or another, the life story of the modern man can be found in the records of public agencies, business firms, or private associations. It begins with his birth certificate, follows him through school, through his work life, and ends with the certificate of his death.

During the American colonial period and later, as the frontier moved westward, there were few scribes. Always it was necessary to find someone who could "write it down," although there was not much that needed to be so preserved—mainly court records and records of land sales. The change came as trade and industry developed. With them came the stenographer, the typewriter, the tape recorder, the accountant with all his machinery, and finally the many systems of filing. Public records were made, but it is only in recent years that interest has grown in the preservation of these. This former lack of interest is shown by the fact that the Federal Archives as a central depository of records was not established until the 1930's.[45]

Today no work place, no small store, no modern farmer, not even a taxi driver, can operate effectively without written records. With reference to employment, the trail of the worker can be traced back job by job, even though he has moved from one place to another. His complete work history is contained in records.

One may ask what this has to do with social control. The answer is found in the attitudes people have toward their records. It begins in school when children learn to be concerned about low marks. These grades follow the individual as he moves from level to level, and when he tries to get his first job he knows that the employer may ask to see his school record. Once he begins to accumulate an employment record, the school record will recede in the background. The individual may also accumulate various records quite aside from his work, in the various secondary organizations to which he may belong.

[45] During the Great Depression the federal government established a Records Project on which relief labor was used to collect and file public records in all types of public administrations. Important historical documents were often found in boxes stored in attics or in basements. Thus millions of birth certificates were found and pasted into large tomes. It was due to this project that this writer was able to get his own birth certificate. The project did much to make public officials record conscious, especially in counties and small cities.

In these connections his name may appear in the newspapers or be mentioned on the radio. He may have a record with the police, in various health agencies, in the welfare agencies. He must apply for various licenses—to drive a car, operate a boat, to go fishing, to own a dog, and so on; each record may be quite aside from his work, but each tells something about him.

It is not always true that a man's off-the-job record is something private and unrelated to his work record. Especially in the United States during the past two decades, as personnel management becomes more of a profession, these off-the-job records are being carefully scrutinized. They may be pertinent to the decision whether the individual will be employed or not, whether he keeps his job or not, whether he will be promoted or not. Such questions as these are asked: What does he do with his spare time? What organizations does he belong to? What about his home life? Has he ever been arrested? What about his reading? Often the answers to these and many other questions can be found in various records, or a record is made on the basis of data furnished by informants.

### *Use and Abuse of Personal Records*

In a secondary society such as ours these personal records are very useful in securing knowledge about community problems and behavior. If we want information about sickness, deaths, births, marriages, about the mobility of people, about occupations, about memberships in organizations, about consumer sales, participation in sports or other leisure activities, it can be had from the public or private records. Such uses of records, however important, do not involve the individual as a person. He is reduced to a mere statistic. But the same records may be used with the attention on the individual as a person.

Those who live in the midst of record keeping tend to become record conscious. This may be a matter of less concern for one who has little reason to worry, but even these must know how to behave toward "the record." Others, for one reason or another, may find it convenient to keep something "off the record" or to put it "in the record." Even the most upright and law-abiding individual must take care in this regard, since what goes into the record today may have a very different meaning a few years hence. One may be haunted by something put into the record long ago which may have been incompletely reported.

From the point of view of the individual, personal records may have a positive value. They can afford him a sense of security in the contractural society. How true this is may be seen in the biographies many write about themselves for *Who's Who*. The criminal, on the other hand, resorts to many devices to conceal his record. We cannot even guess how much of a force personal records are for control in the secondary society, but we cannot dismiss them as unimportant.

One use of records, and it has frightening aspects, is that resorted to by politicians in positions of power. To illustrate, when the Germans occupied Holland during World War II, they collected all personal records. While no one knew what records were being held, everyone knew that these records could be used in devious ways. When the citizens learned that the building in which these records were held has been completely destroyed by air bombing, there was a great release of tension in the country.

Such a use of personal records is fear-instilling. It is a use against which every citizen is helpless. He never knows whether the content of the record is true or false, complete or fragmentary. Here is one of the most demoralizing weapons the dictator can hold, and to safeguard such records is one of the most sacred trusts of government.

## Social Control and Authority

One of the difficulties many people encounter when they try to understand social control in the secondary society is that it is so impersonal, so fragmented, and so without central purpose. There is no central focus of authority. Perhaps they compare it to the apparent order of the village society. If it is pointed out that, as McCormick found in the Indian villages, these villages are really centers of continuous strife,[46] they will cite some phase of village life that holds it together. McCormick noted that although families have their never-ending feuds and neighbors cheat one another, there are always limits; the community does not break from strife within.

But the primary community may become disorganized by the importation of ideas from the outside. Carey observed that the African tribal society, so strongly under primary group authority, is really fragile and becomes demoralized in the face of outside influences.[47] Secondary society, on the other hand, is not rendered less secure by the entrance of the new and novel; rather, it thrives on change.

How the intrusion of a single idea may disturb the equilibrium of primary

[46] William C. McCormick, "Mysore Villagers' View of Change," *Economic Development and Cultural Change*, Vol. V, No. 3, April 1957, p. 358.

[47] Joyce Carey, "Policy for Aid," *Confluence* (The Relation of Advanced to Underdeveloped Countries), Part II, 1955, p. 295.

community life is illustrated by the changes of land ownership and occupancy among the Basoga at the north shore of Lake Victoria in Uganda, as reported by Fallers. Among these people, land was never property in the Western sense, although the right of occupancy might be inherited. It was not something to be sold or mortgaged. It was a social possession. Yet ways of work, rules for holding other property, and conditions essential for acquiring social status and the organization of the family were all oriented to the land, but in another sense. Difficulties arise when such a non-land-owning culture is encroached on by a culture having different ideas about land. This single intrusion threatens the balance in the whole social order. Problems of control arise for which the traditional control system is not prepared to cope. This is the beginning of the Western economy and all it entails and it means that a good share of the Basoga culture must be set aside. Fallers concluded "the landholding system is so intimately involved with the entire structure of Basoga society at the village level that its alteration would necessarily involve far-reaching changes in the social system."[48]

Such change as it comes means taking one step and then another along the way of confusion from the primary type to a secondary type of social control, also away from the focused authority of the primary society and into the secondary society with its diverse competing authorities, each a type of authority within a particular sphere of interest. This transition is described by Friedmann (using an expression of Sir Henry Sumner Maine) as a development "from status to contract":

> The evolution from status to contract, from immobility to mobility, gradually pervaded all phases of life, beyond the fields of commercial and labour contracts. It invaded family relations, and the law of succession. It became the basis of club and union membership. Gradually it penetrated into the law of land tenure, sale and succession.[49]

The secondary society is one of many kinds of organizations enjoying different degrees of power and exercising different kinds of authority in their respective private spheres. Within these and between them are innumerable more or less transient and fluid primary organizations, the points of contact for much of the social control in this kind of society. The focus of authority centers in civil government which tends mostly to stand apart from all this, although participating in one way or other, now and then, here and there. If it

[48] A. L. Fallers, "The Policies of Landholding in Basoga," *Economic Development and Cultural Change*, Vol. III, No. 3, April 1955, p. 281.

[49] W. Friedmann, *Law and Social Change in Contemporary Britain*, London, Stevens and Sons, 1951, p. 85.

is democratic government, it mixes in only as the situation demands, and no more than by common consensus is deemed necessary.

Apparently the strength of the secondary society is found in its lack of any central focus of the different authorities, except that they are all finally subservient to the civil administration. It is not a strict subservience, but permits a wide range of freedom. A change from this to some strict form of regulation could not stop short of dictatorship.

## Stability in the Social Order

In this kind of society, no less than in the primary society, the individual is under pressures to conform to the generally accepted norms. Those individuals who do not so conform comprise what Neumeyer calls the "deviant and conflicting elements."[50] In the rapidly changing community such persons are to be expected. Considering the amount of adjustment that must be made, we may well marvel that the deviant types are not more numerous. This may be good evidence that most modern communities are unusually stable.

Davis, Bredemeier, and Levy add the thought that there are "competing definitions of proper conduct in the same situation." That misbehavior in the eyes of one group may be called normal behavior by another group means that there exist different ideas about what social control should be. Davis and his associates note further that in this as other societies there is the contest between generations. They add that each "stabilizing element also gives signs of potential instability."[51] The total society comprises many subsocieties each of which contains its elements of stability and instability and each of which is in process of change. But the entire social order tends to maintain a high degree of equilibrium, or we may call it residual stability.

As we noted, the village as a primary group is never without conflict, but it maintains an equilibrium. The same can be said of the secondary urban society; it is the scene of competition, even of conflict, but it does not fall apart. It manages somehow to keep itself under control.

## Summary

Social control in the modern urban community is different not only from

[50] Martin H. Neumeyer, *Social Problems and the Changing Society,* New York, Van Nostrand, 1953, p. 419.

[51] Kingsley Davis, Harry C. Bredemeier, and Marion J. Levy, Jr., *Modern American Society,* New York, Rinehart, 1949, pp. 703-704.

that in rural communities, but it is probably different from that found in pre-industrial urban communities. One unique feature of the industrial urban community is that, more than in rural communities or in early cities, the urban populations tend more to carry on their major activities through large secondary organizations. Apparently, too, more than in earlier cities, people are associated with a larger variety of primary groups, with the family in a less dominant position than formerly, while the role of the individual has become more important. These changes in various ways have doubtless figured in modifying the character of social control in the urban community.

Social control is the process by which a society influences the behavior, and usually the thinking, of its members in accordance to rules which to some degree are culturally inherited. The more rural and isolated the society, the more it will be governed by such inherited rules. The more exposed a society is to change-prompting influence, the more difficulty it may experience in conforming to such inherited rules of conduct and belief. Since the urban society is most exposed to pressures for change, and most receptive to such pressures, it is also readier in the adoption of new ways of behaving and thinking and is often least submissive to the inherited codes, especially to those inherited rules which fail to square with the realities of the immediate situation.

The enforcement of controls in the secondary urban society tends to be both more indirect and more lenient than in the rural society, but there is a tendency also in the urban community for enforcement to fluctuate between extremes of strictness and laxness. In times of emergency, controls temporarily tend to tighten and such tightening tends to be expected. Conformity to social controls also tends to fluctuate; in fact, an extreme of laxness in conformity may stimulate a tightening of control enforcement, for a time at least. Control and conformity tend not to follow an even course as is normal in primitive or very rural societies.

In the urban community more than in the rural, there are different areas of control, although they may overlap: family relations, leisure-time activity, work and economic relations, education, religion, and so on. The enforcement of codes in one may be quite separate from that in another, and conformity expectations may differ. Moreover, different sections of the population may subscribe to different standards of enforcement and conformity. Less urbanized people or older people or more religious people may have different expectations about behavior than the more urbanized, the younger, or the less strict in terms of religion. There tend to be social class differences also.

While no more than a third to a half of the urbanites are members of formal organizations—economic, political, professional, social, cultural, and so on—these secondary associations, in a subordinate relation to civil authority, tend to share with civil authority the function of control and order to the extent that they have control and guidance relations with their members. Within this network of secondary voluntary organizations is found a myriad of primary groups, most of them small and many of them short-lived, with which urbanites as individuals have informal contact. They are the tolerant and friendly social nuclei which participate variously and intimately in the social control process.

Civil authority, operating under law, however indirect at times, is the final enforcer of the behavior codes. It is law on a more impersonal plane than tribal or other preurban systems of authority. The civil law is the reserve force for social control, invoked at times with swiftness and firmness, at other times mildly and tardily. Its utility may be as much in its "on-guard" functions as in its occasional operation.

In the less urban societies, people are controlled in large measure by the forces of nature, from which urban people are somewhat detached. The latter live in more of a man-made environment, and they learn to behave in relation to it. Mass society is a large part of the urban environment and urban man conforms to its movements and its ways. The disciplined crowd is also a disciplining crowd, impersonally setting and enforcing standards of conduct.

Urbanism in many ways is a contractual way of life, a natural consequence of its secondary and impersonal character. It is a way of life that functions through the making and keeping of records, and these records have an important secondary control function. This holds whether they are employment records, records of private associations, bank accounts, reports in the press, or the many kinds of public records on which the individual's name is written for one purpose or another. Whether these records are used by the urbanite for his advantage or used against him, they constitute another of the secondary control mechanisms in modern society.

Such are characteristics of social control and conformity in the urban society and it may be added that they pertain to any society to the extent that it is industrialized and urbanized. Some such control and conformity pattern may be expected in urbanism everywhere, whether it is firmly established as in the Western countries or only taking form as in the countries described as less advanced.

CHAPTER 18

# Urbanism and Resources Control

The making of plans for a city a generation ahead may seem an academic, if not a futile occupation; yet it has been going on for about 5,000 years. In that time the nature of the planning process has changed considerably, but the end in view is still the same—to create a place fit for our children to live in. It remains an act of statesmanship, first to envision the kind of a city by which we want to be remembered and secondly to bring it about.[1]

THIS CHAPTER is about planning, but in the widest sense of that term. In order to survive, urbanism must be sustained. Cities and towns must have access to wide areas of supply, as they must also be able to reach wide-flung markets. We use the terms "resources" and "control" as they are pertinent to the well-being of urbanism. The resources of a city include its streets and houses, its work places and its lines of transportation, its public facilities of all kinds and, by no means least, its labor force. There are even more vital resources that lie beyond the city limits: the sources of food and clothing, and many other kinds of supply. To the extent that the outlying area is a market, that is also a resource. Whatever is in the hinterland that may be described as resources for the great city must also be described as resources for the hinterland. The city as a source of services and finished goods is similarly a resource for the village and the small town.

Control, as it applies to resources in this wide meaning, is also a term of global content. It is not concerned merely with the discovery and creation of resources, and having access to resources; it must be concerned also with the continuity of access to these resources. Control must be concerned as well

[1] William Holford, "Plans and Programs," *Annals of the American Academy of Political and Social Science*, Vol. 314, November 1957, p. 94.

with the conservation of resources. Control in this sense is also planning, except that planning is usually associated with organized effort for achieving particular tasks: a road system, a pattern of construction, the removal of slums, a network of sewers and water mains, reconstruction of the urban center, and so on. Control must also be associated with policy making and with operations.

Much that would be included under this meaning of control is also embraced by regional planning schemes. Often, however, the focus of attention in regional plans is more on natural resources than on cities as such. Another term is "metropolitan planning" which centers in the great city but follows the interests of the urban community outward. It is more concerned with gaining access and securing supply, and is less preoccupied with creating and conserving hinterland resources. In this chapter we must be aware of the wider interests as they might be embraced by the term "control" as described above. However, for the most part, our examination of the subject would be more in keeping with the concept "metropolitan planning."

Metropolitan planning is somewhat akin to what Bardet calls "la mission de l'urbanisme," which is concerned with planning but also with other matters which relate to community living. It concerns controls over the environment to make community living safer, healthier, cleaner, more convenient, and more attractive, and it sees the community in wider terms than the metropolis itself.[2]

## The Community and the Situation

In our efforts to understand better the modern urban community, we have come to recognize that it must be seen as a phenomenon in whose development various factors have operated and continue to operate. We can describe it in terms of its chronological growth or history, in terms of its natural or social evolution, in terms of its geography, or in terms of the ecological processes operating within the community. These different factors that have functioned to make each community the kind of a place it is tend to operate, as it were, in collusion. Moreover, there is an on-going relationship between them. The story of what the community has been, what it is, and what it is tending to become is stamped in the total situation.

The United Nations Food and Agricultural Organization has made some studies of the fragmentation of the rural land supply. Fragmentation results for different reasons: The expansion of industrial and urban uses tends to break

[2] Gaston Bardet, *Mission de l'Urbanisme.* Paris, Editions Ouvrières Economie et Humanisme, 1949, p. 368.

land into smaller units, but the increase of population on the land also results in greater fragmentation.[3] When land is taken out of food-producing uses and becomes so much space in the urban community, the tendency toward a more intensive use continues. In this process we are confronted with ecological factors which not only have their peculiar local history, but which are operating now and will continue to operate in unpredictable ways. Erickson speaks of the dominant and subordinated elements in this process of space competition:

> We might say it involves the idea of the survival of the fittest, survival depending not simply upon economic factors, but upon social, political, geographic and technological forces as well. Unfortunately it has been used primarily as a device to explain the pattern of the community rather than its repatterning. That is, it has been used as a static rather than a dynamic concept, demonstrating that all the spatially fixed elements of community structure, such as roads, factories and homes, become integrated into fairly permanent patterns, certain of these functions dominating in one area while others are subordinated.[4]

Erickson reminds us, however, that these seemingly fixed patterns are not permanent, but the community goes on shaping and reshaping itself as if under the guidance of some invisible authority. Thus we may speak of the situation in any community at any particuar time, recognizing that the situation is in process of change. This means that if we would control the situation we must know considerable about its component elements, the different factors that are involved and the relative importance of each. The idea of the situation is conveyed by Pomeroy in considering the planning needs of the dynamic community:

> We need to know much about the physical community, to be sure; we need to know much about the economics of the community; above all we need to know about people; and maybe even most of all, we need to know what we do not know. A slide rule or a comptometer cannot make a mistake, but the person who operates it can. There are unknown factors in all our statistical analyses; and in statistical projections the variables and the unknowns pile up so rapidly as we move into the future that what we get from the slide rule and the comptometer must first be strongly conditioned and then replaced by good judgment.[5]

The situation for a community, if it is to be controlled by planning or

[3] Bernard O. Binns, and others, *The Consolidation of Fragmented Agricultural Holdings,* Rome, Food and Agricultural Organization, 1950, p. 9.

[4] E. Gordon Erickson, *Urban Behavior,* New York, Macmillan, 1954, p. 177.

[5] Hugh R. Pomeroy, "The Planning Process and Public Administration." See Gerald Breese and Dorothy E. Whiteman, Eds., *An Approach to Urban Planning,* Princeton, Princeton University Press, 1953, p. 36.

otherwise, includes a variety of factors as they were, as they are today, and as they may become ten or twenty or more years hence. There is the population which, whether it increases or not, may change in its age, sex, marital, or occupational characteristics. There is the work pattern which does not cease to change. There are the systems of transportation which change in response to population and technology. And there are the geographical, the ecological factors which both influence and are influenced by the factors just named as well as others.

## Misgivings about Planning

Situational control is concerned for the most part with physical properties and with the uses of space. But it may also be concerned with trying to put people here or to prevent them from going there. Space reserved for some uses may not be occupied for other purposes. Planning in general may involve the total scene, and most people tend to approve of planning as it concerns wide areas, but the application of any wide area control must be concerned with many details that make up the total picture. Here it is that the objectives of the plan and the purposes of individuals or groups may come into conflict.

Situational control usually means control under law. There must be a central authority empowered to tell people what to do, but more often what not to do. This may raise issues involving property rights, but social rights may also be involved. Thus Brohi, having in mind the village problems in East Pakistan, declared himself opposed to planning, and he had national planning in mind. He made this objection, but he excepted "planning by freely associated groups." All other planning he regarded as evil.[6] But such a freely associated private group in East Pakistan would have to be a village or group of villages, and they would have little interest in the wider objectives of national planning in a situation where such planning is imperative.

Hayek sees danger in the tendency to promote planning under government. To him it means centralized control for the sake of order and stability. It may mean that the individual is suppressed to a condition of "dependence on a power which deliberately molds and shapes it." As he seems to assume, public control over the environment must lead to a tightening control over people, and the centralized power becomes more centralized. Individuals in such a situation become mere "interchangeable units."[7] He does, however,

[6] A. K. Brohi, "Asia and the Western Man," *Confluence* (The Relation of Advanced to Underdeveloped Countries), Vol. IV, Part II, No. 3, 1955, p. 309.

[7] Friedrich A. Hayek, *Individualism and the Economic Order,* London, Routledge and Kegan Paul, 1949, p. 27.

recognize that government needs to participate in managing the affairs of the community, but there must be some degree of decentralization in decision making and some "two-way communication of knowledge."[8]

There are other reservations people may have about planning. In the past there have been occasional well-informed individuals who favored planning, but they feared to entrust it to the professional planners with their costly schemes. However, these reservations notwithstanding, the idea of planning to control the physical environment of the community goes forward. It is being recognized as needed in order to get the greatest value out of national as well as local resources, and people are coming to accept it.[9]

## History of Town Planning

Many important cities began as perfectly planned places, but there is probably not one that grew according to plan. However wise the planner, he cannot know what the needs of a community will be, or what the place will be like a hundred or fifty years hence, or even a decade later. Yet most plans had long-term objectives. Not infrequently they were the ego creations of rulers who hoped their own would excel the creation of some other ruler. Versailles, started by Louis XIV in 1661, stimulated a great amount of imitation. Thus, as Greer put it, each king was trying to impress the others. In some cases other improvements were added making the place the better wherein "to live and work and play."[10] Such objectives were generally afterthoughts.

In these old planned cities, as in other old cities, the common people had to fit in where they could find room. There were broad avenues and monumental structures which even today impress visitors, but the people crowded into the narrow back streets or toiled in dark and narrow work places. Gallion and Eisner remind us that the streets were "dark and filth-ridden from refuse thrown from dwelling windows." There were cesspools beneath the floors. The "house of the nobleman grew spacious while the typical dwelling of the poor remained cramped or was moved higher into the attic."[11]

Some American cities started with plans. William Penn gave Philadelphia a gridiron pattern in 1682, and James Oglethorpe in 1733 set up a unique

[8] *Ibid.*, p. 84.

[9] At the theoretical level, a very important consideration lies in what Robert K. Merton has called "the unanticipated consequences" of much purposive action. See his *Social Theory and Social Structure*, Glencoe, Illinois, The Free Press, 1949.

[10] Guy Greer, *Your City Tomorrow*, New York, Macmillan, 1947, p. 1.

[11] Arthur B. Gallion and Simon Eisner, *The Urban Pattern*, New York, Van Nostrand, 1950, p. 41.

plan for Savannah. The plan for the city of Washington was laid out in 1791. Buffalo in 1804 and Detroit in 1807 began as planned cities, while the unimaginative plan for Manhattan was adopted in 1811. There have been numerous plans for redesigning cities. Prominent among these are the Burnham plans, one for San Francisco in 1905 and one for Chicago in 1909, and there were still others during the first decade of this century, the period of the "city-beautiful" ideal of which more will be said later.[12]

While the experts would call this an oversimplification, we can think of modern city planning as beginning at the close of the last century and as having two sources of inspiration: the garden-city movement and the "city-beautiful" movement. The inspiration for the first of these is credited to an Englishman, Ebenezer Howard whose book, *Tomorrow; A Peaceful Path to Real Reform,* appeared in 1888.[13] His idea was that the way to solve the slum problems was to vacate the city, in fact to scatter the city to many smaller places where people would have space and beauty. Tunnard quotes Howard, "There are in reality not only, as is so constantly assumed, two alternatives—town life and country life—but a third alternative, in which all the advantages of the most energetic town life, with all the beauty and delight of the country, may be secured in perfect combination." Tunnard ventured the thought that "the countryside would be (the) unhappy partner in this forced marriage."[14]

The Howard idea did lead to various garden-city projects, one being Letchworth Village, near London, which started in 1903, a type of nonprofit organization. There have been many other garden villages, Sunnyside in New York City being an example. However, the coming of cheap and rapid transportation brought the suburban movement which has emptied the cities to the point where decentralization is the urgent problem.

Apparently the city-beautiful idea was imported from France. It was concerned with building impressive urban centers, monumental edifices, grand avenues, great fountains, and the like. It was a sentimenal although sincere false-front concept and it gained wide support from women's clubs and civic improvement groups, but it had no concern for the total environmental problems of the community. The movement, said Walker, to "inspire men's minds" with architectural creations was "patently impracticable."[15] Mumford observed

[12] For a summary of city planning efforts in the past and planning aims, see Thomas Adams, *Outline of Town and City Planning,* New York, Russell Sage Foundation, 1936.

[13] Ebenezer Howard, *Garden Cities of Tomorrow,* F. J. Osburn, Ed. (3d ed.), London, Faber and Faber, 1945.

[14] Christopher Tunnard, *The City of Man,* New York, Charles Scribner's Sons, 1953, p. 236.

[15] Robert A. Walker, *The Planning Function of Government,* Chicago, University of Chicago Press, 1950, p. 15.

that its basic assumption was one that made planning something that could be bought, "that could be plastered on impoverished life," not itself a way of life.[16]

## New Approaches to Planning

Pomeroy makes the point that while the "star-gazing planners" were occupied with designing civic centers that seldom got off the paper, perhaps fortunately, the city engineers were at work putting their mistakes in stone and concrete, below, on, and above the surface.

> If the "city-beautiful people" had not been effective by reason of being somewhat visionary, the engineers probably did more damage to cities by reason of the fact that they *were* effective. They were interested in widening streets and building things, and somehow tended to overlook what may be called the soul of the city, to which the city-beautiful people had in some degree been sensitive.[17]

What the visionary people did do was to stir civic interest. It was not long before the sociologists became interested in the planning problems of the city, as did people with economic training. Public officials, informed about the ways of getting things done, also became interested. That was to be expected, since civic motivated groups were securing legislation for zoning and planning. Even occasional engineers—specialists in water supply or sewer systems or traffic—began to develop an interest in the social implications of their work. Planning came down from its visionary sphere, and much of this has happened since the 1920's in most of the Western countries.

The new and broad approach is exemplified well by developments in town and country planning in Great Britain during the past two decades. The case of Middlesbrough, a city of 138,000 in 1944, is typical, This town was in a period of rapid and disturbing change. The authorities decided that there should be a plan for guiding its development. It was decided to have a social survey to supplement the physical survey. The social study, carried out under the direction of Ruth Glass, was in many respects an ecological one, although that term was not used. It was a study of areas and of the different types of mobility.

### *Social Planning in Middlesbrough*

Not only was Middlesbrough growing, but its "center" had ceased to be at a convenient point in the town. It had remained in the same old place, an

[16] Lewis Mumford, *City Development*, New York, Harcourt, Brace, 1945, p. 10.
[17] Pomeroy, *op. cit.*, p. 13.

area of old buildings, housing, and key institutions. The history and traditions of the town centered here, but the people were moving toward the south. The better-to-do moved first and often farthest. Others followed, making it a sort of procession. Between the center toward the north and the traditionless new-homes area to the south was a mixture of areas, including some blight and having little order. Of the old center and the new border Glass and associates wrote:

> Thus the old part is crowded, an array of brick with hardly any green patches, but it is still the place for the most essential urban functions: work, trade, transport outlets, administration and entertainment. The people are sociable. Here are the slums, and here is also the warmth of the town.
>
> The new part is spacious and barren; not designed to draw people together, but rather to drive them from each other. It is free from the noise and smoke of the old town. The countryside has been allowed to infiltrate here.[18]

Apparently Middlesbrough is one of those cities that grew without much guidance, except as shaped by the influences in the situation. Residential and other uses of space became segregated naturally, as they will probably do so again, but for the present the town is disturbed by growth and inner mobility. Facilitated by improved transit, suburbs are developing. Now the problem is to decide what planning can do to "pull all the parts together" and to revive a sense of community.[19]

Perhaps the city needs a new center, bright, clean, and roomy, somewhere between the north and south ends. This will come, with or without planning, but planning may help guide these processes and thus obviate the need of so much trial and error. Other areas will need a new orientation to such a new central area. The location problems, so the report concludes, are minor ones, although numerous and varied.

> It is not a town the structure of which shows grave gaps and major dislocations. It suffers from comparatively minor ailments only, all of which are the typical legacy of its period of growth. It suffers from social and geographical segregation and from an unequal distribution of amenities throughout the town. The old quarter now needs reconditioning because it was neglected whilst the suburbs sprawled southward.[20]

The approach to planning here was a fairly complete examination of the environmental situation, in which the studies of the engineers and planners

[18] Ruth Glass and associates, *The Social Background of a Plan*, London, Routledge and Kegan Paul, 1948, p. 12.

[19] *Ibid.*, p. 14.

[20] *Ibid.*, p. 15.

were reinforced by the work of sociologists and psychologists. Social and physical planning are complementary, in Gutkind's words, "and both must be kept within the limits of reality, which they should neither force nor displace, but help to clarify in the orbit of things as well as people."[21] Planning in this sense is a slower process than that of the earlier period. It is less certain and also less arbitrary. Decisions, involving the meeting of many minds, are more difficult, but they are likely to be more enduring. Moreover, this approach to planning is not a once-and-for-all matter; it continues. It is not for the city alone, but extends outward.[22]

## The Wide Reach of Metropolitan Control

Whoever tries to ease the traffic problems of the city cannot focus his attention only on the transit lines and streets within the city, for these streets end in roads which go in all directions. These outside roads are important to the life of the urban community. Whoever would solve the water supply problems of the city cannot limit his planning to the conduits and lines that distribute water to users; he must go into the hinterland where the water originates and there controls must be established to insure its purity. This is a part of planning, as it is also a part of planning to be concerned about the disposal of the waste and sewage of the city. The waste disposal problems of the great city may be of very much concern to other cities, towns, and villages, and these may take their troubles to the courts. Because of high taxes and limited space, factories may decide to move from the city to outlying areas. Since they are still included in the metropolitan economy of the city, they must still be of interest in any planning that the city undertakes. For recreation purposes the central city may need to provide play areas, camping areas, and beaches in places beyond its boundaries. This is also part of any planning of the city for metropolitan control.

The planners in Great Britain convinced Parliament that a plan for decentralization was needed for the large cities, and in 1946 the New Towns

[21] E. A. Gutkind and others, *Creative Demobilization,* New York, Oxford University Press, 1944, Vol. I, p. 22.

[22] During the depression of the 1930's Great Britain undertook the task of encouraging the development of industry in areas of acute unemployment. There had been studies by commissions on planning, and these helped give form to the idea of town and country planning. A Town and Country Planning Act was passed in 1943, another in 1944, and finally came that of 1947. These stimulated a variety of community social studies that were oriented to planning. See Ben W. Lewis, *British Planning and Nationalization,* London, George Allen and Unwin, 1952, pp. 141-146.

Act was passed, enabling the building of several new cities of from 30,000 to 50,000 inhabitants. One of these was planned at the village of Stevenage near London. In 1801 Stevenage was a farming village of 1,251 population. The population was about 5,000 in 1946 when the residents learned of the Greater London Plan to take possession and to build a city there. This proud and conservative town that in 1900 tried to keep the telephone out of their community regarded the London Plan as an intrusion. They used every means to oppose the venture, even taking their case to the courts, but to no avail.[23]

The case of Stevenage illustrates how wide-reaching city planning can be, and how, in bringing "the greatest good to the greatest number," the big city can impose itself on the small town. Questions about fairness, as in the case of Stevenage, are answered with the bigger question: How else can such an urbanized, industrialized country make the best overall use of all of its space and natural resources? Such plans center in the great city. Gutkind observed that, although there have always been and still are rural-urban antagonisms, the time has come in Great Britain when agriculture must be accorded the "same rights as industry." This is the trend in some countries and it is guided by the scientific and social thinking of the city man. Nonetheless, quoting Gutkind again, "agriculture must be integrated into the complete structure of our urbanized social and economic life."[24]

## The Rural Stake in Planning

It should not be overlooked that the sentence from Gutkind quoted above expresses the view of the urban man who is determined to be fair with the rural man: nonetheless, *"agriculture must be integrated."* The urban man's concept of the country is usually leisure-oriented and his attention is on the trees, the grass, and the birds. After telling us that only 6 percent of the English are in agriculture, Brogan adds:

> As a consequence, there is no country in the world in which feeling for the soil as a factor of production is as rare as in England, or where knowledge of farming as a way of making a living is so much a specialist knowledge, in which the most romantic and unrealistic views of country life can be advanced with less fear of brutal contradiction from people who know what agriculture, as an economic and social system, involves. This truth should hardly need stating. To

[23] Harold Orlans, *Stevenage, A Sociological Study of a New Town,* London, Routledge and Kegan Paul, 1952. This is a step by step report on the opposition offered by Stevenage.

[24] Gutkind, *op. cit.,* Vol. I, p. 120.

> believe that a people who have for generations lived in towns, of whom only a small proportion has any direct connection with the land, has, in some mystical way, evaded the consequences of this state of affairs, is to believe in miracles.[25]

What Brogan says of British urbanites could be said of increasing numbers of American urbanites. They have no urgent reason for being informed about rural ways of work and life. If planning needs force them to look toward the hinterland for purposes of situational control, the problems are approached with urban thinking. The rural man would have no interest in their kind of planning, unless he is a very urbanized person. An example of rural people carrying on in old ways is given to us by Friedmann in a report on Matera, a community of 30,000 in Southern Italy.

More than half of the Matera people are peasants, and a good share of these for centuries have been living in caves dug into the rock walls of the canyon, and for these they have always paid rent. Under the United Nations Relief Administration and with the cooperation of the Italian Government, a plan was developed for moving a good share of these cave dwellers to a new farming project where each family would live in a house. Friedmann's report relates how a study was made, how the families were persuaded to accept the plan, and how an attempt was made to design the houses to suit their needs.[26] Traditionally the family mule also inhabited the family cave. Thus in the new houses space had to be provided for the mule under the same roof. As with all plans, there was opposition to this one from those who stood to lose financially. Once the project had been completed, "politicians took credit for it."[27]

The story of the Matera project also illustrates how powerful tradition may be in resisting even the most obviously needed plans. Incidentally, not all the resistance to rural plans in Italy can be charged to peasant conservatism. Much of it comes from big land owners whose unused acres are being parceled out to farming projects. Describing such resistance as feudalistic, Nelson observed that this is the end of one of the last strongholds of this kind of feudalism in Europe.[28] The initiative for this kind of planning was urban, not rural.

The planners of these rural projects in Italy are trying to teach these

[25] D. W. Brogan, *The English People*, New York, Alfred A. Knopf, 1948, p. 275.

[26] F. G. Friedmann, "The Impact of Technically Advanced Civilizations on Underdeveloped Areas," *Confluence* (The Relation of Advanced to Underdeveloped Areas), Vol. IV, No. 4, pp. 397-401.

[27] *Ibid.*, pp. 401-402.

[28] Lowry Nelson, *Land Reform in Italy*, Washington, National Planning Association, Pamphlet No. 97, 1956, p. 38.

resettled farmers to join in various cooperatives and to enter into a contractual urbanized type of society. They hope that the transition to the new way of life can be made in about twelve years. Nelson questioned whether a people can be moved in so short a time from backward peasantry to a condition of advanced agriculture.[29] It has taken much longer than twelve years to get the less backward American hill villagers to accept the idea of TVA.

## The Tennessee Valley Authority

Although regional planning is a term with many meanings, there would be no disagreement among the various viewpoint groups about the activities of the Tennessee Valley Authority being so designated. We can add that the work of TVA was oriented, even at the beginning, to rural planning. The program would make a river navigable, but it would also carry on erosion control projects. It would produce low-cost electric power which might be used in city or country, but it would be within the price range of poor farmers. It would manufacture and distribute fertilizers at low cost. Under the project, barren hillsides have been reforested. TVA is now world-recognized as an effective rural planning program. It is looked upon with pride in the region itself, although the idea did not start there and it is hardly likely that the local people could have envisioned such a gigantic regional program.

As seen by Selznik, TVA has become a symbol, but unfortunately too much of a partisan symbol. The name is seldom mentioned except to be warmly applauded or strongly condemned.[30] Lilienthal, onetime director of TVA, is one of its supporters. To him the project has returned a depleted area to "fertility and productiveness."[31] With respect to our interest in the community, it must be said that TVA was not designed as a community planning program. Its purpose was to restore the natural resources of a region, to reclaim and bring under control the wide physical environment. But in performing this task it has served as a frame of reference for different kinds of community planning and it has lent its support to such planning.

Especially for the rural people, TVA has gone beyond its original purpose in giving instruction and leadership. These efforts Selznik calls "an adventure in executing broad social responsibilities for the development of a unified

[29] *Ibid.*, p. 45.

[30] Philip Selznik, *TVA and the Grass Roots*, Berkeley, University of California Press, 1949, pp. 19-20.

[31] David E. Lilienthal, "The People's Stake in Planning," *Proceedings of The American Society of Planning Officials*, Chicago, The Society, 1944, p. 5.

region."[32] Abundant low-cost electric power has encouraged industry in this previously nonindustrial region, which means that the general level of living has been raised. It is certain that no politician in any of the seven states in the TVA region would oppose the program today.

Precisely because TVA is a gigantic regional project of wide horizons, it captures the imagination, and planners are tempted to look at other regions that might be similarly organized and probably one day will be. One importance of such a wide metropolitan plan is that within its region it stimulates many local plans and programs oriented to the purposes of TVA. Such local plans may be urban or rural or both.

## Wider Implications of Metropolitan Control

TVA is important because it has brought under control and is exploiting without waste the natural resources of a region. More people can live there than before and enjoy a higher level of life. Because the rural people fare better, urban people are also advantaged. Such basic controls will be needed as urbanism expands. It is the recognition of these needs that is stimulating the advanced countries to turn their attention as never before to the plight of the underdeveloped countries. Here are resources which are essential to the well-being of the urban industrial countries, if not to the United States, at least to such countries as Belgium, the United Kingdom, and others.

Understandably the Organization for European Economic Cooperation and the Council of Europe have been occupied with a study of the development needs of Africa. Their experts have reported on the vast natural resources that might be developed: ores and woods, animal products, oils and other vegetable products. But roads must be built, towns established, amenities provided, a labor force trained. According to one group of experts, the first great need is electric power.[33]

So long as this new frontier exists, especially in Africa, and is undeveloped, it must be included in the wide planning of all the advanced countries. One group of OEEC experts reported in 1951 that a ten-year development

[32] Selznik, *op. cit.*, p. 11.

[33] *Energy in Overseas Territories in Africa South of the Sahara,* Paris, OEEC, 1953, pp. 36-37. Territories are mainly those of Belgium, France and England. Because coal is lacking in these regions and electric power must be provided, costly projects are necessary. With such power the other developments could move rapidly on private capital. A program for providing the power would have to be a public investment. Other needs are roads, houses, schools, health facilities, and so on.

program in Africa would require 38 billion dollars. However, if the industrial urban countries expect to make the best use of the raw materials in the underdeveloped countries, says another OEEC report, and if they want to expand their markets in these countries, they must also think of the social needs of these countries.[34] The urbanized civilization, for its own well-being, must provide the means and the leadership to raise the living standards, to raise the educational level of the less advanced peoples, and to guide them in the direction of developing their own self-sufficiency; in short, urbanize them.

Africa is but one of the underdeveloped regions of the world where economic and social development is needed to increase productive capacity. The observations of Salz regarding the Ecuadorian Indians applies to other peoples in Central and South America. Before the time and task systems of industry can be established with full efficiency in these countries a great amount of learning and changing will be needed. In a haphazard manner, this process is now under way. Salz says of this new approach:

> Regardless of the interests which prompt support of industrialization, there are some aspects which distinguish this new "industrial revolution" on a global scale from its earlier prototypes of the Western World. Among such aspects are the emphasis on international cooperation rather than on competition; on bringing about a state of general welfare directly rather than leaving it to result gradually from individual benefits; and, above all, reliance on planned and peaceful penetration of the non-industrial or underdeveloped economies rather than dictatorial methods.[35]

Indigenous responsibility for and leadership in such planning develops slowly in some countries. This is indicated for Uruguay in a report by Fitzgibbon. He found that much of Uruguay's wealth and culture are not only centered in the Montevideo area, but much of the public money for works and services is expended there. More than half of the country's high schools and its only university are in Montevideo, and 90 percent of the newspaper readers are in this metropolitan area. Of 45 radio stations in the country, 25 are in the capital. Almost all of Uruguay's museums and libraries are in Montevideo, as are two-thirds of the hospital beds. Some sections of the country have no doctors nearer than 60 miles. Of 254 Uruguayans in *Who's Who in Latin America,* 236 lived in Montevideo and another eight in 1954 were abroad,

[34] *Investments in Overseas Territories in Africa South of the Sahara,* Paris, OEEC, 1951, p. 79.

[35] Beate R. Salz, "The Human Element in Industrialization; A Hypothetical Case Study of Ecuadorean Indians." *Economic Development and Cultural Change,* Vol. IV, No. 1, Pt. 2, October 1955, p. 2.

leaving only 10 in the rest of the country.[36] The backland areas are lacking in public works. Large sections of the population are landless and live a mobile existence without housing. The morale and morals of the backland peoples are described as low.[37]

Even the United States is not without its underdeveloped areas. Shimkin has reported on such an area near Fort Yukon, Alaska, where the people live mainly on trapping. In recent years the supply of fur-bearing animals has diminished, while the population, thanks to American medical service, has increased. The meat-providing moose and caribou have also decreased. While the village which was studied has 650 inhabitants who derive their living from 17,000 square miles of trapping range, they have passed the point of diminishing returns. Although a hard-working people, they apparently lack the imagination, and certainly lack the means, to undertake other kinds of activity. Shimkin concluded that the leadership and the ideas must come from the outside. Other resources must be found and other kinds of work.[38]

## Metropolitan Control and Civil Authority

Final responsibility for leadership in any planning scheme must be the civil authority. It is in fact this authority, operating on an international scale, that is taking the initiative in promoting programs for aid to the underdeveloped countries. The TVA program, described above, was initiated by the federal government, but it was put into operation with the cooperative support of the several state governments. The American road system is a joint planning and building program of the federal and state governments, and so to some extent is the control of national parks and forest preserves, even of water control areas. State governments in many cases share responsibility with local government for road building, bridge construction, maintenance of water sheds (providing water supply for cities), protection of streams against pollution, and maintenance of parks and play areas.

Without civil authority, planning must be a voluntary arrangement. Voluntary arrangements, however, rarely succeed in the secondary society, although the idea is often naïvely defended, especially since planning concerns

[36] Russell H. Fitzgibbon, "Uruguayan Regional Problems," *Social Sciences*, Vol. 29, No. 2, April 1954, pp. 80-81.

[37] *Ibid.*, pp. 78-79.

[38] D. B. Shimkin, "The Economy of a Trapping Center," *Economic Development and Cultural Change*, Vol. III, No. 3, April 1955, pp. 219-240. We are told by Shimkin that American experts are now at work on the problems of the trappers in Alaska.

the control over economic values and competitive relations. Voluntary arrangements mean that individuals and groups willingly forego advantages. Says Lynd, "groups in American life tend to grind competitively against each other as harsh and unimaginative extensions of the individual competition on which the society is founded."[39] This characteristic of secondary groups is something which must be recognized as a reality, and it is such recognition that generally results in placing planning control under civil authority.

Civil authority at the national level operates over wide areas, and in relationships with the states the national authority is only indirectly exercised. For example, the state builds roads but part of the cost is paid by the federal government; however, the federal aid may be withheld if the road construction falls under generally accepted standards. A somewhat similar relationship exists between states and local communities in planning control. Lewis reported that local planning authority in Great Britain was usually inactive prior to the establishment of a national authority. Once established, it stimulated more effort on the part of local authority.[40] Even when central authority does take the initiative for planning, much of the responsibility for implementation rests with local government. This means, as Jones noted, that local life "is affected for good or ill by every action that local governments take, or do not take." This holder whether the local community has or does not have a planning program.[41]

Municipal authorities in a large city, and this is especially true of American cities, cannot address themselves to wide area planning without finding the way blocked by various surrounding authorities, some of them overlapping. Merriam, Parratt, and Lepawsky found 1,642 such authorities in the Chicago metropolitan region. These included 978 semi-independent school districts, 204 towns and cities, 190 drainage districts, 165 townships (a rural form that continues to survive), as well as various districts for the control of forests, parks, sanitation, and health. There are also county governments and three state governments. Not only do these authorities seldom cooperate, but sometimes the activities of one are contrary to those of others. For purposes of planning and environmental control, such a wide area needs a TVA-like authority, as does every metropolitan area of large size.[42] While the study just

[39] Robert S. Lynd, "Groups and Social Status." See Robert M. MacIver, Ed., *Civilization and Group Relationships,* New York, Harper & Brothers, 1945, p. 99.

[40] Lewis, *op. cit.*, p. 140.

[41] Victor Jones, "Local Government Organization in Metropolitan Areas." See Coleman Woodbury, Ed., *The Future of Cities and Urban Redevelopment,* Chicago, University of Chicago Press, 1953, p. 481.

[42] Charles E. Merriam, Spencer D. Parratt, and Albert Lepawsky, *The Government of the Metropolitan Region of Chicago,* Chicago, University of Chicago Press, 1933, p. 9.

reported was carried out in 1933, Walker in 1950 reported that the situation was about the same, except for minor changes.[43]

## Ecology and the Master Plan

The term "master plan" is somewhat in ill-repute with modern planners because three or four decades ago it was used to designate great dreams that could be put on paper. Pomeroy recalls that what could not be put on maps was then not called planning; "It was not nice to talk about slums, and they probably were the result of sin, anyway."[44] Today the master plan has come to mean the vision a community has of its future. It is wide-reaching in space and far-reaching in time, but not necessarily exact. It is concerned with general policy for land use and population distribution, with the major channels of movement, and with the distribution of physical facilities. It aims "to express the determination of the community as to what it wants to be" in terms of goals and standards, to be applied as development takes place; all these are aspects of the master plan.[45]

The idea of the master plan, as many use the term, is quite similar to that of the metropolitan region, as that term is used here. Usually the master-plan concept is associated mainly with the city itself and often no more than incidentally with the outlying areas. While the term may not be used, growth and change as recognized in the master plan tend to be seen in relation to the ecological processes of the city. While most developments induced under such a plan would affect the ecological processes, knowledge of these processes can be helpful in planning. There is a general recognition, for example, in transportation planning, and we quote Norton:

> The fact is, however, that the larger cities of America are no longer single municipalities. They have become metropolitan areas made up of dozens or even hundreds of separately governed cities, towns and villages. Transportation planning for metropolitan areas of necessity involves many municipalities. This planning is carried on by the Federal agencies, by state highway departments, by independent highway, bridge, and port authorities, by railroads and by bus companies all operating across many municipal boundaries. Local planning, however intelligent it may be, cannot of itself add up to metropolitan planning. The result is that as we have virtually no metropolitan land-use planning in the United States today, metropolitan transportation developments, such as express

[43] Walker, *op. cit.*, p. 260.
[44] Pomeroy, *op. cit.*, p. 18.
[45] *Ibid.*, p. 19.

> highways, are sometimes a detriment rather than a benefit to the many communities through which they pass.[46]

What Norton says of transportation applies to any other aspect of planning for the modern community: Control is dispersed and is seldom effectively applied. As a group of writers said of New York, it has suffered from rapid growth, from a rigid street pattern, from the evils of *laissez faire,* and has been little helped by sporadic plans. The city is called most "unsuitable in the number and distribution of its population, most opportunistic in land use," but still "most dynamic in the pulse and variety of its living ways."[47] But New York today has its master plan and a general idea of what it expects to be in the future.

Inherent in the master-plan concept is the recognition that the urban community is complex and not maneuverable as planners used to hope before they became aware of the forces that really influence urban behavior. The new approach is to understand these forces that they may be somewhat guided, and so, as seen by Gallion and Eisner, the master plan is a guide welding the "sociological, economic and geographic properties of the city into a structure." It sets goals and standards, and calls for decisions, not only before but while development is under way.[48] Moreover, it recognizes that democratic planning is many-sided and that dealing with planning problems is a long-term and continuous operation. This is indicated in the three-dimensional approach suggested by Wirth and Shils who see the modern community as:

(1) A unit having a functional position in a larger—regional, national, or world—economy of men, goods and services;

(2) An integral social structure or mechanism with its own forces of equilibrium and its own social and psychological characteristics; and

(3) A physical apparatus, which men use for the attainment of their biological, economic and cultural ends. About the apparatus:

> The way in which this apparatus is used by given individuals and the particular parts of the apparatus which they use depend on their economic and social status, i.e., on the claims which they can effectually exert on others in the form of income and recognition. Certain types of apparatus are manipulated only by specially qualified persons, as for instance, certain types of machinery,

[46] C. McKim Norton, "Metropolitan Transportation." See Gerald Breese and Dorothy E. Whiteman, Ed., *An Approach to Urban Planning,* Princeton, Princeton University Press, 1953, p. 84.

[47] *New York Panorama,* American Guide Series, New York, Hastings House, 1938, p. 12.

[48] Gallion and Eisner, *op. cit.,* p. 249.

lathes, locomotives, elevators and trolley cars, while others, such as buildings for work and residence, are, to some degree at least, utilized by everyone who works and lives in the city, either as an individual in his own right or as a member of an organized group such as a corporation or family.[49]

## Paper Plans

In 1905 San Francisco adopted the plan of Daniel Burnham, and the following year occurred the earthquake which laid most of the urban center in waste. The opportunity was ideal for a planned reconstruction but, as Pomeroy observed, the people were too occupied with building to think about the plan.[50]

The San Francisco experience is not unique. The story has been repeated in other cities, even in Chicago where a plan for future development was also adopted in the "city-beautiful" period. While the original plan, fortunately, has not been followed, a planning concept has been kept in mind and it has been followed in some respects.

Planning in Russia is total, in that it takes account of all phases of situational control that seem to need planning. It includes the rural as well as the urban; in fact, there was a systematic program for eliminating the old peasant way of life. That part of the plan aimed at making the farmer over in the image of the industrial worker, as reported by Mitrany.[51] Planning is not only ideological, it is mandated.

The plan for Moscow as fixed in the 1920's set the population limit at about 5,000,000. Moving into or from the city was placed under firm control. Controls were set for the different land uses. Apparently, as seen by Perkins, the resolution to limit the size of the city has not been enforced. Moscow's rate of growth has been fairly similar to that of other world cities. It climbed from about 2,500,000 when the plan started to 4,137,000 in 1939.[52] Presumably Russia should be the planner's paradise. There can be no contests about property rights, and civil authority has full power to put any land to any use. Under such conditions the rebuilding of the heroic city of Stalingrad began. The plan for the city was hailed as a great achievement, but the plan on paper in many respects, according to Perkins, was not followed in the

[49] Louis Wirth and Edward Shils, "Urban Living Conditions," *Urban Planning and Land Policies,* Vol. II, Washington, National Resources Committee, 1939.

[50] Pomeroy, *op. cit.,* p. 12.

[51] David Mitrany, *Marx against the Peasant,* London, George Weidenfeld and Nicolson, 1952, p. 223.

[52] Maurice F. Perkins, *City Planning in Soviet Russia,* Chicago, University of Chicago Press, 1953, p. 38.

construction.[53] Persons or groups having objections did not offer criticism. On the contrary, they continued to praise the plan, now and again suggesting "improvements." It seems that suggestions for improvement under dictatorship operate as does criticism under democracy.

Other Russian cities also had plans; the grand central areas were to be constructed first to inspire the people. Actually, factories and houses have been given first priority, and the amenities neglected. "As a result," says Perkins, "a number of cities appeared as conglomerates of scattered settlements, lacking public utilities and landscaping, while the centers remained untouched."[54]

When the planner, after making his study of the lay of the land and other pertinent facts about the city, puts his plan on paper, he is projecting a still picture to be realized at some future time. His colored drawings of buildings, fountains, trees, street layouts, and so on indicate what the city will look like. Not only is the plan a still picture which says, "This is how it will be," but it carries the idea, "This is how it will continue to be." However, such a still picture drawn in 1958 visualizing a city as it will be in 1980 and for decades after ignores the basic facts of urban living and change. We need only to ask what problems certain American cities would have today if they had faithfully carried out broad plans designed at the beginning of this century. Thanks to various interventions, these plans were not carried out.

It is encouraging to note that experts in this field are beginning to speak of dynamic planning; that is, planning that is flexible in design and responsive to change. As Fry put it at a meeting of architects, "We would be better employed in searching for the rules that govern nature and nature's creature, man; for if this city is to be built it will be built not by one man but many, and not by architects only."[55] The idea is growing that planning can succeed only if it becomes increasingly a shared activity; the different organized groups in the community must have a part in it. This is an idea that was emphasized in a widely read book by Wootten who argued that such control is safest in the hands of the man and woman on the street.[56] Progress would be slow, but the results would be more lasting.

The plan of a city must reflect the work and life of the people and it must

[53] *Ibid.*, p. 82.

[54] *Ibid.*, p. 82

[55] Maxwell Fry, "The Idea and its Realization." See J. Tyrwhitt, J. L. Sert and E. N. Rogers, Eds., *The Heart of the City*, London, Lund Humphries, 1952, p. 89.

[56] Barbara Wootten, *Freedom under Planning*, Chapel Hill, University of North Carolina Press, 1945, p. 100.

be a facility for the people in their day-to-day activities. It must relieve them of the major frustrations of urban living.[57] We can, however, be unrealistic in in speaking about democratic planning and "the people participating in planning." The so-called "man on the street" is helpful mainly in a negative way. He can say what he does not want in a plan, but he is not usually able to say how the plan should be. The expert makes the plans, and if he "goes to the people" it usually means that he talks to organized groups. The leaders of these organized groups may be able to make positive proposals, but they can do it more easily if a plan is brought before them.

## The Central Urban Area

More than any other part of the community, planning concerns the urban center. More designs are made and books written about this area than any other section. It offers the greatest challenge to planners. Its character reflects that of the city, but it belongs not to the city alone but to anyone who may come from anywhere. If we speak of Hamburg, Paris, London, or New York as being international cities, that description applies to the square mile or perhaps three or four square miles which make up the "urban core." It is the point of special markets and the "front office" for wide areas.

Here is where space is most precious, buildings climb higher, competition for position is sharper. The problem of planning is to make the center more accessible for more people, for here it is that the lines of communication and transportation from all areas meet. It is not only the central place for work and work direction, it is also the center for fun, even for culture. Sert would bring in more "plants, water, sun and shade, and all the natural elements friendly to man," and exclude the noise and fumes of motor cars.[58] Desirable as such a change might be, how much would it encroach on the paramount work functions of the urban center?

If a fourth to a third of the people of a city enter the central area on a single work day, that usually means a great amount of congestion; in fact, different kinds of congestion at different times of the day—the workers who come and go, the customers who come and go, and the pleasure seekers. There may be thousands of specific reasons for people going to the heart of the city.

[57] H. M. Kallen, "City Planning and the Idea of the City, Considerations Especially about New York," *Social Research,* Vol. 23, No. 2, Summer 1956, pp. 186-198. The article is a good statement of this viewpoint.

[58] J. L. Sert, "Centres of Community Life." See J. Tyrwhitt, J. L. Sert and E. N. Rogers, Eds., *The Heart of the City,* London, Lund Humphries, 1952, p. 11.

The planner needs to know how to classify these reasons, and if he can do it wisely he may find that certain functions performed at the center of the city might be moved to another location. That is being tried in the rebuilding of the war-damaged city of Coventry, England. There will be space for stores, for offices, for culture and beauty, but industry and housing will be excluded.[59] But Coventry is a small city, 300,000, and the sorting task is comparatively simple. It would be a staggering task in a city like London, and there the effort might be very unwelcome.

## Transportation and Access

Whether the planners, however wise and however informed, are able to decide what functions should be located at the urban center and which should be excluded, has yet to be demonstrated. It is being done in a small way, even in large cities, by zoning ordinances, but not in a way to provide more free space and less congestion. There is, however, a sort of control due to economic and ecological forces. The sorting forces are negative: the excluding effect of high rents and land values and the inconvenience of congestion.

This does not mean that there is no need for the planner who may find ways of excluding certain functions from the urban center or who discovers that functions may be excluded by some effective control of the traffic. That is being done with good effect in most large cities of the West. What happens when it is not done is suggested in this item on cities of the East, written by Versluys:

> Traffic in these cities is one of the major problems. Bullocks and cows may quietly cross the main roads; rickshaws, buses, handcarts, bicycles, cars, carriages, trucks, tramways and innumerable pedestrians help to create unsolvable traffic problems. The roads, overcrowded and narrow already in the city centre, become still more so by protruding wooden or bamboo constructions of innumerable stalls put in front of the more permanent shops. Racks of goods are hung against the walls, the roads are full of people offering their cheap articles displayed right on the pavement, or on small folding tables, or carrying them on a tray or in their arms.[60]

Western cities have passed far beyond that stage of congestion. They have in fact moved from one type of congestion to another. Not more than a century ago in New York and other American cities cows, pigs, and chickens

[59] D. E. E. Gibson, "Coventry, England," see J. Tyrwhitt, J. L. Sert, and E. N. Rogers, Eds., *The Heart of the City*, London, Lund Humphries, 1952, p. 135.

[60] J. D. N. Versluys, "Introduction," *Implications of Industrialization and Urbanization* (Five Studies in Asia), Calcutta, Unesco Research Center, 1956, p. iii.

were very common even in congested areas. Much space was needed for the many horses. Administrative action ridded the cities of cows, pigs, and chickens, but it was technological change that ridded the city of the horse, the same technological change that enabled people to escape the congested urban centers to the suburbs. Traffic congestion during the final years of horse-drawn transport was more frustrating than much of the automobile congestion today in the same cities.

Of all the specialists in the planning field, perhaps the most effective in their work have been those who have dealt with the problems of metropolitan transportation, although they have been occupied mainly with relieving congestion. However, as Norton says, they are better able to plan transportation facilities for cities if they know what and where future needs will be.[61]

As already mentioned, transportation problems are not merely urban, for the streets within and roads beyond are a continuous network. Toynbee, with good reason, describes road building as an interest that has always been primarily urban. This is true since the building of the Roman roads over which tribute came to feed urban multitudes. With the rise of industry, roads were needed to connect cities.[62] Decades had to pass before the rural people got interested in the same good roads.

## Urban Decentralization

Unanticipated by planners and quite without their help, cities have begun to spread out. While slums have not vanished, the old congested slums are disappearing; this was not planned either. In some cities, areas of land once occupied by residences are falling into disuse and programs of redevelopment are being promoted. This trend from centralization to decentralization has brought forward many problems that were not imagined previously. The city tends to lose its old identity and becomes a central unit in a cluster of towns and cities embracing, as Gulick reminds us, up to 18,000 square miles and anywhere from five to twenty million people. This area is a metropolitan social and economic unit without any overall metropolitan organization; here are many governments, but no overspreading civil authority. This is a planning problem which Gulick argues not only must be met, but can be met.[63]

Seen from the point of view of the central city, decentralization means

[61] Norton, *op. cit.*, p. 82.

[62] Arnold J. Toynbee, *A Study of History,* Oxford, Oxford University Press, 1954, Vol. VII, pp. 80-107.

[63] Luther Gulick, "Metropolitan Organizations," *Annals of the American Academy of Political and Social Science,* Vol. 314, November 1957, pp. 57-65.

considerable tax losses because much property is not earning and hence the amount received in taxes is lowered. On the other hand, the expenses of the central city are not lowered in proportion. The many people of stature who earn their money in the city but do not live there deprive the city of needed leadership. This leadership talent lives in the suburbs. People living in the suburbs who enter the city at will and use its services make no contribution for their support. On the other hand, the outlying communities are not quite secure because each is in the midst of changes over which it has no control, and the individual suburbs lack the willingness or ability to cooperate with other suburbs on mutual problems. The vexing in-between problems such as drainage, water supply, fire protection, parks, and many others get little or no systematic handling.[64]

## Planning and the Housing Problem

Almost every urban community in the industrial countries is confronted with a housing shortage, and has been over a long period. The problem is not one that is solved merely by building more and more houses. It concerns location: where houses may or must be erected and where they should not or must not be erected. It concerns kinds of housing to meet different needs and the kinds for different locations, since in some locations certain kinds of housing are not wanted. It concerns also the number of houses of different kinds needed in different locations. Moreover, planners need to know in advance what part of such housing will be constructed with public funds and what part might be provided by private enterprise.

With respect to housing, planners are confronted with many variables which make this phase of environmental control highly uncertain. For this reason such an informed specialist at Klutznik declared, with respect to housing needs, "Anyone who claims the capacity to see twenty-five years ahead and who stands on it as conclusive and undebatable would be a fool."[65] One variable which he mentions concerns the possibility of housing renewal in decayed urban areas which will attract families back to the city. It is likely to happen, but when?[66]

64 "The New America, Suburbia, Exurbia, Urbia," *Newsweek*, April 1, 1957. This article contains a great amount of stimulating information about the suburban development, especially around Los Angeles. See also such recent books on the suburbs as A. C. Spectorsky's *The Exurbanites* and John Keats' *The Crack in the Picture Window*.

65 Philip M. Klutznik, "The Provision of Shelter," *Annals of the American Academy of Political and Social Science*, Vol. 314, November 1957, p. 45.

66 *Ibid.*, p. 43.

Another variable which is hard to predict concerns the changing housing needs of the family at different stages of the family life cycle. Agle estimates that the family cycle includes four stages: from marriage to the first children (about 6 years), the family-rearing period (about 18 years), the home-leaving period for children (about 6 years), and the period when the parents are alone (10 to 20 years). Each stage calls for a different type of housing.[67] The nature of this family cycle is different for different social and economic classes, and it may be influenced by the number of children in the family.

Still another variable is population change within a given area. In an area of single family houses, the number of children may drop, say from about three per family to less than two per family in the course of two or three decades; the number of dwellings may remain the same but the number of inhabitants may be fewer. Or the population may remain about the same but the number of households or spending units may increase. This may mean that the level of living of the population is rising. Or it could be that the houses are declining and the population is changing to one of lower income level. There may be an exodus of younger families in an area if some feel uncomfortable in the presence of so many older people. Older people may migrate from an area of single dwelling after their children have married.

As the family has its life cycle so the different kinds of residence have their life cycles. Grebler estimated that the usefulness of a house in an American city tends to be from 20 to 30 years, after which its worth as an investment diminishes.[68] The aging of a house may depend on factors other than physical deterioration. Obviously, the investment attitude toward a family-owned house would be less than toward a rented house. The family house may be kept in repair and not be called old short of 40 or 50 years. The usual attitude of people who rent houses is to demand as much in modernity as they can pay for and when they can afford it they move to better-appointed dwellings. Thus as houses decline and become outmoded, they experience a succession of occupants. This is also called the "filter-down" process, which some real estate people consider as a satisfactory arrangement for housing the poor who usually bring up at the rear of the succession of occupants.

Housing demand is especially sensitive to technological change. A rapid rate of technological change offers a challenge to planners, and it makes pre-

[67] Charles K. Agle, "Housing and Urban Redevelopment." See Gerald Breese and Dorothy E. Whiteman, Eds., *An Approach to Urban Planning*, Princeton, Princeton University Press, 1953, p. 54.

[68] Leo Grebler, *Production of New Housing*, New York, Social Science Research Council, 1950, p. 11.

diction about future needs and future methods of meeting needs somewhat of an exercise in guessing. Nonetheless, planners must make guesses because the need for housing is both urgent and continuous.

## Making Communities Tolerable Places

Hyder and Tobin, after their study of Milwaukee, made some observations about the planlessness of American cities, charging that growth has been chaotic and without control. Most of these cities have as a result been confronted with the "encroachments of spreading blight," and they need "replenishing for unity, cohesion and livability."[69]

Sinclair complained that 70 years were needed in London to realize certain transit reforms. He added that London has 355 rival bodies with "stationary powers" fighting for "the privilege to rule ten million people." Their jurisdictions overlap and "their duties are ill-defined." They operate under rules and precedents, "the mistakes and forgetfulness and ignorance of men now dead."[70] As Sinclair sees it, the city of London has changed and grown without much guidance from the constituted authorities. Yet it must be said that, with all the alleged ineffectiveness of these agencies, London managed to organize itself admirably during the World War II bombings and has since evolved a plan for regional development that seems to work.

To mention one more criticism, Saarinen is of the opinion that many of the growth errors and much of the apparent planlessness of towns and cities are due to the lack of vision on the part of our forefathers. Had they laid the right foundation, most of our present problems would be "much simpler and easier." The moral is that this generation must not make the same mistakes. If we plan wisely, "the problem of our children will be easier, and future generations will be grateful."[71]

The observations quoted above are all too familiar, and that of Saarinen contains two fallacies: first, that if our forefathers had taken thought in building communities, we, their children, would have fewer problems; second, that if we take thought in our building plans our children will have it easier. Obviously our forefathers in 1850 could not have foreseen the automobile, electric power, and steel-structured buildings which were forces in community life in

[69] K. Leo Hyder and Howard J. Tobin, *Proposals for Downtown Milwaukee,* Chicago, Urban Land Committee, 1941, p. 5.

[70] Robert Sinclair, *The Big City, a Human Study of London,* London, Reynal & Hitchcock, 1938, p. 19.

[71] Eliel Saarinen, *The City, Its Growth, Its Decay, Its Future,* New York, Reinhold, 1945, pp. 250-251.

1950. While we are better informed about the factors involved in our problems of environmental control, we too can only guess about future population changes, economic changes, or technological changes. However, we are infinitely better off than were our forefathers because we are more aware of the uncertainties in planning and we recognize, as few of them did, that planning must be based on systematic knowledge.

Leo N. Tolstoy considered towns as "places where mankind has begun to rot, and unhappily the rot is spreading." Peterson said of our present urban civilization that it spreads "deterioration and degradation, with human erosion keeping pace with the erosion of our pregnant fields."[72] Gallion wrote in 1953, "The meanness of our urban environment is intolerable," but he argued that there is much we can do about it; so much is possible, he declared, that it "piques the imagination."[73] If Tolstoy were to see the city about which Peterson and Gallion complain, he would be astonished that so much progress has been made. On the other hand, Peterson and Gallion see other goals that must be achieved to make the modern community a tolerable place.

With all the faults of urbanism, most people who have choices turn to the urban rather than to the rural way of life. And there are many who would not call the rural way of life ideal. On the negative urban stereotype, Baker wrote:

> While cities are open to some blame upon all of these scores, they can proudly boast of some uniquely urban contributions to society, notably the fostering of cultural activities and of intellectual and social freedom. Also ignored by many is the fact that slums are by no means confined to cities. Decaying rural areas are often deficient in both material and cultural advantages.[74]

While it is worthwhile to make efforts to eliminate the negative elements in community life, whether urban or rural, and to resort to planning for any purpose that it can serve, we must also try to be realistic. Churchill does not believe that most urbanites would be happy in some of the planned communites that are being built; in fact, many are not. "A community of like-minded souls dwelling in well-ordered harmony" is to him a dull place and not compatible with the variety and opportunity of urban life.[75]

[72] Elmer T. Peterson, Ed., *Cities Are Abnormal,* Norman, University of Oklahoma Press, 1946, p. 24.

[73] Arthur B. Gallion, "Civic Design and Democracy." See Coleman Woodbury, Ed., *The Future of Cities and Urban Redevelopment,* Chicago, University of Chicago Press, 1953, p. 76.

[74] Gordon E. Baker, *Rural versus Urban Political Power,* New York, Doubleday, 1955, pp. 1-2.

[75] Henry J. Churchill, "What Kind of Cities Do We Want?" See Coleman Woodbury, Ed., *The Future of Cities and Urban Redevelopment,* Chicago, University of Chicago Press, 1953, p. 45.

Making the community a tolerable place in which to live and work must permit variety. The planner's urge for order could very well be stifling to progress. The change process means that some elements in the environment will show decline, since others show dominance. The task of planning is to try for balance, to guide the change processes but not suppress them for the sake of order. Whether there is advance or not must be seen by (1) long-term comparisons, and (2) wide-area comparisons. What advances, if any, have been made between 1858 and 1958? Have there been advances in environmental control, not only in the consuming city but also in the producing hinterland?

During this century technology has advanced beyond measure, giving man more power over his environment; the problem is how to use that power. Roads and streets are more serviceable and cleaner. Facilities for work are safer, cleaner, and more efficient. Living quarters have improved. Water supplies are safer and waste disposal is less a health hazard. More attention is given to the conservation of natural resources. Urban sanitation is more under control. Buildings are safer from fire hazard and better supplied with light and air. Public functions are housed in better buildings and services are better provided with equipment.

Most of the stimulation, the initiative, the know-how, and the financing for these changes are urban, and it is in the urban community where the urge is most felt to carry these developments farther.

## Summary

Since urban communities are dependent on their hinterlands for sustenance, for their survival and growth, they need to find ways and means at least of sharing in programs for the development and control of the resources essential to their well-being. Such a concept of control for the use and conservation of resources, both human and natural, includes different types of planning as applied to metropolitan areas as well as to the central cities. We may speak of it as planning for situational control. It recognizes that the major urban community is a unit in various relationship networks: social, economic, and political. These relationships are influenced by both geographic and ecological factors both in the urban community and outside.

In this chapter attention has been centered mainly on planning and control within cities and to some extent beyond their borders, keeping in mind that cities are functionally identified with ever-widening areas. The evolution of city planning was sufficiently reviewed to make clear that planning is both an old interest and a changing one. Much of the learning in this field has taken place during the past half century. This process of learning has been increasingly stimulated by the rapid growth of the cities in recent decades. It has been stimulated, too, by rapid technological advances, but most of all, it has been stimulated by the entry into the field of various specialists: economists, sociologists, geographers, ecologists, engineers, and others. It is no longer the province of the architects and various uninformed social reformers.

Thus the reform groups in this as in other areas of disorder, especially in the United States and Great Britain, did stimulate demands for planning and they did attract the interest of specialists who have been able to make practical contributions. The service of the reformers should not be undervalued. Even though their proposed solutions may have been unrealistic, they did stir interest.

Some people oppose planning because they fear that, since planning must be under public control, it will encroach on the rights of the individual. They fear the concentration of decision making in central offices. Others favor planning in general, but oppose particular schemes that they fear might intrude upon their economic interests. Perhaps they are the ones whose activities need to be brought under planning control for the community good. Still others may be hesitant because they suspect the judgment of the planners. Democratic planning must expect such different viewpoints.

Planning for the great city, and the control that must sustain it, is not one thing, but many. It may concern housing, the location of industry, the width of streets, creation of parks and playgrounds, routing of traffic, ridding the city of smoke, development of the urban center; or it may be concerned with problems outside the city—drainage, water supply, the organization of police and fire protection. Whatever may be concerned, planning involves disturbing old relationships and habits and establishing new ones. Thus planning is always controversial and usually proceeds by compromise.

Especially in those great cities that have wrestled or even fumbled with the idea of planning and control for years, master plans may have evolved. The master plan is a vision, partly on paper, of what the city hopes to be like at some future date. It is a guide for moving toward a goal, or goals. It may change with experience, but that means redefining the goals. As a very general blueprint of the future, continually held before the people, the master plan is an educational device. Whether there is a master plan or not, the goals of planning can hardly be attained by voluntary action. They must have the sanction of law. Civil authority must have the final responsibility, but civil authority needs the cooperation of voluntary organizations.

Planning is not only the proper responsibility of civil authority; it is generally the work of cooperating levels of government. Some work is essentially national (or regional), like the establishment of TVA or other control over natural resources. Other planning is the peculiar responsibility of state government, or local communities may be empowered by states to exercise certain specific planning functions. In some respects these authorities may overlap, with the result that not much, if anything, gets done. There are other types of planning which lie only indirectly within the sphere of public authority, like that of lumber companies adopting a policy of nonwaste or replanting in the exploitation of their private forest preserves.

Our interest is mainly in metropolitan planning and we have given that term a wide scope. It applies to cities and their hinterlands, and it may also apply to groups of cities and their joint hinterlands within a region. For example, fifty or a hundred cities may have to cooperate in conserving and using the water supply of a region. Metropolitan planning, to some extent, may be thought of as global, since it must be concerned with utilizing the resources of wide areas, even regions, that have already in fact a social and economic unity, although this unity may be concealed by a maze of political subdivisions which usually operate to frustrate rather than facilitate wide-area situational control. Whether we think in terms of single metropolitan areas or a regional cluster of related metropolitan areas, our attention is primarily on clearing the way for a more orderly and less wasteful urbanism.

However much is said about voluntary action, and without undervaluing its importance, we must recognize that metropolitan or regional planning cannot be achieved by that approach. It would be a miracle if voluntary action could bring about such a regional utility for planning and control as the TVA. The need is for interstate, regional, or national civil authorities or quasi-public bodies such as the European Coal and Steel Community, which does not operate in a wider area than that embraced by the TVA but is the type of authority that can initiate a great variety of wide-area planning. With the expansion of urbanism such metropolitan and regional public authorities become more and more imperative. We can be sure that they will take form, but probably not until after much fumbling.

# Names Index

# Subject Index

For Product Safety Concerns and Information please contact our EU representative GPSR@taylorandfrancis.com
Taylor & Francis Verlag GmbH, Kaufingerstraße 24, 80331 München, Germany

www.ingramcontent.com/pod-product-compliance
Lightning Source LLC
LaVergne TN
LVHW011102300926
849160LV00008B/252
*9780415860529*